ASTRONOMY

ASTRO

NOMY

Harper & Row, Publishers
New York   Hagerstown
London   San Francisco

Frank N. Bash
Consulting author
University of Texas

A LEOGRYPH BOOK

Project editor: DANIEL SCHILLER
Sponsoring editor: WAYNE SCHOTANUS
Consulting editor: DILIP BALAMORE
Design: PEDRO A. NOA
Art direction: DEBORAH DALY
Manufacturing supervisor: STEFANIA J. TAFLINSKA
Production manager: EILEEN MAX
Composition: SOUVENIR, BY THE CLARINDA COMPANY
Printer and binder: HALLIDAY LITHOGRAPH CORPORATION

Library of Congress Catalog Card No.: 76-52524
ISBN 0-06-043853-3

CONTENTS

CHAPTER ONE:
SKY AND COSMOS

CHAPTER TWO:
THE STARS IN SPACE

CHAPTER THREE:
LIGHT AND ITS USES

CHAPTER FOUR:
RADIATION, ATOMS, AND SPECTRA

CHAPTER FIVE:
WEIGHING AND MEASURING THE STARS

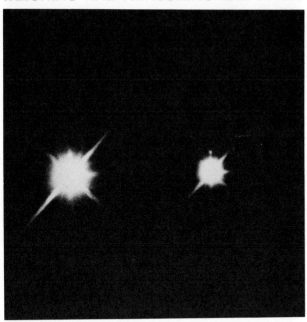

CHAPTER SIX:
STELLAR STRUCTURE

CHAPTER SEVEN:
LIVES AND DEATHS OF THE STARS

CHAPTER EIGHT:
THE MILKY WAY

CHAPTER ELEVEN:
THE SUN

CHAPTER TWELVE:
MOTION IN THE SOLAR SYSTEM

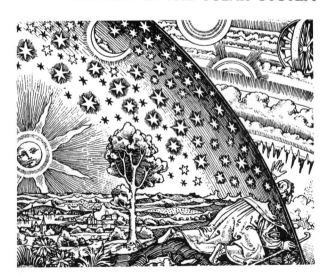

CHAPTER THIRTEEN:
PLANETS AND OTHERS

CHAPTER FOURTEEN:
ORIGIN AND HISTORY OF THE SOLAR SYSTEM

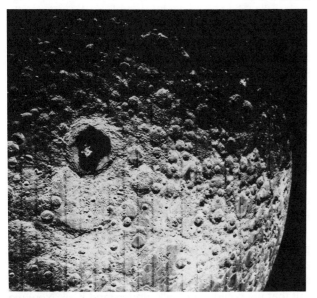

PREFACE

This is the obligatory section which every author writes, but few people ever read. He tells you, and perhaps justifies to himself, why he wrote the book. He has to be explicit because few of us ever succeed in making the reason obvious in the text. Of course, that may be true because most of the reason is hidden inside the author and only a small part is in a composite, imagined reader.

First, although the plan and emphasis of this book are mine, others did much of the work. Even though they didn't want recognition, I must acknowledge the great skill and hard work of Dan Schiller and Dilip Balamore. If the book is successful, they are in part responsible.

I got involved in this project after teaching a one-semester introductory astronomy course for non-science majors to several thousand students over 10 years. It always seemed to me that the available texts either treated the students like miniature future astronomers or were so surgically non-mathematical that the surgeon also removed the logical structure.

I do have a mental picture of an average student in my class. That student is a member of the group of mathematical walking wounded. He became convinced in early grade school that mathematics consists entirely of a memorized set of absolutely arbitrary rules with no logical structure at all. Having forgotten all those rules, he is convinced that he cannot do any mathematics. He is interested in astronomy through having read articles about black holes, quasars and horoscopes. Sometimes, I think that he expects me to appear for the first lecture complete with a black cloak and a peaked hat covered with moons, stars and astrological symbols, and lecture for a semester on constellations and their mythology, including in the mythology, quasars and black holes. He is quite surprised when I confess that I can recognize about four constellations and that I speak physics. Because, I imagine, science (especially physics) conjures up a whole new set of goblins. He feels distrust, some fear of the consequences of past scientific discoveries, and I think, subconsciously, a vague sense of dullness and irrelevance. A feeling that science is a sort of SUPER GAME which exists only in the classroom, independent of nature, and which has been invented by a group of people who have made the rules so obscure and difficult that they have in effect elected themselves a sort of super-elite: the only ones who can play. In addition, scientists have some magic recipe called the "scientific method." I imagine this leading to a picture of a room full of disheveled, white-haired men sitting at tall stools (like a scene out of Dickens), writing with fountain pens, dully applying the scientific method, and occasionally shouting "eureka." All failures are due to not being sufficiently careful in applying the "method."

I have always felt that it was important for my students to see that science is successful because it deals with simple problems—problems in which only a few variables are involved. Science is also quantitative; it always works with measurements. Finally, science attempts only to describe, not to explain. One of my worries is that the success of science is causing it to be applied where it is not applicable, in areas where too many variables, or ones that cannot be quantified, are involved.

The process of doing science is often entirely irrational. Ideas come while driving your car as often as by any rational intensive process. Correct descriptions have been made as the result of mistakes, or in spite of contrary evidence. Ideas come whether or not they have any practical application. But in science, these ideas must be cleaned up, quantified, made subject to test before they are communicated, and this, I suppose, makes the ideas appear always to have been logically arrived at.

Of all the sciences, astronomy especially reveals that this search for truth is not inexorable. For example, it was suggested in the third century BC that the earth revolves about the sun, yet we remember Copernicus for this discovery in the sixteenth century AD. It is interesting to inquire about what happened to that idea in the intervening 1800 years, and to discover that Plato and Aristotle are largely responsible for its demise. They felt that they could discover, by pure thought, some "self evident" properties of the universe. They and everyone else with the same idea have been wrong. We have to look.

It is fascinating to follow the development of our perception of nature. At the time of Isaac Newton, about 1700, it seemed clear that the universe was like some kind of huge mechanical system with lots of parts, but that we could eventually understand it all. Now we have a physical law which describes limits on what we can know. We now can predict with high precision what will happen when light strikes a group of atoms, but in the process we have had to abandon any description of what light actually is, and we cannot say what will happen to a given individual atom.

In this book we have not exhaustively covered the field of astronomy. We concentrate on stars and attempt to show you what we know about them, and especially how we know it. We want you to see your terrestrial experiences, how they have been codified into physical laws which describe them quantitatively, and how they have been applied off the earth. We want you to see how we arrive at the awesome statements that the carbon in your body was produced in a star; that if you look far enough you can see the universe being born. We want you to see how we have unravelled the evolution of stars, even though the shortest lived star lasts longer than all of recorded human history.

I am a professional astronomer who deeply loves his subject, is continuously in awe of the beauty of nature and the beauty of its logical structure. I enjoy teaching astronomy because, like every astronomer I have ever met, I am evangelistic about my subject. I recognize that I am not often logical and that even my love of the people that I love and the love of my subject isn't logical. I couldn't care less about astronomy's practical benefits. I am grateful that society has a structure which allows me to be paid to study it, if only as a purely intel-lectual endeavor, like art, history or philosophy. Astronomy certainly isn't everything, but it's fun, challenging, and beautiful.

The greatest success of this book would be for you to see that whereas celestral objects are beautiful, human comprehension of some logical structure in the universe, and the ability to extend terrestrial experience to celestial objects, are both as beautiful.

Probably this poem, which is one of my favorites and also the one which Bertrand Russell said expressed his feeling about science, conveys it best:

THE INFINITE

Dear to me always was this lonely hill
And this hedge that excludes so large a part
Of the ultimate horizon from my view.
But as I sit and gaze, my thought conceives
Interminable vastnesses of space
Beyond it, and unearthly silences,
And profoundest calm; whereat my heart almost
Becomes dismayed. And as I hear the wind
Blustering through these branches, I find myself
Comparing with this sound that infinite silence;
And then I call to mind eternity,
And the ages that are dead, and this that now
Is living, and the noise of it. And so
In this immensity my thought sinks drowned:
And sweet it seems to shipwreck in this sea.

—Giacomo Leopardi (1798–1837)

Translated by R. C. Trevelyan, in *Translations from Leopardi*, Cambridge University Press, 1941

A NOTE ON THE USE OF THIS BOOK

Although this book begins with stellar astronomy, and treats the solar system only after having dealt with galactic astronomy and cosmology, its use need not be confined solely to courses organized in the same way. It should be possible to employ this book in a course that begins, in the more traditional and historical way, with the solar system. The sequence of chapters for such a course would be 1, 12, 5 (first half only), 13, 14, followed by either 11 or 2, depending on the instructor's preferred organization. In this ordering, some instructors may wish to supplement the material in Chapter 12 with the discussion of parallax on pp. 24–27.

A NOTE ON UNITS

For the most part, this book employs units of the cgs system, but they are not used with dogmatic rigidity. In one or two places, mks units have been allowed to appear where they were more convenient; and in many examples from everyday experience, British units (e.g. 40 mph; the 200-inch telescope; 800° F) have been used for the sake of their familiarity. A discussion of units, and a table of equivalents and conversion factors, appears in the appendix, which also contains an outline of the limited mathematical apparatus used in the text.

CHAPTER ONE

SKY AND COSMOS

Astronomy was once the study of the heavens; today, it is the study of the universe. The purpose of this chapter is to introduce you to both of these realms. The chapter text is intended to orient you with respect to the visible sky. It presents a brief account of the motions of the celestial bodies, and of the temporal phenomena associated with them: day and night, the month, the seasons, the year. It also explains these observations in terms of the earth's motions and those of the other bodies in our solar system.

The movements of the heavenly bodies have been known since the dawn of history, and the correct interpretation of them had begun to emerge hundreds of years before the birth of Christ. By the seventeenth century the dynamics of the solar system were pretty well understood. There is very little in this chapter that would come as a surprise to a well-educated person of 1776, or even 1676. Yet since that time, our picture of the universe beyond the solar system has changed dramatically. In 1776, the most remote astronomical object whose distance was known with any degree of certainty was the planet Saturn, which at its farthest from us is some 1.6 billion kilometers away. This may seem a huge number at first—but anyone studying astronomy must soon get used to far larger numbers. Today, we have observed objects that are about 50 trillion times more distant than Saturn.

In the past few centuries, each generation has seen the scale of the universe revised upward, and we cannot be sure that this will not happen again—perhaps more than once—in your lifetime. The Portfolio in this chapter is intended to orient you with respect to the universe as a whole, to the extent that we understand it today.

3

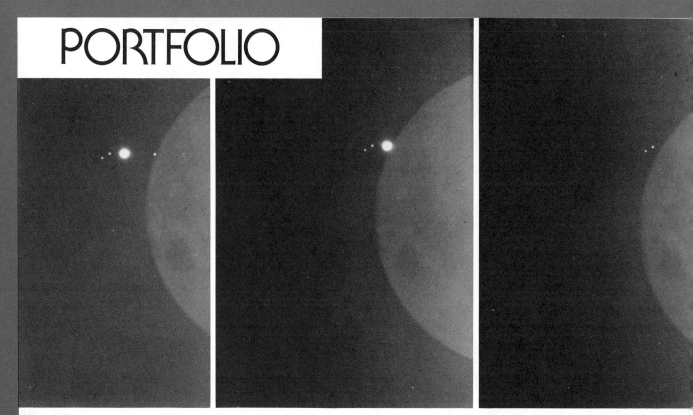

Above, the moon progressively occults Jupiter and its four major satellites. Below, a close up of Jupiter and one of its satellites from a camera aboard the Pioneer 10 spacecraft.

scale of the universe

The universe is very large and very empty. A region of space a million miles in diameter contains only a pound of matter on the average. But matter in the universe is not uniformly distributed. It is extremely lumpy, concentrated into entities that we call planets, stars, and galaxies.

The earth, a planet, is a spherical ball of rock and metal, 12,750 km in diameter. It circles the sun, a sphere of hot gas whose diameter is some 1,392,000 km. The sun is itself a rather unremarkable star, just one of roughly 100,000,000,000,000 stars that make up the stellar system known as the galaxy. This system is 100,000,000,000,000,000 km across. Our galaxy is, in turn, just one among the billions of galaxies photographed by our telescopes.

Clearly, astronomers deal with very large numbers. They can be handled as easily as small ones, but there is difference between manipulating large numbers and grasping their meaning. We talk easily of millions and billions, but seldom stop to think how large these numbers are. An average person takes about 10 days to count a million and 25 years to count a billion. The problem is worse when we try to visualize *distances* of a million kilometers.

To get some feel for the scale of the universe let us consider various scale models.

In our first model the earth is reduced to a ball 10 cm in diameter, the size of a grapefruit. The moon is a plum 3 meters away, just across the room. The sun is a 11 m (36 foot) globe, 1.2 kilometers away, about 15 city blocks. The outermost planet, Pluto, is 47 km away, in the next county. But the nearest star, Proxima Centauri, is 314,000 km away, far larger than any distance on earth. We, obviously, have to construct another model to deal with stellar distances.

In our second model the solar system is reduced to 10 cm. The earth is then .00001 cm across, the size of a bacterium. The sun is dust mote, barely visible to the eye, 1/8 cm away. The nearest star is at 1/3 of a kilometer, 4 city blocks. The center of the galaxy if at a distance of 2400 km, halfway across the United States. The nearest external galaxy is too far away to far away to fit in this model.

In our third model the entire galaxy has shrunk to 10 cm. The neighbouring stars within the galaxy are then only .0002 cm apart. They can be told apart only with a powerful microscope. Our nearest galactic neighbour is about an arm's length and the most distance object observed is at 10 km (6 miles). We can jog to the edge of the observed universe in an hour.

At right, M 31, the great galaxy in Andromeda. This galaxy, which must be counted as one of our nearest neighbours, is nevertheless so far away that its light takes 2 million years to reach us. Two small, elliptical companions can be seen near the galaxy. The bright line at the bottom of the picture was caused by a fireball that streaked across the sky during the long exposure required for this photograph.

In the photograph below, we see some of the hundreds of galaxies that belong to a cluster in Corona Borealis. Light from this cluster takes 1.3 billion years to reach us.

When you look up at the sky, on a clear, dark night, the stars appear to be attached to the inside of a huge, inverted bowl spanning the heavens. This impression is so strong that it has become almost permanently embodied in our language. We speak of "the dome of the heavens," of the "canopy" of stars that "arches" overhead. It is easy to imagine that the heavens are really a sphere, half of which is below the horizon and therefore hidden from our view. In astronomy, as in everything else, we have to start with what we see. So we shall start with this sphere—the **celestial sphere.**

The Celestial Sphere

The ancients were convinced that this sphere had a physical existence, and that it was of enormous radius. For our present purposes, however, the celestial sphere is merely a geometrical fiction, and its size is not important. We shall soon see that most astronomical objects are so far away that quite sophisticated techniques are needed to measure their distances. At this point we are concerned only with the directions of objects, not their distances from us. The stars on the celestial sphere form a permanent pattern; they do not seem to change their positions with respect to each other. Very careful observations reveal that the stars do move, but so slowly that it takes thousands of years for the changes to be noticeable to the unaided eye. We can therefore regard the stars as fixed in their places on the celestial sphere.

The stars appear fixed to the inside of an enormous, imaginary dome called the celestial sphere.

The celestial sphere as a whole, however, appears to change its orientation with respect to an observer on earth. Since the stars define the celestial sphere, we can use them as points of reference. As we watch the sky for a few hours, we see that the constellations have moved. Some have set below the western horizon, and others have risen in the east. The entire celestial sphere seems to revolve about the observer. We know now that this is due to the rotation of the earth about an axis passing through the north and south poles. To the observer, who does not sense that the earth he stands on is spinning, it seems as if the celestial sphere is spinning instead. The axis about which the celestial sphere rotates is the same as that of the earth's rotation. It is defined by two points in the heavens which do not move—the north and south **celestial poles,** located just above the corresponding poles of the earth. Halfway between the poles is a circle called the **celestial equator,** which corresponds exactly to the earth's equator (Figure 1–1).

The celestial poles and celestial equator are projections of the earth's poles and equator onto the celestial sphere.

The earth rotates once about its axis every 23 hours and 56 minutes. The celestial sphere, therefore, appears to turn in the opposite direction at the same rate. If a star rises at 9 PM today, in other words, it will rise again 23 hours and 56 minutes later—that is, at 8:56 PM. This period of time is called a **sidereal day,** or "star's day."

Angles

Astronomy has always been a science of measurement. Indeed, for its first few thousand years there was little else it could be. The earliest astronomers could do no more than measure the positions of the sun, moon, and planets, and try to predict their future positions. Such measurements had a very practical significance to people in ancient days, for astronomy was a highly utilitarian science—an aid to timekeeping, navigation, and agriculture. Sailors steered by the stars, priests foretold the future on the basis of celestial phenomena, farmers studied the heavens to know when to plant and when to harvest. None of these uses would have been possible without a system of marking and measuring positions on the celestial sphere. Even today, our sophisticated telescopes and spectrographs would be useless without some precise way of locating the objects that we study.

But how do you go about measuring distances in the sky—on an imaginary, invisible, infinitely distant sphere? The answer is not obvious at first. Children, for example, generally try to use the methods with which we are most familiar: measuring distances in inches or feet. You may have heard children try to point out a particular star by saying, "it's about three inches above that tree," or say that the moon is "about an inch wide." But a moment's thought will show you that this method is of no use. You could, of course, mark off celestial distances with a ruler—but only if you know exactly how far from your eye the ruler is supposed to be. Two stars that seem to be an inch apart on a ruler held a foot from your eye would seem to be 440 feet apart if the ruler were a mile away.

The method used by astronomers since ancient times—one that avoids this problem—is to measure the *angle* between two objects or locations on the celestial sphere, imagining your eye to be at the apex of the angle. A circle contains 360°. Thus the angle from one point on the horizon to the opposite point is 180°, and the angle from a point directly above your head to (the **zenith**) to any point on the horizon is 90°—a **right angle.** Each degree is subdivided into 60′ (minutes of arc, or simply **minutes**), and each minute is further subdivided into 60″ (seconds of arc, or **seconds**). You may find this system, based on the number 60, a bit odd at first (though it is familiar from our system of timekeeping). It is very convenient, however, because 60 can be evenly divided by so many numbers. It is therefore very easy to measure off such fractions as 1/8 of a circle (45°), or 1/15 of a degree (4′).

Sizes and distances on the celestial sphere can be deceptive. Many people think of the sun or moon as about the size of a quarter held up at arm's length. In fact, however, both sun and moon have diameters of about 31′, or slightly more than 1/2°. To exactly cover the moon, you would have to hold a quarter about 2 3/4 m (9 feet) from your eye. If your arms are not that long, you can still use a quarter for measuring distances on the celestial sphere. Held at an average arm's length—about .7 m, or 27 inches—a quarter will have an angular diameter of about 2°.

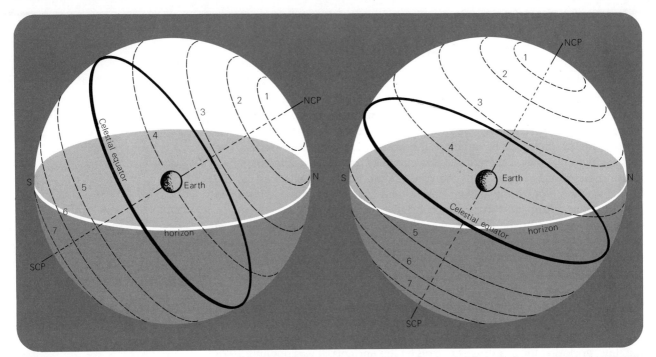

Figure 1-1
The celestial sphere, as it appears to observers at two different latitudes on earth. In the panel at left, the observer, whom we imagine to be standing on the little globe at center and looking up, is located at 30° north latitude (approximately the latitude of New Orleans). In the panel at right, he is located at 60° north latitude (approximately the latitude of Anchorage, Alaska, and Oslo, Norway). For the observer at 30°, only stars 1 and 2 are circumpolar, and never set. For the observer at 60°, star 3 is also circumpolar, and star 4 spends much more time above the horizon than it does for the observer at lower latitude. But the observer at 60° sees star 5 for a much briefer period of time than his southern counterpart, and can never see star 6 at all, though it rises and sets for the observer at 30°. All told, the observer at 60° sees less of the sky than the one at 30°. At the equator, all the heavens are visible; at each pole, only 50 percent.

It would be very convenient if we could describe positions precisely on the celestial sphere, just as we do on the surface of earth by means of latitude and longitude. The former is measured from the equator, which is geographically convenient; the latter from the Greenwich meridian, which was politically convenient in the heyday of British naval power. On the celestial sphere there are several alternative ways of choosing the principal circles from which positions can be measured. They are discussed in greater detail in the Appendix.

Figure 1–1 represents the celestial sphere as seen by a person at latitude 30° north, and another at latitude 60° north. The north celestial pole is 30° above the horizon for the first observer, and 60° above the horizon for the second. If we can locate the position of the celestial pole, we can find, not only the direction of north, but also the latitude of our location on earth. One way of doing this is to take a very long exposure photograph of the stars—a star trail (see page 2). The celestial pole is the point about which all the stars seem to revolve.

In the northern sky, there is a fairly bright star very close to the north celestial pole. This star, called Polaris, serves as a convenient guide for navigators, since it does not move. There is no conspicuous star near the south celestial pole, and so navigators in the southern hemisphere, who cannot see Polaris, have to make do with stars that move as the celestial sphere rotates. Hawaiian legends tell how their ancestors came from the south sea islands. They had much difficulty sailing in the southern hemisphere. Then they crossed the equator, and a star became visible above the northern horizon that did not move. They could navigate with ease.

Stars and Constellations

Early peoples, seeing the apparently unchanging pattern of stars in the sky, gave fanciful, mythological names to groups of stars that seemed to form some recognizable shape. These star pictures are called **constellations.** In the northern sky, for example, we have Orion, the hunter; Leo, the lion; Gemini, the twins; and many more. Different cultures, naturally, often gave different names to the same group of stars. Thus the constellation which we know as the Big Dipper was the Plow to the ancient Britons, and Ursa Major, the great bear, to the Greeks and Romans. Most of the constellations with which we are familiar today were named by the Greeks, though they themselves inherited many of them from earlier civilizations. (In the skies of the southern hemisphere — invisible from Greece — many constellations were not named until modern times. Consequently we have such odd constellations as Antilia, the air pump.)

As time went on, the concept of constellations gradually changed, until today we think of them not as groups of stars, but as areas of the sky. There are 88 modern constellations, and they cover the sky completely. Originally, the boundaries of even the modern constellations wandered rather aimlessly among the stars, enclosing the star groups named by the ancients. It was not until 1928 that the presently accepted boundaries were established. Modern sky maps generally do not show the old figures of men and beasts that the ancients associated with the constellations, but they do retain the old names.

We also retain the proper names of the bright stars that have come down to us from antiquity. Even in highly sophisticated scientific journals, these stars are referred to by their old names. Most of these are of Arabic origin, for the Arabs were the greatest astronomers of the middle ages, and preserved Greek astro-

The constellation Orion: left, as it appears in a long-exposure photograph (not all the stars are visible to the naked eye); right, in an old star atlas.

The modern boundaries of the constellation Orion, together with the names and designations of some of the more prominent stars. Many others are listed in various catalogs.

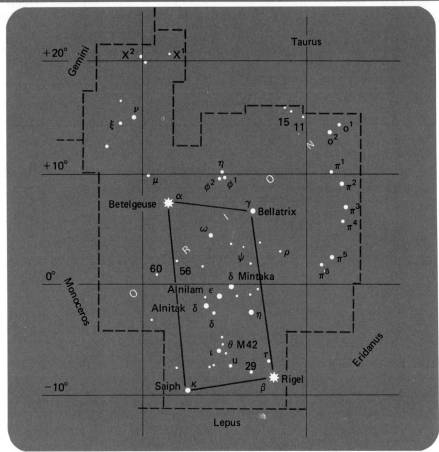

nomical knowledge at a time when Europe had largely lost this heritage. Thus the star that marks the tail of Cygnus, the swan, is called Deneb, from the Arabic *dhanab al-dajaj*—"tail of the hen." Betelgeuse, the bright red star marking the shoulder of Orion, means something like "the armpit of the central one" in Arabic. Others are Greek; Sirius comes from the Greek word for "scorching."

As the study of astronomy progressed, astronomers needed a simpler and more universal method of designating stars. In 1603, Johann Bayer suggested in his atlas a method of designating the brighter stars that is still in use. He gave Greek letters, in alphabetical order, to the stars of each constellation. Following the Greek letter he used the Latin possessive form of the name of the constellation. In general, he started with the brightest star, which he called alpha; the second brightest star was called beta, and so on.

As investigations in astronomy proceeded, it became obvious that the Greek alphabet was not large enough to provide designations for all the stars. Approximately a century after Bayer, the English astronomer John Flamsteed suggested a more extensive system. After the Greek letters had been exhausted, the fainter stars were numbered, moving across the constellations from east to west. This system included only stars that were visible to the naked eye. Today, large telescopes far beyond the dreams of ancient and even early modern astronomers provide photographs of literally millions of stars, all of which must be designated by numbers in various catalogs.

To an observer at either latitude in Figure 1–1, many stars will appear to rise and set. Stars on the celestial equator will remain visible above the horizon for half a sidereal day. Those north of it will remain visible longer, and those south of it for a shorter time. Some stars, close enough to the north celestial pole, will never set. The ancient Egyptians called these stars "the stars that know no destruction." Today we call them **circumpolar stars.** Which stars are circumpolar depends on the latitude of the observer. For the observer at 30° north, stars 1 and 2 will never set; for the observer at 60° north, star 3 will also be circumpolar, though the first observer will see it rise and set. A person at the equator will find no circumpolar stars; a person exactly at the pole will find that all the stars are circumpolar, so that none of them ever rise or set.

The Sun

The earth rotates once every 23 hours and 56 minutes. Why then is the day taken to be 24 hours long? It is because the interval between one noon (the time when the sun is at its highest point in the sky) and the next is 24 hours. The reason for the discrepancy between the day measured by the stars and the day measured by the sun is the earth's annual revolution about the sun in nearly circular orbit.

We can understand the effect of the earth's revolution if we consider an observer on earth for whom the sun is directly overhead at a particular moment, and then imagine what he will find 23 hours and 56 minutes later. In that time, the earth will have completed one rotation on its axis, and the observer will be oriented exactly the same way, with respect to the stars, as he was before. But the position of the earth with respect to the sun has changed. The earth has moved about 1° in its orbit around the sun, and thus has to rotate about 4 minutes longer to bring the sun exactly overhead again for our observer (Figure 1–2). In a year of 365 days, the earth actually rotates 366 times.

The sun seems to move among the fixed stars, rising 4 minutes later each day.

Because of the earth's revolution about the sun, the sun is not fixed on the celestial sphere, as the stars are. It does not share fully in the rotation of the stars, but lags slightly behind, moving about 1° eastward with respect to the stars each day. Thus from day to day and from month to month, the sun appears to lie in the direction of different stars (Figure 1–2). In a full year, it makes a complete revolution around the celestial sphere. The constellations through which the sun moves on its annual journey around the sky are the 12 signs, or houses, of the **zodiac.** When we say the sun is "in" a particular constellation, therefore, we are specifying its progress along this path.

The motion of the earth around the sun also changes the appearance of the night sky. Stars which cannot be seen in April, for example, because they are above the horizon during the daytime, will rise progressively earlier as the year goes by. In October, they will be rising at night, and be visible in the night sky. Thus some constellations can be seen in the summer and fall, and others in the winter and spring. Year after year, though, the same stars and constellations return, each in its proper season.

We do not ordinarily use the sidereal day for time-keeping, since we live according to a day defined by the sun. Everyday life is geared to when the sun rises and sets, rather than when a particular star crosses the horizon. Astronomers, however, are usually more interested in star-time than sun-time. Observatories have special clocks which keep sidereal time.

Seasons

Having examined the celestial sphere, we can easily understand why we have seasons on earth. The apparent path of the sun against the background of the fixed stars is known as the **ecliptic.** (The constellations of the zodiac thus straddle the ecliptic.) The plane of the ecliptic and the plane of the celestial equator are tilted at an angle of $23\frac{1}{2}°$ to each other. To put it another way, the earth's axis of rotation is not exactly perpendicular to the plane of the earth's orbit, but makes an angle of $66\frac{1}{2}°$ with it (Figure 1–3). As a result, in the course of a year, the sun's motion on the celestial sphere takes it from $23\frac{1}{2}°$ north of the celestial equator to $23\frac{1}{2}°$ south of the celestial equator, and back again. On March 21, it is on the celestial equator, going north. On September 23 it crosses the celestial equator going south. These are the times of the **vernal** (spring) **equinox** and **autumnal equinox,** respectively. "Equinox" means literally "equal night," and at the time of each equinox, the lengths of the day and night are equal everywhere on earth: each lasts exactly 12 hours.

At the spring and fall equinoxes, the sun crosses the celestial equator, and day and night are equal in length all over the world.

When the sun is at the northernmost point of its path, daytime in the northern hemisphere is long compared to night; the sun is higher in the sky, and its rays fall more vertically. The result is summer in the northern hemisphere. At these times, the sun is lowest in the southern skies, and the southern hemisphere experiences its winter. Thus the seasons of the year are reversed in the two hemispheres. Points at which the distance of

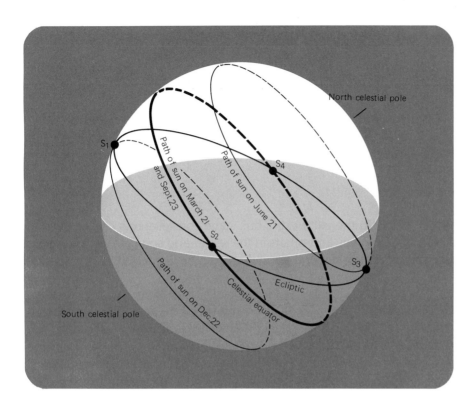

Figure 1-3
The sun's annual path around the celestial sphere, and its daily path through the heavens at four different times of the year. On December 22, when the sun is at the winter solstice (S_1), the day is short and the night long in the northern hemisphere. On June 21, when the sun is at the summer solstice (S_3), the opposite is true. At the spring and fall equinoxes (S_2 and S_4), day and night are equal in length.

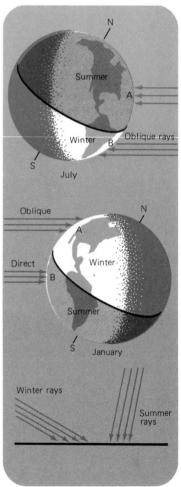

the sun from the equator is greatest occur midway between the vernal and autumnal equinoxes, at the **summer** and **winter solstices.** They fall about June 21 and December 21 each year.

If there were no inclination between the ecliptic and the celestial equator, the earth's position in relation to the sun would be that of a perpetual equinox, and we would enjoy an eternal spring. In actuality, though, the seasons are not quite so simple; the atmosphere, weather patterns, and the irregular distribution of land and oceans also play a role. Because of these factors, the hottest days of the year in the northern hemisphere lag behind June 21, and the coldest behind December 21.

The Calendar

The earth's revolution around the sun defines another set of relations with which we are all familiar—the calendar. A calendar is a device for keeping track of the passage of time by dividing it into convenient units based on the earth's rotation and revolution. It takes the earth 365.242 days to complete its orbit around the sun. As you might imagine, a calendar with 365 days would soon go out of step with nature. In order to bring the year back into accord, we have **leap years** in which one extra day is added to the February of every year which is divisible by 4. This is slight overcompensation because the earth's orbital period is 365.242

rather than 365.250. The mistake amounts to about 1 day only every 360 to 400 years. This is further corrected by making the final year of a century (i.e., 1600, 1700, 1800, 1900) a leap year only if it can be divided by 400. Thus 1600 was a leap year, whreas 1700 was not. This new calendar has an error of only 1 day every 3000 years.

The Moon

Another object that is not fixed among the stars is the moon. The moon travels around the earth in a nearly circular orbit with a radius of 384,400 km. On the celestial sphere, it seems to follow a path near the ecliptic, completing a full circle, so that it returns to the same position among the stars every $27\frac{1}{3}$ days. This is its period with respect to the stars, or sidereal period. In this time, however, the sun has moved. To return to the same position with respect to the sun, it must travel once around the celestial sphere, and then slightly more than two additional days to catch up. The entire process takes about $29\frac{1}{2}$ days, the moon's **synodic period.**

The synodic period is related to the phenomenon of the moon's **phases** (Figure 1–5). When the moon lies in the same direction in the

Figure 1-5 (right)
The phases of the moon. The view is from above the moon's orbit; the inserts show the appearance of the moon from earth at each point in its orbit.

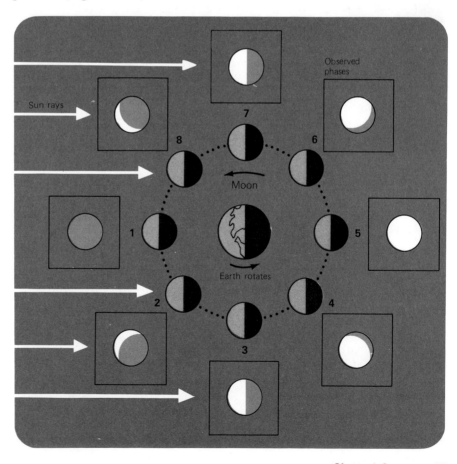

Figure 1-4 (left)
In July, the sun's rays are relatively direct in the northern hemisphere, which then experiences summer. In the southern hemisphere they fall at a steep angle, so that less radiation strikes each square meter of the earth's surface. It is then winter. In January, the situation is reversed.

Eclipses

Eclipses have long had a powerful impact on the human imagination. An eclipse of the sun can be one of the most awesome experiences of a lifetime — and, for the unprepared, one of the most terrifying. In 585 BC, it is recorded, a battle was raging between the armies of the Medes and the Lydians. In the middle of the fight, the sun began to disappear — an omen so powerful that the terrified soldiers on both sides laid down their arms, and a peace treaty was concluded on the spot. Yet many people find it equally astonishing that so dramatic an event can be predicted with such minute accuracy. Even in ancient times, astronomers had little trouble in forecasting eclipses, almost to the hour, many years in advance.

Eclipses of the sun are possible only through a fortunate accident of geometry. The sun, which is almost exactly 400 times larger than the moon, is also almost exactly 400 times farther away. As a result, the two bodies, seen from the

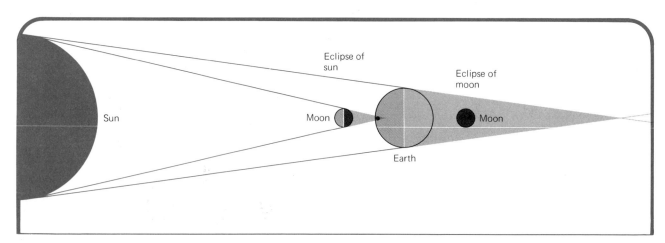

Sun

Eclipse of sun

Eclipse of moon

Moon

Moon

Earth

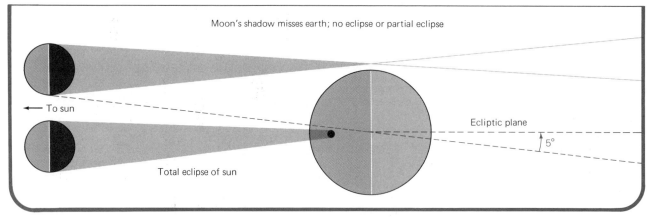

Moon's shadow misses earth; no eclipse or partial eclipse

To sun

Total eclipse of sun

Ecliptic plane

5°

earth, have practically identical angular sizes. When the moon passes between the earth and the sun — as it does once each month, at the time of the new moon — it may cause an eclipse of the sun by neatly covering the sun's disc for a few spectacular minutes.

If eclipses are produced so simply, why are they rare — so rare that you have probably never witnessed one? The odds of your having done so are in fact rather slim. For any given place on earth, in any one year, the chances of an eclipse being visible are about 1 in 360, on the average. If you don't travel much, you will probably pass your whole life without ever seeing a total solar eclipse.

There are several answers. For one thing, the plane of the moon's orbit is tilted about 5° to that of the earth's orbit. Thus on many of its revolutions, the moon passes above or below a line joining the earth and sun. Its full shadow misses the earth, producing no eclipse, or only a partial one. Then too, even when the moon is in position to cause a total eclipse — about once every 18 months, on the average — its shadow on the earth is never more than about 300 km wide. As the earth turns under it, the shadow sweeps across our planet's surface in a path that may be thousands of kilometers long. As a result, observers in different locations see the eclipse at different times. Nowhere, however, does the phase of totality last more than about 7 minutes.

If the moon were appreciably more distant from the earth, we would never see a total solar eclipse, for the moon would not cover the entire disc of the sun. As it is, variations in the earth-moon and earth-sun distance, resulting from the fact that neither the moon nor the earth moves in a perfectly circular orbit, sometimes produce an eclipse that is **annular,** rather than total. At these times, the moon is somewhat more distant from the earth than usual, and can cover only the central portion of the solar disc, leaving a bright ring or annulus.

Besides being beautiful, eclipses are useful to scientists. Before the invention of such devices as the coronograph, they provided the only opportunity for studying the sun's atmsopheric layers, and they still give astronomers their best chance to observe the inner corona. The bending of light rays in a gravitational field (Chapter 7) predicted by the theory of relativity, was confirmed during an eclipse in 1919. Sun-grazing comets, too, are often found during eclipses. When far from the sun, they are often too faint to be noticed. Near the sun, they are at their greatest brilliance, but except during solar eclipses, they are likely to be lost in the sun's glare.

An eclipse of the moon takes place when the moon moves into the earth's shadow, so that it is cut off from the sun's illumination. There are actually fewer lunar eclipses than solar, but they are much more likely to be observed. The moon may take up to 6 hours to complete its passage through the earth's shadow, and the event can be witnessed from every point on the earth's night side. Thus your chances of seeing a lunar eclipse are much better than your chances of seeing a solar eclipse; in any given year, they are greater than fifty percent.

During a lunar eclipse, the moon is not completely dark, but glows with a coppery-red color. The effect is caused by sunlight scattered through the earth's atmosphere. To an observer on the moon, the dark disc of the earth would seem to be surrounded by a thin, glowing, reddish ring.

sky as the sun, it is said to be new. At the time of the **new moon,** the sun illuminates the half of the moon that is turned away from us, and we see only the dark hemisphere. As the moon moves slowly eastward in the sky, away from the sun, it becomes a **waxing,** or increasing, **crescent.** When it is separated from the sun by 90°, half the face we see is bright and half is dark. It is then said to be at **first quarter.** When the moon reaches a position opposite that of the sun in the sky, so that it crosses the meridian (the north-south line that passes through the zenith point) exactly at midnight, we see the entire illuminated half. The moon is then said to be **full.** After the full moon, the moon **wanes** (decreases), going through the same sequence of phases in reverse until it reaches the new moon once again. These phases were first explained by Aristotle, about 350 BC. The period of the moon's phases was the basis of the calendar month, which has approximately the same number of days as the moon's synodic period.

The moon takes 27.3 days to complete its orbit about the earth, but 29.5 days to complete its cycle of phases.

As seen from the moon, the earth also shows phases. The phase of the earth from the moon is the complement of the phase of the moon seen from the earth; when the moon is new the earth is full, and vice-versa. An interesting situation occurs when the moon is in the crescent phase. As seen from the moon, the earth is nearly full and is shining down with a brightness 75 times that of the full moon. This lights up the part of the moon *not* illuminated by the sun, so we on earth see it faintly illuminated. This phenomenon is called "the old moon in the new moon's arms," and was explained by Leonardo da Vinci in the fifteenth century.

The Planets

Besides the sun and moon there are other apparently permanent objects that do not have fixed positions. These objects, which at first sight seem to be stars, move in complicated patterns over the celestial sphere. The ancient Greeks knew five of them, and called them "wanderers" or planets. Since then, three other planets have been discovered, and thousands of minor planets or asteroids have also been observed. In addition, the earth itself has been recognized to be a planet.

The motion of the planets on the celestial sphere is very complex, and elucidating the patterns was difficult for early astronomers. Most of the intricacies of planetary motion that perplexed astronomers of long ago can be easily understood once we realize that all our observations are being made from a moving body, the earth, which both rotates on its own axis and revolves about the sun. The modern view is that the sun is in the center of the planetary system, and that all the planets, including the earth, revolve about it in nearly circular orbits. Mercury and Venus occupy orbits smaller than the earth's and are called **inferior planets.**

Mars, Jupiter, Saturn, Uranus, Neptune, and Pluto, the **superior planets,** revolve in orbits larger than that of the earth. The larger the planet's orbit, the more slowly it moves, and the longer it takes to complete its trip around the sun.

The various laws of planetary motion are discussed in Chapters 5 and 12; here we will present only some useful terminology. When a planet appears nearest another object on the celestial sphere (for example, a star, another planet, or the sun), the two bodies are said to be in **conjunction.** When a planet is in conjunction with the sun, it is said to be in **superior conjunction** if it lies beyond the sun. This is possible for any planet. Inferior planets may also lie between the sun and the earth at times. This is known as **inferior conjunction.**

The angle between a planet and the sun, as seen from earth, is called **elongation.** We can easily see that the elongation of Mercury and Venus has to be less than 90°. In fact, Mercury can never be seen more than 28° from the sun, and Venus never more than 47°. When a planet lies in the opposite direction from the sun, it is said to be in **opposition.** Its elongation is then 180°. This can only happen for superior planets.

EXERCISES

1. A star rises at 10:00 PM on January 1. When will it rise on a) January 4? b) January 30? c) July 1?
2. If you wish to observe as much of the celestial sphere as possible, at what latitude should you live?
3. There are places on earth where the sun does not set for days on end. Where would you find such places? At what season would this phenomenon occur? Explain your answers.
4. Why is the solar day longer than the sidereal day?
5. Imagine a planet in an orbit like the earth's, but whose sidereal day is ¼ of a year long. How many sidereal days are there in its year? How many solar days? How long is each solar day?
6. Answer the questions in the previous problem assuming that the planet rotates in the direction opposite that of its revolution in orbit.
7. Why are planetary motions seen on the celestial sphere quite complex?
8. Why do we have seasons? How would they change if the inclination of the ecliptic to the celestial equator was 40°?
9. Explain how a "full earth" is visible from the moon at the time of the new moon on earth, and vice versa.
10. Is it possible for planets to show phases? Which ones might you expect to have them, and why?

THE STARS IN SPACE

Today we know a remarkable amount about the stars—more, probably than we know about the living cells that make up our own bodies, or the workings of our brains. A star is in many ways a relatively simple object in a very complicated universe. Yet as recently as the nineteenth century, philosophers could assert confidently that knowledge of the stars would remain forever out of reach. Fifty years ago we did not know how stars produce the energy that enables them to shine for millions or even billions of years. Sixty years ago we did not know whether there were any stars outside our own galaxy. A hundred and fifty years ago we had not determined the distance of even a single star. It had not yet been conclusively proven that the stars were suns like our own.

Indeed, from that moment of prehistory tens of thousands of years ago, when human beings first began to study the heavens, until the seventeenth century, practically nothing was learned about the true nature of the stars. The reason for this slow progress is simple. In order to formulate a scientific theory, and to test that theory once it has been formulated, evidence is needed. Since we cannot perform experiments with the stars, as biologists or chemists or psychologists can with the things they study, all our evidence must come from simple observation. Unfortunately, even with the most careful naked eye observation, the number of things we can learn about the stars is disappointingly small.

If you live in a place where the air is clear and the sky dark (that is, away from a city), and have the patience to look at the stars carefully night after night, you can easily put yourself in the place of the early astronomer. To appreciate the limitations of pretelescopic astronomy, try making a list of what you can observe. It will not be a very long list:

1) A large number of stars seem distributed at random over the face of the sky, which is also crossed by a broad hazy band (The Milky Way). Their number is hard to estimate; there are too many to count easily, in any case.

2) Faint stars are much more numerous than bright ones. No individual star, however, seems to vary in brightness from day to day or year to year. (There are a few exceptions, but they were apparently not noticed in ancient times.)

3) Though most stars appear white, some of the brighter ones are tinged with various colors. The brightest stars look larger than the faint ones, and twinkle more noticeably.

4) Most important of all, the stars seem never to change their positions with respect to each other. The same pattern appears unaltered night after night and year after year.

Particularly in this last respect, the appearance and behavior of the stars is markedly unlike that of the sun, moon, and the five planets visible to the naked eye. The stars seem to form a permanent, unchanging backdrop against which these seven bodies trace their intricate paths. The motions of these bodies had been studied by people of many civilizations since the dawn of history. Greek astronomers in particular had, by the second century BC, collected a very large body of careful observations. On the basis of this evidence they were able to propose various theories about the solar system, perform calculations, and make predictions which could then be checked against observation. From these efforts to explain the movements of the sun, moon, and planets, the first scientific astronomy emerged.

Though the models of the solar system devised by Greek astronomers were imperfect by the standards of our present knowledge, they were at least grounded in careful observation and measurement, and therefore represented steps toward truth. It was the absence of such concrete, quantitative data that hampered the attempt to learn about the stars. Most of the available evidence was purely negative: the stars did **not** move with respect to each other, they did **not** vary in brightness. With so little to build on, early astronomers could not formulate meaningful theories. They could only engage in speculation, inventing theories that could be neither supported nor refuted.

Without more evidence, the early astronomer could not interpret even the simple observations he could make. Were the stars that appeared brightest, for example, truly more luminous than the others? Or larger? Or were they merely closer to us? This is one of the most basic questions in astronomy, but there seemed no way to answer it. The as-

We can learn much less about the stars than about the sun, moon, and planets through naked-eye observation.

tronomer's problem is similar to that of a person lost in a forest at night. Ahead of him he glimpses a faint flicker of light. But how can he tell if it is a candle 10 meters away, a lighted window 100 meters away, or a bonfire a kilometer away?

This analogy suggests something about the kinds of evidence that astronomers needed to move beyond mere speculation. If the physical nature of the stars were known — candles, or bonfires, or suns — it would be possible to form some idea of their distance. Conversely, if their distance could be estimated, one could at least make reasonable hypotheses about their nature. In other words, knowing **where** the stars are would represent a great step towards learning **what** they are, and vice versa. But the earliest astronomers knew neither.

The Search for the Stars

The ancients never realized that one of the keys to the riddle of the stars was literally staring them in the face each day. We know today that our sun is a star, and much of our understanding of the stars has come from the study of the sun. Unfortunately for early astronomy, it required too great a leap of the imagination to identify the hot, brilliant, blazing body we see by day with the faint points of light visible in the night sky. In fairness to the astronomers of ancient times, there was no real evidence to point the way; the idea would have been merely another speculation. But there was probably another reason for their failure even to consider the possibility. The Greeks had an estimate for the distance of the sun from the earth. It was much too low — some 8 million kilometers, compared to the correct value of 150 million — but it was still a huge figure. If the stars were suns, they would obviously have to be almost inconceivably distant to appear as faint as they do. Accepting the stars as suns would have meant accepting a universe far larger than most Greek philosophers were willing to imagine.

Early astronomers did not realize that the stars are suns like our own.

The Greeks did, however, realize that the solar system might provide the key to the stars in a different way. By about 250 BC they had devised two major models of the solar system. Both were fairly successful in explaining the apparent motions of the sun, moon, and planets. We are not concerned here with the details of these rival systems, or the history of the 2000 year struggle between them, topics which are dealt with later in this book. One crucial difference between the two, however, had the greatest importance in the story of man's search for the stars. In the **geocentric theory,** popularized by the astronomer Ptolemy about the middle of the second century AD, the earth was considered to be fixed, while the other bodies of the solar system moved around it. The alternative theory, known as the **heliocentric theory** and associated with the name of Aristarchos, proposed that the sun was the unmoving center of the solar system. The other bodies, including the earth, were thought to travel in paths around the sun.

If the earth was immobile, as Ptolemy and his followers maintained, that fact would tell us nothing about the stars; they would remain as much a mystery as ever. But if the earth moved as Aristarchos claimed, the apparent positions of the stars should shift slightly as a result. Moreover, measurement of that shift could provide a way of calculating the distance of the stars.

PARALLAX **Parallax** means a change in the apparent direction of a distant object because of the observer's motion. The phenomenon of parallax is a constant part of our every day experience, though we do not usually stop to analyze it. We all know that when we move from one place to another, objects around us seem to change their positions. The amount of this shift decreases with the distance of the object. Looking out the window of a moving car, for example, we see nearby objects such as telephone poles along the highway seeming to streak by us, while a mountain range in the distance seems hardly to change until we have covered many miles.

You can demonstrate this effect for yourself without moving at all, for the human body has a remarkably efficient parallax mechanism of its own. Hold a pencil at arm's length and close your left eye. Observe the pencil's position against the far wall; then close your right eye and open your left at the same instant. Notice how the pencil seems to jump against the background. If you move the pencil closer to your eyes, the amount of the "jump" increases; if you move if farther away, the jump seems smaller. You can see from this experiment that the left eye and the right eye always record slightly different images because of their separation. The brain interprets this difference to determine the distance of various objects.

What the brain does automatically, a surveyor or astronomer can do by using some relatively simple geometry. If a surveyor wishes to measure the distance to a remote or inaccessible object (say a tree growing on the far rim of a distant canyon), he records its direction from two different locations; this gives him two angles of a triangle. He then measures the precise distance between the observation points; this distance becomes the **baseline** of the imaginary triangle (Figure 2–2). He can then draw a scale model of the triangle, and simply measure the distance on the drawing. In practice, though, a surveyor would most likely calculate the distance with the aid of trigonometric tables, which give the numerical relationships among the sides and angles for certain classes of triangles.

It is clear from these examples that the angular shift, or parallax, depends on two factors: the distance of the object being observed, and the length of the baseline. A distant object will generally have a small parallax, but that parallax can be made larger by increasing the baseline. Since small angles are difficult to measure accurately (especially with the

Figure 2-1
These two photographs were taken from points about 6 feet apart. In comparing them, notice how the child in the foreground seems to have moved with respect to the more distant trees in the background. This shift is called parallax, and it is entirely due to the change in the camera's position; the child's location was the same when both pictures were taken. The second child, farther away, shows a proportionally smaller parallax. (The trees are distant enough to have very little parallax, and can thus be used as a fixed frame of reference.)

The nearer an object, the more it seems to shift its position when seen from different locations. This effect is called parallax.

naked eye), a remote object must be observed from widely separated places in order to have a measurable parallax.

Greek astronomers were familiar with the geometry of parallax. They realized that if the earth completed a full revolution about the sun every 12 months, as Aristarchos claimed, an astronomer viewing the stars in January and July would be making his observations from points separated by the full diameter of the earth's orbit. This the Greeks estimated to be about 16 million kilometers—enough of a baseline, they reasoned, for some stellar parallax to be detectable, if the earth did indeed move.

With only the naked eye and relatively crude devices for the measurement of angles, Greek astronomers were nevertheless able to determine stellar positions to an accuracy of about 10′ (i.e. ⅙°). This is about ⅓ the apparent diameter of the full moon in the sky. But when they searched for stellar parallaxes, they could find none. Of course this did not rule out the possibility of the stars having parallaxes smaller than they could measure, but this the Greeks were very reluctant to consider. It is not hard to calculate how distant the stars would have to be to have par-

Figure 2-2
Surveyors make use of the geometry of parallax to find the distance of remote objects. The object is observed from two different points. Once the baseline and the two angles are known, the distance can be found from a scale drawing, or from mathematical tables.

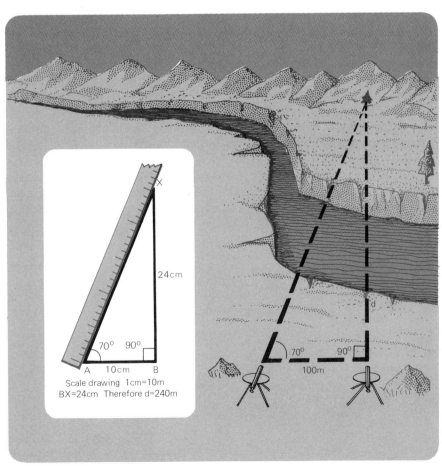

The Calculation of Parallax

The calculation of parallax can be done in several ways, but the basic principle is quite simple. And if the distance we are trying to measure is very great compared to the baseline being used — which is certainly the case for astronomical measurements — the arithmetic involved is also very simple. To understand the basis of parallax calculations, consider the family of triangles shown in Figure 2–A. Each has the same base, but their heights are different. It is obvious that the taller the triangle, the smaller is the angle at its peak (i.e. opposite the base).

For a triangle whose height is equal to its base, this angle is 45°. When the height is twice the base, the peak angle is 26½°, and when the height is four times the base, the peak angle is only 14°. A very important property of such triangles is the fact that, for small peak angles (less than about 5°), the size of the angle is very nearly inversely proportional to the height:base ratio. Thus if we know that a triangle with a 1° peak angle has a height 57.3 times its base, we can calculate the height:base ratio of a triangle with a peak angle of only ⅙° (10′) : $57.3 \times 6 = 343.8$, to a very close approximation.

A triangle with a 1° peak angle is too tall and skinny to be drawn to scale on this page; if its base were only an inch, its height would still be almost five feet. But the triangles used in astronomical calculations are far skinnier than this. In Figure 2–B, the earth is shown at two points in its orbit six months apart (B and D). The sun is at C, and the star whose parallax we are measuring is at A. Triangle CAB is the skinny triangle that is used to find the distance to the star. If we were to try and construct this drawing to scale, representing the distance between the earth and sun, BC, as one inch, the nearest star would have to be drawn over four miles away.

When observed from opposite sides of the earth's orbit, the star at A seems to shift its position against the more distant background stars. The angle of parallax is defined as half of this total shift. It can be shown by geometry that the total

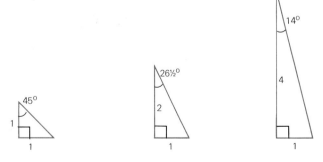

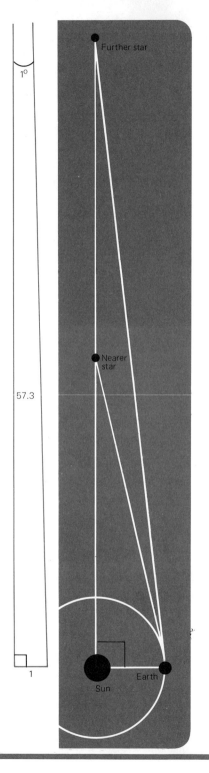

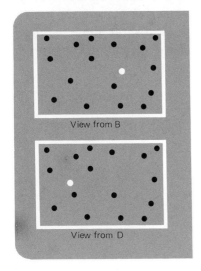

View from B

View from D

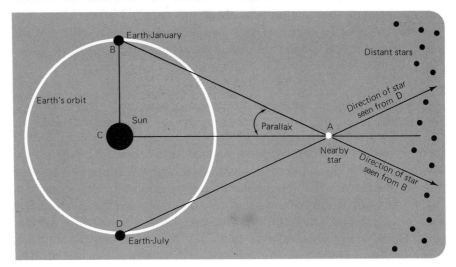

angular shift is equal to angle DAB, and that half of this, the star's parallax, is therefore equal to angle CAB. Thus careful measurement of the star's parallax gives us the peak angle of our triangle. The next thing to do is to look up, in a mathematical table, the height-to-base ratio of a triangle with this peak angle. This will tell us the relationship between the star's distance, AC, to the distance between the sun and earth, CB. If the latter is known, simple arithmetic will give us the distance we are looking for:

$$BC \times AC/BC = AC$$

As an example, let us repeat the calculation done by Greek mathematicians over 2000 years ago. They wished to find how far the stars must be to show no parallax as great as $10'$ of arc ($\frac{1}{6}'$). A triangle with a peak angle of $10'$ has a height about 344 times its base. If the stars had parallaxes smaller than $10'$, their distance had to be even greater than 344 times the distance of the earth from the sun. This distance the Greek astronomers did not know with any accuracy. The best estimate they had was about 8 million km. The nearest star, then, was at least

$$(3.44 \times 10^2) \times (8 \times 10^6) = 2.75 \times 10^9 \text{ km}$$

or nearly 3 billion km away.

This seemed an impossible distance to most early astronomers. In fact, however, the stars are much farther still. The largest known parallax belongs to one of the stars of the Alpha Centauri system. It is $.76''$, or $.0002°$. If we set angle CAB equal to his figure, the height of our triangle will be some 270,000 times its base. We know today that the base is not 8 million km, as the ancients thought, but about 150 million. Thus the distance to Alpha Centauri is

$$(2.7 \times 10^5) \times (1.5 \times 10^8) = 4.05 \times 10^{13}$$

—41 trillion km, or about 25 trillion miles.

allaxes smaller than 10′ (see Box). Mathematicians made the calculation, and the results were unbelievably large: nearly 3 billion kilometers! Thus for the next 17 centuries most people preferred to believe that the earth did not move around the sun. This seemed to them much more plausible than accepting the fact that the universe was so large.

The Greeks realized that if the earth moved, the stars should show some parallax. But they could find none with the naked eye.

The Quest for Stellar Parallax

But the earth **does** move about the sun. The heliocentric theory, out of favor for so many centuries, was revived by Nicolaus Copernicus in 1543. By the middle of the seventeenth century it was supported by such a mass of evidence that it could no longer be doubted. The absence of stellar parallax could therefore mean only one thing: the stars were in fact enormously distant, so distant that their parallaxes were too tiny to measure. By this time observational technique had improved considerably from the days of Greek astronomy. Angles in the heavens could be measured with ten times their former precision, to about 1′ of arc ($1/60°$), yet still no parallaxes could be found. In order to show no detectable parallaxes, the stars had to be at least ten times further away than the ancients had calculated, or about 30 billion kilometers. And this figure represented only the minimum possible distance; the actual distance of even the nearest star would not be known for another two centuries.

Once the heliocentric model of the solar system was accepted, and the implications of the failure to find stellar parallaxes understood, astronomers had their first real inkling of the true vastness of the cosmos. This knowledge led to the formulation of new theories about the stellar universe. Just a few decades after the publication of Copernicus' work one of his followers, the Englishman Thomas Digges, suggested that the stars were scattered throughout the infinite depths of space surrounding the solar system. Until that time, most people had imagined them as occupying a spherical shell outside the orbit of Saturn, so that all were roughly the same distance from earth. Even Copernicus himself had never abandoned that view.

Digges' speculation was carried further by the Italian philosopher Giordano Bruno. He conceived the notion that the stars were actually suns like our own. The idea had occurred to others during the previous century (it appears briefly in the journals of Leonardo Da Vinci, for example), but no one had made any attempt to popularize it. Bruno did not stop here; he made the remarkable imaginative leap of suggesting that the stars might have planets like our earth, populated by races of beings like ourselves. (These strikingly modern ideas were considered impious by the religious authorities of the day, and were among the heresies for which Bruno was burned at the stake in 1600).

Neither Digges' idea of an infinite space filled with stars nor Bruno's identification of the stars with the sun could be proven in the sixteenth

The Inverse Square Law

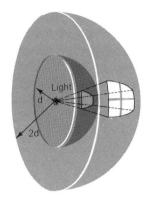

The inverse square law is basically an effect of geometry. To understand it we need know nothing about light itself beyond the fact that it travels in straight lines. Imagine a small lamp placed at the center of a large transparent globe. All the light from the lamp passes through the globe. If we add up the amount of light passing through each square meter of the globe's area each second, the total will be the amount of light emitted by the lamp per second. Suppose you look at the lamp by placing your eye very close to the surface of the globe. Only a fraction of the total light coming through the globe enters your eye. The size of this fraction is just the ratio of the area of your pupil to the total area of the globe.

Now imagine that a larger transparent globe is placed around the original one. Suppose that its radius is twice that of the smaller globe. Thus every point on its surface is twice as far from the lamp as the surface of the small globe. The area of a sphere increases with the square of its radius, so the larger globe has 2×2, or four times the area of the small one. Evidently all the lamp's light will also pass through the larger globe. The same amount of light, in other words, is spread out over an area four times larger than when it passed through the inner globe. Each square meter of the more distant surface must therefore be receiving only $1/4$ the amount of light that fell on a square meter of the surface closer to the lamp. If your eye were placed near the surface of the outer globe, and your pupil size didn't change, only $1/4$ as much light would enter your eye.

The intensity of light falls off inversely with the square of the distance.

century—nor for long afterwards. But they were beginning to seem more and more reasonable. If the stars were really as distant as men were coming to think, then they must be immensely bright to be seen at all.

This line of thought led some astronomers in the seventeenth century to try a new way of calculating stellar distances. Suppose, they reasoned, we assume that a bright star is in reality about as luminous as the sun. The distance of the sun was by then fairly well known. The way in which the brightness of light diminishes with distance was also known. This relationship had been clearly formulated by the great English physicist Isaac Newton, who had shown that the brightness of a luminous body varied inversely as the square of its distance from the observer. In other words, doubling the distance between observer and light source causes the light intensity to drop to $1/(2 \times 2)$, or $1/4$ of its previous value. If you triple the distance, the light will be nine times fainter, and so on (see Box).

By comparing the apparent brightness of a star and the sun, then, it should be possible to determine how much further away one was than the other. The method is entirely sound in theory, and is in fact used today. But when it was first tried it did not meet with success. At the time there were no reliable instruments for measuring brightness. The comparison of the brilliant sun, seen in daytime, with a star seen at night many hours earlier or later, could not be carried out accurately enough using only the naked eye to give meaningful results.

In the meantime, the search for stellar parallaxes went on. Galileo had been the first man to study the heavens through a telescope (in 1609), and throughout the seventeenth and eighteenth century larger and better telescopes were coming into use. The telescope made it possible to measure far smaller angles in the heavens than an observer could ever hope to detect with the naked eye. Moreover, the telescope soon helped astronomers to determine that the baseline they were using in their search for stellar parallax—the diameter of the earth's orbit—was actually far larger than had been realized in ancient times. In 1671 a team of investigators obtained a fairly accurate figure for the distance of the planet Mars by telescopic measurement of its parallax from two different observatories on earth. (Mars is close enough so that a baseline of a few thousand miles on the earth's surface is sufficient to give a measurable parallax.) From this figure they were able to calculate the dimensions of the solar system. They found that the distance of the sun from the earth —the **Astronomical Unit** (AU) in modern terminology—was not some 8 million km, as the best Greek estimates had suggested but about 140 million km. (This was a reasonably accurate result, only about 7 percent lower than the accepted modern value.) It seemed merely a question of time and technology before the long-sought stellar parallaxes were detected.

As telescopes of better quality and larger size became available, astronomers redoubled their efforts in the search for stellar parallaxes. Success finally came to several men almost simultaneously in the 1830s. The first to publish his results was Friedrich Bessel, who determined the parallax of the faint star 61 Cygni in 1838. At nearly the same time the Englishman Thomas Henderson, working at the Cape of Good Hope in South Africa, measured the parallax of Alpha Centauri, a bright star of the southern hemisphere.

Stellar Distances

The parallax of Alpha Centauri is only about $3/4''$ of arc, or $1/4800$ of a degree—about the angular size of a golf ball seen from 7 miles away. Yet this is the *largest* parallax ever recorded—which means that Alpha Centauri is the star nearest to us, other than our own sun. Using the method shown on p. 27, we can compute its distance. It comes out to about 41 trillion kilometers (4.1×10^{13} km), or 25 trillion miles. We can see from this figure that the mile and the kilometer are not very practical units for measuring astronomical distances. Instead astronomers generally use either the light-year or the parsec.

The **light-year** (ly) is the distance travelled by light through a vacuum in one year. Since light travels at 3×10^5 km/sec, a light-year turns out to be 9.5×10^{12} km. Alpha Centauri can therefore be said to be about 4.3 light-years from us. This figure means that, if anything drastic were to happen to Alpha Centauri—if it exploded, for instance—we would not even learn about it for some 52 months, for it would take that long for light from the castastrophe to reach us. Nor is there any other way the news could reach us more quickly, for nothing can travel more rapidly than light. The light-year is very useful for expressing distances of remote objects such as other galaxies. It tells us how long the light that we see has taken to reach us, and thus exactly how up to date is the information it is bringing.

The **parsec** (pc) is defined as the distance at which a star will appear to have an annual parallax of 1″ of arc ($\frac{1}{3600}°$). Look again at the diagram on p. 26. A long triangle whose apex angle is 1″ has a height 206,265 times its base. Since the base is the distance of the earth from the sun, 1.5×10^8 km, the height of the triangle, corresponding to the distance of the star, is $(206,265) \times (1.5 \times 10^8)$ km. This is about 31 trillion km, or roughly 3¼ light-years.

1 parsec = 3¼ light-years
  = 19 trillion miles
  = 31 trillion km

We have seen that the more distant a star, the smaller its parallax. If one star is three times as far as another, it will have exactly ⅓ the parallax. Another way of saying this is that the parallax of a star is inversely proportional to its distance. Defining the parsec in the way stated above makes it possible to express this relationship in a very simple mathematical form:

$$D_{(\text{parsecs})} = \frac{1}{\pi_{('')}}$$

where D is the star's distance and π (the Greek letter pi) is its observed parallax in seconds of arc. This enables astronomers to convert parallax measurements directly to distance measurements without complicated calculations. The parsec is therefore the most convenient unit for expressing distances to the stars.

The distance of a star, in parsecs, is the reciprocal of its parallax expressed in seconds of arc.

Even with the largest telescopes, stellar parallaxes cannot be measured with an accuracy greater than about .005″. Attaining that degree of precision is itself a remarkable achievement, for .005″ is roughly the angular size of an average bacterium (diameter $\frac{1}{10,000}$ of an inch) seen from a distance of 30 ft. If a star has a parallax of .01″ its distance will be $\frac{1}{.01}$, or 100 pc. However, due to the uncertainties of measurement, the real parallax must be considered to be .010 ± .005″. In other words, the true value could be anywhere between .005″ and .015″. This means that the distance lies between $\frac{1}{.005} = 200$ pc and $\frac{1}{.015} = 67$ pc, a rather large uncertainty.

Thus parallax measurements are not very useful for determining stellar distances beyond about 40 pc or so. But the importance of parallax as a distance indicator extends, indirectly, far beyond that range. There are various other methods of determining astronomical distances, some of which are described in later chapters. But many of them are calibrated by using parallactic distances to nearby objects. Parallax is thus one of the foundations of our scale of astronomical distances. It is the ruler that we use to measure our yardsticks.

Knowledge of stellar distances is our key to the astronomical universe. Once we know the distance of a star we can begin to learn many other things about it. Two of its most important properties follow almost immediately: its luminosity, and its velocity across our line of sight.

Stellar Motions

More than a century before the first stellar parallax was determined, evidence had been found that the "fixed" stars did in fact move—though very slowly by human standards. The English astronomer Edmond Halley made the discovery in 1718 while comparing the positions of a number of stars in the sky with their positions as recorded on Greek star maps dating from the second century BC. He found that at least three of the brightest stars seemed to have changed their positions significantly with respect to their neighbors over the intervening 1800 years. The discrepancy was too great to be considered an error on the part of the ancient observers. The star Arcturus, for example, one of the most prominent stars in the Spring sky, had moved by more than a degree—twice the diameter of the full moon. The Greek astronomers could not have made a mistake of that size. It seemed much more reasonable to conclude that the stars did move—but too slowly to be seen with the naked eye in the course of a human lifetime.

The change in position of a star in the sky with time is called its **proper motion.** It is important not to confuse proper motion and parallax. Parallax is entirely the result of the earth's yearly orbital motion about the sun. It does not represent an actual motion of the star in space. The apparent motion caused by parallax retraces the same path over and over again each year, back and forth along a short line or in a tiny ellipse. Proper motion is the result of a star's velocity with respect to our sun. Though it may be small, over the years it adds up; that is why Halley was able to make his discovery. With large modern telescopes, it is no longer necessary to wait centuries to measure proper motions; many can be measured over the course of a few years or even months. The largest known proper motion (10.5"/yr) belongs to the inconspicuous star known as Barnard's star (after the American astronomer who discovered it), and there are several dozen others with known proper motions of more than 1"/yr. Arcturus, for example, has a proper motion of 2.28"/yr. Thus over the 18 centuries between the time the Greek star maps were made and the time Arcturus was observed by Halley, it had moved 2.28"/yr × 1800 yr = 4101" = 68.4' = 1° 8.4'.

The proper motion of a star is related to its velocity relative to our sun, but two other factors determine the exact amount of the proper mo-

Figure 2-3
The proper motion of Barnard's star is evident on these photographs taken 22 years apart. The angular motion shown is about $\frac{1}{16}°$—about $\frac{1}{8}$ the diameter of the full moon.

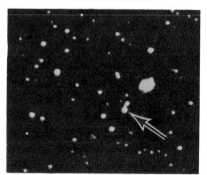

1894

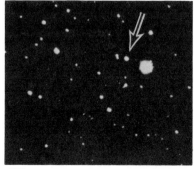

1916

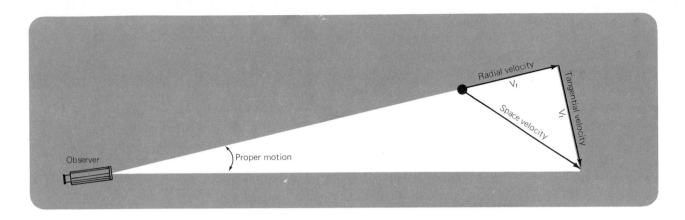

Figure 2-4
Tangential velocity, v_t, is a star's velocity across our line of sight. If we know the star's distance, we can find v_t easily from its proper motion—its angular movement across the sky. But the star's space velocity, its "real" velocity with respect to ourselves, cannot be found unless we also know its radial velocity, v_r, toward or away from us. This can only be determined from study of the star's spectrum.

The stars are too distant for their proper motions to be visible to the naked eye in the course of a human lifetime.

tion we observe. One is the direction of the star's true spatial motion with respect to us. Only in rare cases will a star be moving straight across our field of view, perpendicular to our line of sight. Most stars will be moving at some angle to our line of sight. For convenience we can divide the star's space velocity, into two **components:** its motion toward or away from us, called the **radial velocity** (V_r), and its motion across our line of sight, called the **tangential velocity** (V_t). These two can be represented by arrows lying at right angles to each other as in Figure 2–4. The length of each arrow is proportional to the velocity it represents. Thus if V_t is 60 km/sec and V_r is 36 km/sec, we might find it convenient to represent V_t by an arrow 60 mm in length and V_r by one 36 mm in length. We can add these together to get the true motion of the star, **V,** but in doing so we must take into consideration the fact that they are not in the same direction. They must therefore be added in space.

A quantity that has both magnitude and direction, such as the distance between two points, or the velocity of a body, is called a **vector.** We can see from the examples already given that when the components of a vector are known, its magnitude and direction are determined. They can be found graphically, simply by drawing the components on a scale diagram and measuring the size and angle of the resultant. They can also be found mathematically. The square of the magnitude of the vector is equal to the sum of the squares of its components. For a star, $V^2 = V_r^2 + V_t^2$; thus in the example originally given, $V^2 = 60^2 + 36^2 = 4896$; $V \simeq 70$. The exact direction of the vector can also be found using trigonometric tables. Even without such tables, however, we can solve certain simple cases. For example, if V_t and V_r are equal, then the angle the space velocity **V** makes with our line of sight is 45° (Fig. 2–5).

From these diagrams it is evident that what we see as proper motion is the result only of the tangential component, V_t, of the star's velocity with respect to us. A star moving directly toward or away from us would have no tangential component, and exhibit no proper motion, no matter how fast it was moving. (Similarly, people are sometimes told to fear a tornado only if it appears to be stationary. If it shows no motion across your line of sight, it must be moving either directly away from you or directly **towards** you.) So we cannot determine either the magnitude or direction of a star's true space velocity without knowing its radial velocity (toward or away from us). We shall see in the following chapter, however, that there is a very convenient way of measuring the radial velocity of most stars by analyzing their spectra. Thus we can for the moment consider the radial velocity to be, in principle, a known quantity.

To measure a star's true tangential velocity, we must know one other fact: its distance. A bird 500 feet overhead and a jet plane a mile overhead may seem to cross a particular region of sky in the same time; yet we know that the plane is in reality moving at several hundred miles per hour, while the bird is flying at only a fraction of that speed. The more distant an object, the more slowly it seems to move across our field of view. Two stars with the same tangential velocity will exhibit proper motions that are inversely proportional to their distance. If we were three times as far from Barnard's star, for example, we would find its proper motion to be only 3.5″/yr. Thus it is not surprising that the most distant stars have proper motions too small to measure even over very long periods of time.

In one respect this fact is actually helpful to the astronomer. A photograph of almost any region of the sky will contain the images of thousands of stars. If all of them had large proper motions, comparing photographs taken months or years apart would be like comparing two aerial photos of the New Year's Eve crowd in Times Square half an hour apart. Many of the same people would be visible in both pictures, but they would have changed their positions enough to make identification of any one person difficult. When an astronomer compares two photographs of the same region taken at different times, however, he can expect that most of the stars will not have moved at all. They will form a fixed pattern which he can use for reference in measuring any proper motions that do show up. Very distant stars are used as reference points for the measurement of parallax in exactly the same way.

Even so, however, finding the few stars that have moved slightly in a large star field would be almost impossible without an instrument known as the **blink microscope** or blink comparator. This is a device which allows an observer to view two different photos of the same area of the sky in rapid succession. If any star has changed position between the time when the first photograph was taken and the time of the second, it will appear to jump back and forth when the pictures are flashed in alternation. The principle is exactly the same as that of the motion picture; when two similar images are projected in rapid succession, the eye interprets the change as motion. Using a blink microscope is exactly like viewing two frames of a movie, photographed days or months apart instead of 1/24 of a second apart as in an ordinary motion picture.

To find the space velocity of a star, we need to know its proper motion, its distance, and its radial velocity.

Figure 2-5
Some famous right triangles. The sum of the squares of the short sides always equals the square of the hypotenuse (the side opposite the right angle).

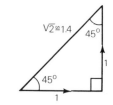

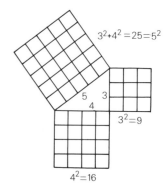

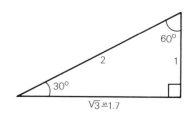

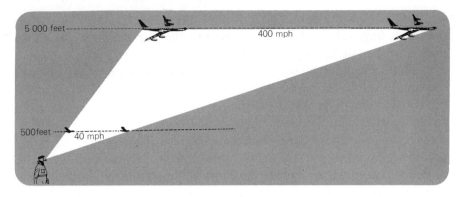

Figure 2-6
When we measure proper motion, we are measuring an angle. The size of that angle depends not only on the velocity of the moving object, but also on its distance. If the airplane in this figure were flying at the same speed as the bird, its proper motion would be much smaller, because it is more distant. (The drawing is not to scale.)

Figure 2-7
A blink microscope, used for comparing 2 photographs of the same region taken at different times. When the photos are viewed in rapid succession, any object that has changed its position appears to jump, and any object that has changed in brightness appears to flicker.

Once the proper motion of a star has been detected and measured, a knowledge of its distance enables us to calculate its tangential velocity. If we also know the star's radial velocity from study of its spectrum, we can determine both the magnitude and direction of its velocity.

Frames of Reference   So far we have been discussing the motions of other stars with respect to the sun. We have taken the sun as a fixed point of reference. There is no reason why we cannot do this, and it is convenient for many purposes. It would seem logical to assume, however, that our sun has its own motion through space like any other star, and our measurements of the velocities of the stars around us reflect in part our own motion. We may see a particular star approaching us at 40 km/sec, for example; is there any way of knowing how much of that velocity is ours and how much is the other star's? It is tempting to think that if we knew the sun's motion through space we could allow for it in our measurements of stellar velocity and so find the "true" velocity of each star.

But is it possible to measure motion in empty space? This is a basic question in physics. The answer, implied by Newtonian mechanics and confirmed by Einstein's theory of relativity, is no. There is no such thing as absolute motion or absolute rest for bodies moving in straight lines with uniform velocity. We can speak meaningfully only about **relative** motions. We can say, for example, that the relative velocity of Arcturus with respect to the sun, or the sun with respect to Arcturus, is about 120 km/sec. This means that if we take the sun as our fixed frame of reference, we can consider Arcturus to be moving at that velocity. But we can just as legitimately take Arcturus to define our frame of reference, and say that the sun is moving at that velocity. We cannot say which of these statements is "really" true. Both are equally true, equally valid. And it would be just as valid to use any other star as our standard of "rest."

In studying the stars, it is convenient to be able to calculate their motions with respect to our stellar neighborhood as a whole. This means defining a frame of "rest" so that the average motion of all the stars in our vicinity is zero. We can then measure the motion of any individual star — including the sun itself — with respect to this frame. In order to do this we must determine the sun's motion, not with respect to absolute space (which is impossible), but with respect to the population of stars that makes up our stellar neighborhood.

This can be done if we know the radial and tangential velocities of the stars in our vicinity. You have probably noticed how, when you look out the window of a rapidly moving car or train, objects ahead seem to move apart as you approach them, while objects to the rear seem to move closer together as you draw away. Similarly, the stars in the direction towards which the sun is moving will appear to be spreading out, and those lying in the opposite direction will appear to be drifting closer together. Thus if we look at the proper motions of a large number of stars, we should find that on the average they tend to converge towards one point in the heavens, and diverge from the diametrically opposite point. The point of divergence is the point in space towards which the sun is moving.

The sun's motion can also be deduced from studying the radial velocities of neighboring stars. Stars ahead of us tend to have radial velocities towards us, and stars behind us tend to be receding. The average radial velocity in each case is simply the sun's own velocity with respect to its neighbors as a group. If we observe a large number of stars at right angles to the sun's path, they should have, on the average, no net radial velocity towards or away from us. They should, though, exhibit the largest tangential velocities, in the direction opposite that of the sun's motion. The average magnitude of this velocity will again be that of the sun through the stellar neighborhood.

All of these observations have been carefully made, and the results indicate that the sun is moving towards a point in the constellation Hercules, not far from the bright summer star Vega, at a rate of about 20 km/sec. This is fairly typical for our part of the galaxy, where the average velocity is about 25 km/sec. There are exceptions, however; one star less than 4 pc away is moving at 292 km/sec. Such "high-velocity stars" belong to a special class which will be discussed in a later chapter.

What of the motion of the sun about the center of the galaxy, which was described in Chapter 1? We could, of course, use the galactic center as our frame of reference, but the results would not be very convenient. For one thing, it is not easy to measure velocities with respect to the center of the galaxy. In fact, we cannot even see the center of the galaxy with an ordinary optical telescope. More importantly, most of the stars in our neighborhood are moving in relatively similar orbits at high speeds (roughly 250 km/sec) about the galactic center. It is the small differences in the speeds and directions of those paths that give rise to the stellar motions we observe in our neighborhood.

It is much easier to deal only with these smaller relative motions, and ignore the larger movements around the galactic center that all the stars share. To take an analogous situation, suppose you are waiting for dinner

There is no such thing as absolute velocity with respect to empty space. Motions can only be measured with respect to some physical frame of reference.

Figure 2-8
When we look forward from a moving train, we see trees and telephone poles along the track seeming to move apart as we approach. Looking backward, we see them appear to move together. This principle is used to determine the direction of the sun's motion among the nearby stars.

to be served in the lounge of a 747 jetliner traveling toward Chicago at 600 mph. You observe the stewardess coming toward you. Which is a more useful statement to you—that she is approaching you at 3 ft/sec, or that she is approaching Chicago backward at 598 mi/hr? Both statements are equally true, but it is certainly more convenient for most purposes to refer motions to ourselves or to our immediate environment.

Measuring Stellar Brightness

One of the most important physical characteristics of any star is its **luminosity,** or its rate of radiation of electromagnetic energy. We have already seen that the inverse square law tells us how the brightness of a star varies with our distance from it. If the distance of the star is known and its apparent brightness can be measured, it is then easy to calculate its true luminosity.

Measuring the apparent brightness of a star using only the naked eye is quite difficult, however. The first attempt to do this was made by the Greek astronomer Hipparchos about 150 BC. He charted the positions of nearly 1000 stars, and divided them into six classes on the basis of their brightness. The brightest stars he designated first magnitude; the faintest that could be seen with the naked eye, sixth magnitude. Astronomers today still use Hipparchos' terminology, which is why the brightest stars are referred to as ''first magnitude stars.'' But Hipparchos' estimates of the relative brightnesses corresponding to his six classes were badly in error, and the classes have been redefined for our modern magnitude system (Appendix 4).

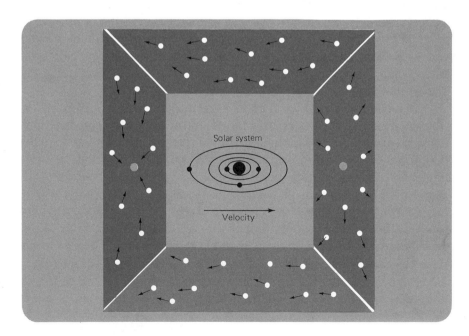

Figure 2-9
By observing the average radial and tangential velocities of the stars around us, we can determine the sun's velocity with respect to the stellar neighborhood.

The first successful attempt to measure the brightness of various stars was carried out during the eighteenth century by the English astronomer William Herschel (1738-1822). Herschel fitted his telescope with an adjustable diaphragm, like that of a camera or the iris of the human eye. This allowed him to vary the light-gathering area of the instrument. Each time a star was observed, the telescope aperture was readjusted so that all the stars were reduced to an identical level of brightness: just barely visible through the telescope.

The brighter the star, the more the telescope aperture had to be closed down to achieve this common level. From the relative size of the aperture used to observe each star, the brightness ratios of the various stars could be determined. Suppose, for example, that two stars appear equal in brightness when one is viewed with the telescope aperture adjusted to 7 inches and the second with an aperture of 5 inches. The light-gathering area of the telescope is proportional to the square of the diameter of the aperture. Therefore the ratio of the two areas is $7^2 : 5^2 = 49 : 25$, or nearly two to one. Thus the second star, to appear as bright as the first with only half the light-collecting area in use, must actually be twice as bright.

Brightness can be measured by varying the light-gathering area of an optical instrument with an adjustable iris diaphragm.

Today more accurate methods can be used. If two stars are photographed on the same film under the same conditions and using the same exposure, the brighter of the two will produce a larger and darker image. The two images can be compared by eye, or measured with greater precision by an instrument called a densitometer. The advantage of this method is that photographic plates are inexpensive and can record the images of many stars at the same time. The most sensitive measurements are done with electronic devices which translate the light of the star directly into an electric current that can be measured with extreme accuracy. A series of stars whose brightnesses have been carefully determined and do not vary are generally used as standards of reference. The brightness of any other star is measured in relation to one or more of these known standards.

Once the apparent brightness of a star and its distance are known, its true luminosity can be calculated by use of the inverse square law. One convenient way of expressing the true luminosity of stars is in terms of the brightness they would appear to have if observed from a standard distance: 10 pc. Suppose, for example, that our sun were moved to a distance of 10 pc from earth. It would then be about two million (2×10^6) times further away than it is at present. Its brightness would be diminished by the square of that number: $(2 \times 10^6)^2$, or 4×10^{12}. The sun would therefore be about four trillion times fainter. It would appear to the naked eye as a dim star, not easily visible except on a clear, dark night.

If the sun were 10 pc away, it would seem about 160 times fainter than the bright southern star Canopus. Canopus, however, is about

Knowing the brightness of a star and its distance, we can find its true luminosity using the inverse square law.

Star	Parallax ('')	Distance (pc)	Proper Motion (''/yr)	Velocity (km/sec)	Luminosity (sun=100)
Proxima Centauri	.763	1.31	3.86	—	.005
α Centauri A	.741	1.35	3.68	34	145.
α Centauri B	.741	1.35	3.68	31	40.
Barnard's Star	.552	1.81	10.34	139	.044
Wolf 359	.426	2.35	4.70	55	.00174
Lalande 21185	.397	2.52	4.78	103	.53
Sirius A	.377	2.65	1.33	19	2300.
Sirius B	.377	2.65	1.33	—	.19
Luyten 726-8 A	.368	2.72	3.36	53	.0064
Luyten 726–8 B	.368	2.72	3.36	—	.004
Ross 154	.345	2.9	0.72	11	.04

Table 2-1
The eight stellar systems within 3 pc of the sun.

30 pc from us. To compare its true luminosity with that of the sun we must allow for its greater distance. If Canopus were seen from the standard reference distance of 10 pc, three times closer than it actually is, it would appear 3^2 or 9 times brighter. It would then be $160 \times 9 = 1440$ times as bright as the sun. This is the true ratio of the luminosities of these two stars. Several of the brighter stars in our night sky are in reality about 50,000 times as luminous as the sun.

Our Neighborhood

Our sun's nearest neighbor is about 1.3 parsecs away — a typical figure for this region of the galaxy.

We can get a good idea of the population of a typical stellar neighborhood if we look at the stars within 3 pc of the sun. In this region, whose volume is 113 cubic parsecs, there are 12 known stars. (There may be additional very faint ones as yet undetected.) Not all of these are solitary, however. The majority belong to multiple star systems, groups of two or more stars that travel through space together, bound by their mutual gravitational forces. In fact the closest "star" to us, Alpha Centauri, is actually a triple system. The nearest of the three stars is not the bright, sun-like Alpha A, nor the slightly less bright, sun-like Alpha B, but the dim third member of the family known as Proxima Centauri. In addition to this triple system, there are two double stars in our 3 pc neighborhood. Some of the properties of these eight stellar systems are listed in the accompanying table. Notice that most of them are considerably less luminous than the sun. The density of stars in this vicinity is about .11 stars per cubic parsec, or .07 systems per cubic parsec. Nearest neighboring stars tend to be about 1.3 pc apart, just like the sun and its nearest neighbor Alpha Centauri. The average velocity is about 50 km/ sec (in this respect our neighborhood is somewhat atypical, since the usual velocity in this part of the galaxy is only about half that figure).

As a result of these random stellar motions an observer on earth 100,000 years in the future will see many of the same stars, but the patterns they form in the skies will be very different. The familiar constellations of today will be distorted beyond recognition.

Since the stars are moving about in random directions through space, do they ever collide? We cannot say it is absolutely impossible, but the odds are enormously against such an event. The stars are very tiny in relation to the huge volumes of space which separate them. It has been estimated that they occupy no more than 10^{-21} of the space of our galaxy. The probability of two stars colliding is thus extremely low. If there were only two fish swimming at random through all the oceans of the earth, they would be more likely to collide than two stars in our galaxy.

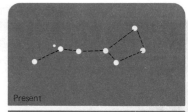

Present

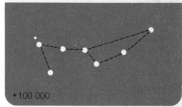

+100 000

SUMMARY

In ancient times, astronomers could count the stars visible to the naked eye, record their positions, and estimate their apparent brightnesses. But they could learn virtually nothing of their distances or their true nature. The evidence available from observation was hardly sufficient even to formulate plausible theories.

Greek astronomers had devised two chief models of the solar system. In the model perfected by Ptolemy, the earth was considered to be the unmoving center of the solar system. According to Aristarchos' theory, however, the earth was in motion, and completed a journey around the sun each year. The Greeks realized that if this were true, the stars might be expected to change their apparent positions slightly as a result of the earth's motion. This effect is known as parallax. The greater the star's distance, the smaller should be its parallax. No stellar parallaxes were observed by the astronomers of that age, who took this fact as evidence that the earth did not move. Actually, stellar parallaxes do exist, but they are far too small to detect without a telescope.

Astronomical distances are generally expressed either in parsecs (pc) or light-years (ly). A light-year, the distance travelled by light in a vacuum in one year, is equal to 9.5×10^{12} km. When distance of an object is given in light-years, the figure tells us how long the light we see from that object has taken to reach the earth. This is very useful in certain branches of astronomy. The parsec is convenient in a different way. It is defined as the distance at which a star will exhibit a yearly parallax of $1''$. Since the parallax of a star is inversely proportional to its distance, defining the parsec in this way enables us to use the simple formula: $D_{(pc)} = 1/\pi_{('')}$ for converting parallax measurements (π) into stellar distances (D). One parsec equals about $3\frac{1}{4}$ ly, or some 31 trillion km.

The term "fixed stars" reflects the long-held belief that the stars do not move with respect to each other. But in 1718 Edmond Halley found that several bright stars apparently had changed their positions over the previous 1800 years. The motion of a star across our line of sight is called its proper motion. Proper motions are very small, but unlike annual parallaxes, they add up over time, and can be measured in observations made years or even months apart with a large telescope. A star's velocity through space can be thought of as consisting of two components: its tangential velocity (V_t), across our line of sight, and its radial velocity (V_r), towards or away from us. If we know a star's distance, we can find its tangential velocity simply from its observed proper motion. Its radial velocity, however, must be found from spectroscopic measurements, described in the next chapter. Once both components are known, the speed and direction of the star's actual motion can be found.

There is no such thing as absolute motion or absolute rest. In order to give a meaningful statement of velocity, it is necessary to specify the frame of reference against which it is being measured. Stellar motions are generally measured with respect either to the sun, or to our own stellar neighborhood as a whole, defined so that the *average* velocity of the stars in this vicinity is zero.

If the distance of a star is known, we can find its luminosity (the rate at which it emits radiant energy into space) by measuring its apparent brightness and applying the inverse square law, which states that the intensity of light from a source falls off as the square of the observer's distance from the source. The luminosity of a star is generally expressed in terms of the apparent brightness the star would have if seen at the standard distance of 10 pc.

EXERCISES

1. What can you learn about the stars by observation with the naked eye? How many stars do you think you can see on a clear night? Since it is obvious that the stars rise and set, what is meant by the term "fixed stars"?
2. Can we calculate the distances of the stars by observing how bright they appear? What are the difficulties involved in this method?
3. Give some examples of the role played by parallax in every day life. Can you name any optical devices that make use of this principle?
4. Why did the Greeks fail to determine stellar distances by the method of parallax? Did any factors besides observation enter into their conclusions?
5. What is a parsec? Why is it a convenient unit for astronomers?
6. Find the distance of a star whose parallax is .02″ in parsecs, kilometers, and light-years.
7. Two stars have parallaxes of .1″ and .005″, respectively. Which star is closer to us, and how much? If the stars are equally luminous, how much brighter will the nearer one appear than the farther one?
8. Star A has a parallax of .2″ and star B has a parallax of .04″. Star B appears 3 times as bright as star A. Which of the stars is more luminous, and by how much?
9. The radial velocity of a star is 5 km/sec and its tangential velocity is 12 km/sec. What is the space velocity of the star?
10. Why is it easier to detect proper motion than parallax? What else do you have to know besides its proper motion in order to find the tangential velocity of a star?
11. A star at a distance of 100 pc has a radial velocity of 25 km/sec. By what percentage does its distance from us change after a) 100 years, and b) 1 million years? (1 year $= 3 \times 10^7$ sec.)

CHAPTER THREE

LIGHT AND ITS USES

When Hipparchos compiled his star chart about 150 BC, he could record nothing but the positions of a few hundred stars on the dome of the heavens, and the apparent brightness of each. The distances of the stars, their motions, their true luminosities were all unknown to him. Indeed, in his day they seemed unknowable.

In the last chapter we have shown how astronomers, using only a few simple laws of geometry, have determined the distances of the nearby stars, adding a third dimension to Hipparchos' two-dimensional universe. We have seen how it is possible to measure the velocities of the stars, though Hipparchos would probably have been astonished to learn that they move at all. We have seen also how it is possible to translate our measurements of stellar brightness, so crudely estimated by Hipparchos, into precise determinations of true stellar luminosity by the use of the inverse square law. We would appear to know, then, quite a bit about the cosmos, or at least our own particular corner of it.

Yet in one important sense, all this information does not represent a great advance over what Hipparchos knew, for it does not reveal very much about the *nature* of the stars. None of the knowledge described in the previous chapter could have been acquired without the use of the telescope. Yet even in the most powerful telescopes, the stars remain mere pinpoints of light. Using only a telescope, it is impossible to find out much more about what the stars really are than could Hipparchos with nothing but his two eyes. In order to learn about the stars, it is necessary to study the nature of the light we receive from them. Properly analyzed, these faint glimmers that have travelled trillions of miles to reach us carry a remarkably rich and intricate message.

The naked eye observer can catch the merest hint of this message. We mentioned in the last chapter that many bright stars seem to be colored. Antares, in the heart of the Scorpion, and Betelgeuse, in the arm of the hunter Orion, are red; Rigel, in the foot of Orion, and Spica in Virgo, are blue. In reality, colors other than white are found among all stars. With the naked eye, we see these colors only in the brightest stars because of a peculiarity of our visual sense: the cells in our eyes which respond best to faint light happen to be color-blind. But noticing that the stars have colors is a little like glimpsing a newspaper headline over someone's shoulder in a bus. The story told by starlight is a complex one, of which color is only a part (though a very important part). To decode all the information that starlight carries it is necessary first to understand many things about light itself. We shall see in Chapter 4 that it eventually involves learning many things about the atom as well.

"We all **know** what light is; but it is not so easy to **tell** what it is," remarked the English writer Samuel Johnson two hundred years ago. Light is not a simple phenomenon, and so the question "what is light" cannot have a simple answer. For several centuries, two rival theories about light were widely debated among scientists. One held that light was a wave, the other that light consisted of particles. Today we know that both views hold an element of the truth. Light has both wave properties and particle properties. In certain situations, it behaves as if it were a wave; in other situations its behavior can best be described by imagining it as a stream of particles. Both models are needed for a complete account of the observed facts. We will start with the wave model, since it is the more useful in describing how light behaves in those optical instruments used by astronomers: the telescope, the camera, and the spectrograph.

LIGHT

Waves

Nearly everyone has had some experience with waves. Water waves are the most familiar, and provide a good introduction to wave motion. If you drop a stone into a pond, ripples — small waves — will radiate out over the surface of the water. If a cork is floating nearby, it will bob up and down as they pass, but it will not move along the surface. This is because the water is not moving horizontally — only up and down, perpendicular to the direction of the wave's advance.

This fact can be demonstrated even more clearly by tying one end of a rope to a fixed object such as a fence, and flicking the free end up and down quickly. A wave will pass along the length of the rope. Yet if you mark any point on the rope with a dab of paint, you will see that it jumps up and down when the wave pulse passes, but at no time does it move toward your hand or toward the fence. Evidently, then, no matter is carried by a wave. Yet in both of the examples mentioned, **something** seems to be moving across the surface of the pond, or down the length of the rope. What is it that we are seeing?

Most waves that we are familiar with are disturbances in material media. Light, however, does not need a material medium, and travels fastest in vacuum.

A wave can best be thought of as a continuously moving disturbance. It is not a physical object but an event, or series of events — something that happens in or to the region through which it moves. We recognize this fact when we speak of "a wave of fear" or "a wave of uncertainty" spreading through a crowd. Fear is not a material object but a psychological disturbance that can spread from one person to another.

In the case of water waves or a rope wave, the disturbance is taking place in a material medium, and takes the form of a distortion of that medium. A region of the water's surface or of the rope is displaced from its normal position, upward or downward. Though it returns almost immediately to its original position, it has meanwhile pulled the next adjacent region out of line. In this way the disturbance propagates through the medium. We shall soon see, however, that there are some waves that can travel through space without a material medium. These waves represent a disturbance of certain properties of space itself. Light is a wave of this kind.

So far we have been speaking of a single wave pulse — a unique disturbance that travels through the medium only once. When we speak of a wave, however, we usually mean **periodic wave** — a series of disturbances that repeats itself regularly over and over again. Suppose, for example, that you keep twitching the end of the rope up and down — or better, build a machine to do it. The result will be a continuous series of wave pulses following each other in a regular rhythm. Any point in the path of such a wave will be subject to the same cycle of disturbance again and again as long as the wave persists.

FREQUENCY AND WAVELENGTH Figure 3–1 is a picture of a simple rope wave of the sort discussed above at a single instant of time. The regions where the disturbance of the rope is at a maximum are called crests and troughs. The distance between one crest and the next, or one trough and the next, is called the **wavelength** of the wave. It is usually represented by λ, the Greek letter lambda. As the wave advances, every point in the medium through which it is moving will become part of the successive crests and troughs. At one instant a particular point will be riding a crest, the next it will be down in the bottom of a trough, and a moment later it will be part of the following crest. The entire cycle will be repeated over and over. The number of such cycles — crest to trough and back to crest again — recorded by a stationary observer each second is

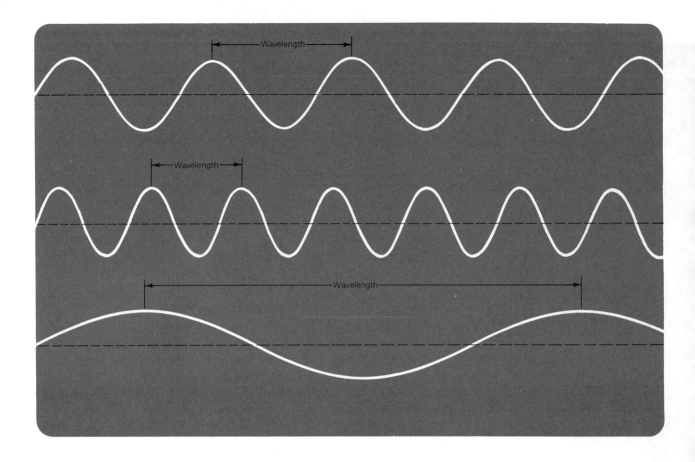

called the **frequency** of the wave, abbreviated f. It is measured in cycles per second, or hertz (Hz).

If the frequency and wavelength of a wave are both known, it is easy to find the velocity of the wave. Suppose, as an analogy, we want to determine the speed of a passing railroad train. If we had an observer at the next station with a synchronized stopwatch, we could measure how long it took the train to cover the distance. But this is inconvenient, and furthermore requires that we know the distance to the station with some accuracy. If we know how long a single car of the train is, however, we can accomplish our objective in a much simpler way. All we need do is stand by the track and count how many cars pass us in a minute. If the cars are 20 m long, and 30 of them pass us each minute, we can see that the train must be advancing at 600 m/min, or 10 m/sec.

We can apply the same method to determining the velocity of a wave. The frequency of the wave tells us how many crests will pass a fixed point each second. The wavelength tells us how far apart are these crests. Multiplying the two figures will give us the velocity at which the wave is traveling. Put in mathematical form, with the velocity of the wave represented by v, this relationship is $v = f\lambda$.

Figure 3-1
A periodic wave is a regular, continuously moving disturbance. The distance between one point of maximum disturbance and the next is the wavelength of the wave. The greater the frequency of the wave, the shorter the wavelength.

The frequency of a wave is the number of vibrations completed each second. The wavelength is the distance between successive points of maximum disturbance. The product of the two is equal to the velocity of the wave.

Roemer and the Velocity of Light

Because light travels so quickly, it was long thought to be instantaneous. Galileo questioned this assumption, and tried to measure the velocity of light directly. His method was to flash a beam of light to an assistant stationed some 2 km away and try to record its travel time. Not surprisingly, the attempt failed, for light can travel such a distance in about .000007 sec.

The speed of light is of great importance to astronomers, and so it is fitting that the first plausible measurement of this quantity came from the astronomical realm. By 1675 several of the moons of Jupiter had not only been observed, but their periods had been accurately timed. Tables had been compiled, giving the exact instant when each satellite should disappear into Jupiter's shadow on its path round the giant planet. But the young Danish astronomer, Ole Roemer, found curious discrepancies between these tables and his observations. Sometimes the eclipses were inexplicably early, at other times late.

Ole Roemer

Roemer concluded that these irregularities were the result of the earth's changing position in its annual orbit about the sun. At certain times of the year, when the earth was far from Jupiter, the light from the satellite had farther to travel, and took longer to reach us. At such times an observer on earth would see the eclipse (or anything else that happened in the vicinity of Jupiter) a few minutes late. Using Roemer's reasoning and observations, it was possible to calculate the velocity of light. The value obtained from Roemer's data, 212,000 km/sec, though low by nearly 30 percent, gave astronomers their first reasonable estimate to work with—a remarkably good one, for its time.

Today, the velocity of light in a vacuum is considered one of the most fundamental constants in nature, and has been determined with an accuracy of about one part per million. The most recent measurements give a figure of 299,792.5 km/sec. For most purposes, the convenient figure of 3×10^8 m/sec, or 3×10^{10} cm/sec, is sufficient to remember.

The velocity of light in a vacuum (abbreviated c) is about 3×10^{10} cm/sec, or 186,000 mi/sec—fast enough to circle the earth more than 7 times in a second. Thus we can rewrite the equation just given, for light in a vacuum,

$$c = f\lambda$$

This means that each wavelength of light has its own unique frequency, and each frequency its own unique wavelength, given by the above equation. The two are inversely proportional; the shorter the wavelength, the higher the frequency. Thus we can describe light by specifying either its wavelength or its frequency. In practice astronomers generally use the wavelength. The wavelengths of visible light can be conveniently expressed in nanometers (nm). One nm is equal to 10^{-11} cm—that is, a billionth of a meter, or about 4 hundred-millionths of an inch.

Our eyes are quite good at perceiving and discriminating among different wavelengths of light. The human brain interprets differences of wavelength as the sensation we call color. Red light has the longest wavelength, and violet the shortest. In between are the other colors of the rainbow: orange, yellow, green, and blue. Together these colors constitute the **visible spectrum.** The wavelengths of visible light range from roughly 400 nm (4×10^{-7} m) for violet to 700 nm (7×10^{-7} m) for red. We can use the formula given above to find the corresponding frequencies. That of violet light, for example, is

$$f \times \frac{c}{\lambda} = \frac{3 \times 10^8 \text{ m/sec}}{4 \times 10^{-7} \text{ m}} = 7.5 \times 10^{14} \text{ Hz}$$

or some 750 trillion cycles per second.

The velocity of light in a vacuum—about 3×10^{10} cm/sec—is one of the fundamental constants of nature.

Different wavelengths of light are perceived as different colors. Red light has the longest wavelength, violet the shortest.

Optics

When traveling through space, or through any homogeneous medium such as water or clear glass, light propagates in straight lines. Rays of light can be made to change their direction, however, by mirrors, lenses, and prisms. It is even possible to break up a beam of light and send its component wavelengths in different directions. Most of our knowledge of the astronomical universe is obtained with the help of certain instruments—telescope, camera, and spectrograph. These devices make use of a few simple laws, which describe how the paths of light rays are affected when they bounce off or pass through different substances. The study of these laws regarding light and its properties is called **optics.**

REFLECTION   Everyone is familiar with the images formed by mirrors or other smooth surfaces, such as that of a still body of water. The return of light rays by such a smooth surface is called **reflection.** A beam of light bounces off a smooth surface in exactly the way that a ball bounces off a flat wall. The angle that the reflected ray makes with the surface is

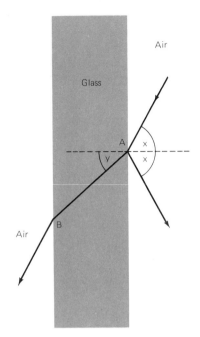

Figure 3-2
When light is reflected from a smooth surface, the original and reflected ray always make identical angles with the reflecting surface. When light passes from one transparent medium into another, in which its velocity differs, it is bent, or refracted.

always the same as the angle at which the original ray struck the surface (Figure 3–2). This effect is independent of wavelength; light of all wavelengths is reflected in exactly the same way. Only if the reflecting surface is extremely smooth, however, will the rays of light be reflected so uniformly that a mirror image is formed.

REFRACTION When light passes from one transparent medium to another—from air to water, for example, or from glass to air—its direction is generally changed. This bending of light rays is called **refraction.** Refraction takes place because the velocity of light, which is greatest in a vacuum, decreases by different amounts in various physical media. In water, for example, its speed is about .75 *c,* while in a diamond it is only about .42 *c.*

Figure 3–2 shows a beam of light passing from a medium in which its velocity is high (such as air) to one in which it is appreciably lower (such as glass) at an angle x. The first part of the beam of light to cross into the new medium, at point A, slows down. The adjacent parts of the beam, which have not yet reached the slower medium, continue to move at their original velocity, and gain on the light already in the slower medium. The result is that the entire beam will swing round and alter its direction. The effect has often been compared to a marching band or an army column making a turn.

On leaving the slower medium, this effect will be reversed. The light that entered the slow medium first will reach the fast medium first, and pull ahead. The rest of the beam, still bogged down in the slow medium, will fall behind. This lag will cause the beam to veer in the opposite direction. If the two boundaries are parallel, however, the changes in direction will be equal and opposite. The beam will emerge from the second medium traveling in the same direction as it was before it entered. This is why we see the same scene through the glass of a window pane as we see when the window is open.

THE LENS Suppose, however, that we pass light through a piece of glass whose sides are not parallel, such as a prism. A prism is wedge-shaped; a ray of light is refracted once on entering, and a second time, in the same direction, on leaving. The ray is not bent back to its original course, as it would if the sides of the glass were parallel, but deflected towards the base of the prism. If we then place two prisms base to base, as in Figure 3–3, they will cause two rays of light, diverging from a point source, to meet again at a point beyond them. Pairs of rays striking the prisms at different angles, however, will converge at different points. If we want *all* the rays from a single point to converge at a single point on the other side, we must make the sides of the prisms, not flat, but curved. If we do that we find that we no longer have two prisms; what we have is a simple **lens.**

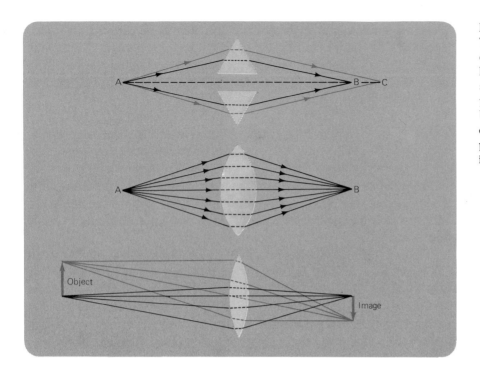

Consider, next, a solid object rather than a point source of light. Each point on the surface of this object is reflecting the light falling on it in all directions — that is why it is visible. We can see from Figure 3 – 3 that the lens focuses all the light from *each* of these points at a corresponding point on the other side. The result is an image of the object. If we place the lens at one end of a light-tight box, and a photographic film at the other, we have a camera. When the lens is opened for a fraction of a second by a mechanical device known as a shutter, the image is recorded on the film.

An advantage of such an arrangement is that we can easily vary the amount of light reaching the film by changing the size of the lens opening. This is usually accomplished with a mechanical diaphragm, similar to the iris of the human eye. When we reduce the diameter of the lens opening by half in this way, we do not lose half the image. Rays from all parts of the image still strike the smaller opening, and are focused to form the image; only the total amount of light gathered by the lens is reduced.

Notice that the image formed by the lens in Figure 3 – 3 is upside down, and reduced in size. It can easily be enlarged, however, with an eyepiece containing one or more powerful magnifying lenses. When a magnifying eyepiece is used to examine the image created by an objective lens, the result is a telescope.

THE TOOLS OF
ASTRONOMY

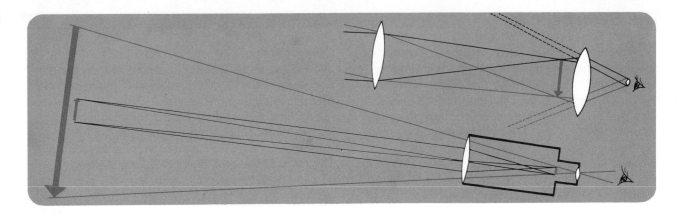

Figure 3-4
A refracting telescope makes use of a magnifying eyepiece to enlarge the image created by a convex objective lens. The image formed by such a telescope is inverted—not a problem for astronomical uses. The more powerful the eyepiece used, the greater the magnification. Extremely high magnifications, however, sacrifice the brightness of the image without increasing its clarity.

The objective lens or mirror of a telescope gathers light from an object and focuses it to form an image. An eyepiece is used to magnify the image.

The Telescope

A telescope that is constructed entirely with lenses is called a **refracting telescope** or **refractor.** In the seventeenth century it was pointed out that a mirror of the proper curvature could also gather light and form an image. It should be possible, therefore, to construct a telescope using a mirror as the principal image forming component, or objective, instead of a lens. The first person to actually build such a **reflecting telescope** or **reflector,** was Isaac Newton, who made one with a two inch mirror in 1668. Astronomers soon found that reflectors could be made more easily and cheaply in larger sizes than refractors, but these instruments were not entirely satisfactory. Their mirrors had to be made of metal, which did not reflect light very efficiently, tarnished easily, and tended to sag under its own weight. Thus for a long time after Newton's day, refracting telescopes were used for most serious astronomical research.

In the middle of the nineteenth century, however, glass mirrors coated with a thin film of silver were developed. These greatly increased the practicality of the reflector, and the tide began to turn in its favor. Further technological advances in the 1920s and 30s made it feasible to build ever larger reflecting telescopes. Aluminum replaced silver as the coating material for mirrors, providing a more durable reflective surface that practically eliminated tarnishing. At about the same time, Pyrex replaced ordinary glass as the primary mirror material. This greatly reduced the problem of expansion and contraction with changing temperatures.

The crowning achievement of these new techniques was the great Mount Palomar telescope, until recently the largest in the world. The surface of the main mirror, 200 inches in diameter, was ground and polished to within a millionth of an inch of its mathematically correct shape over its entire area of more than 200 square feet. The entire process took seven years, during which some 5 tons of glass were removed from the original cast disc. In 1975, an even larger reflector, with a 236 inch mirror, was put into service in the U.S.S.R. The glass for the mammoth mirror, cast in a furnace specially adapted for the purpose, took two years to cool.

The Herschel Family

Throughout history, there have been a sprinkling of families whose members have made outstanding contributions to our civilization: the Bachs in music, the Curies in physics, the Huxleys in biology and literature, the Adams family in American politics and historiography, the Jameses in literature and psychology. Perhaps not so well known is the Herschel family—William, his younger sister Caroline, and William's son John. These remarkable scientists shaped the course of astronomy through nearly a century of passionate, patient study of the heavens.

If you look at the index of any good history of astronomy, you will find the name of Sir William Herschel followed by perhaps a dozen or more citations, for Herschel made important contributions to virtually every area of astronomy: he was a skilled and innovative technician, a superb observer, and a brilliant theoretician. William Herschel was born in 1738 in Hanover, Germany, the son of an army musician. Following his father's footsteps, William began playing the oboe in the Hanoverian Guards at the age of 14. When the French invaded Germany and took Hanover in 1757, William went to England with only a few dollars in his pocket. Eventually he became a successful musician in his adopted country,

rising from music copyist to teacher, performer, conductor, and composer. In 1766, he took the "agreeable and lucrative" position of organist at the fashionable resort of Bath.

Herschel's inquisitive mind soon led him into other areas of study: first mathematics, and then astronomy. In 1773, curious about the objects visible in the night skies, he built his first small telescope. Soon afterward he sought, with the help of his sister Caroline, to build a more powerful instrument. Since the opticians of the day could not provide mirrors as large as he desired, he decided to grind his own. At first he was unsuccessful—his initial attempt at casting the molten metals in his basement "foundry" ended when the mirror cracked upon cooling. But Herschel's persistence paid off in those early days, as it would throughout his career. Eventually he produced mirrors even better than those at the Greenwich Observatory.

The first years were difficult for Herschel. After putting in long hours teaching or conducting, he would rush off to his laboratory to work on his telescopes. Indeed, he was so engrossed in his work, the story goes, that he barely took time to eat. It is said that Caroline had to feed him as he polished the mirrors for his instruments. Another story has it that

William Herschel.

Caroline Herschel.

the Herschels would set up William's telescope outside the concert hall where he was conducting or performing, so that he could rush out at intermission and steal a few precious minutes of observing.

Whether it is true or not, the anecdote is in character, for Herschel devoted all the time he could to the study of the stars. Night after night, he peered through his telescope, surveying section after section of the sky. Many contemporary astronomers would find this sheer drudgery, but for Herschel these hours were filled with excitement and opportunity. This was the golden age of telescopic astronomy, an age when new objects were found at the eyepiece of the telescope, not on photographic plates or spectrograph tracings. A hundred and sixty-five years after Galileo first turned his telescope on the heavens, there were still many objects to be discovered that could be glimpsed with even a small instrument. A diligent observer might be rewarded with the sight of something never before seen by human eyes, and a single important discovery could make a person's reputation overnight.

It was also the age of the amateur. The profession of "scientist" hardly existed; everywhere discoveries were being made by self-trained people using home-made equipment. A century earlier, a Dutch janitor named Anton van Leeuwenhoek, with instruments of his own making, became the first man to explore the microscopic world. Now Herschel, along with other gifted and diligent amateurs of his generation, was doing the same thing for the heavens.

On Tuesday evening, March 13, 1781, Herschel's long hours of toil had their most dramatic

reward. Among the myriad stars of the constellation Gemini he sighted a bright object that, unlike a star, seemed to show a disc. Comparing observations on successive nights, he found it to be moving. "Being struck with its uncommon magnitude," he wrote, "I compared it to H Geminorum and the small star in the quartile between Auriga and Gemini, and finding it so much larger than either of them, suspected it to be a comet." It was "without beard or tail," but this was not unusual for a comet still far from the sun. Further observation, however, showed that, unlike a comet, the object had a nearly circular orbit. It was a planet!

Herschel's discovery shook the world, for this was the first new planet discovered in recorded history. Herschel wished to name it **Georgium Sidus**—Star of George—after King George III of England, but he was persuaded to give the planet a name from classical mythology, such as the other planets bore. Accordingly, he called the planet Uranus, the personification of the heavens in Greek

Herschel's 40-foot telescope.

mythology. Herschel's find brought him much fame, and his gesture to name the new planet after the King did not go unnoticed. George III appointed Herschel his Royal Astronomer at a comfortable salary of 200 pounds a year. His new position and wealth enabled him to devote full time to his astronomical investigations. (Herschel's sole duty in his official post was to periodically show members of the royal family interesting sights in the heavens.)

The next decades, between 1782 and 1802, were busy and fruitful ones for the Royal Astronomer and his sister. During this time he proved himself to be one of the most tenacious and systematic observers the world has ever seen. He surveyed the entire sky four times over, and with the help of Caroline, compiled three catalogs that included some 2500 galaxies, star clusters, and nebulas. Realizing, in those days before photography, that no person could map the entire sky, he developed the technique that he called "star gauging." This involved counting the stars of various magnitudes in some 3400 representative regions of the heavens. It was as a result of this selective sampling procedure that Herschel proposed his theory of the shape of our Milky Way Galaxy. The sun, he believed, lay at the center of a huge stellar system having the shape of a lens. In certain areas of the galaxy, however, there appeared to be no stars at all. Of one of these regions Herschel commented, "Surely, there is a hole in the heavens." We know today that these dark lanes are areas obscured from our view by interstellar dust. But Herschel's idea about the shape of the galaxy was basically correct. Only in placing the sun at the center, rather than out toward the edge, was he mistaken.

Herschel's observations of the nebulas enabled him to make yet another discovery: that some of these patches of "cloud" or "luminous fluid," as they were called at the time, were actually aggregates of numerous stars. From these observations, he concluded that the nebulas were distant "island universes"—or galaxies—like our own. Later, however, he voiced reservations about this idea. It is unfortunate that he did so, for he had had a major insight, one which was not fully confirmed until more than a century later.

Herschel's contributions to the study of the stars earned for him the title of father of stellar astronomy. His attempts at measuring the distances of the stars ended in failure. But, as so often happens in science,

he made an even more important discovery in the process: the orbital motion of binary stars. This was the first evidence that the same law of gravitation at work in our own solar system also operates in distant parts of the universe. As he observed, "I was like Saul who, journeying out to find his father's asses, stumbled upon an entire kingdom instead." The discovery once again illustrates Herschel's careful and methodical nature. Only by repeating his observations of double stars 20 years after first studying them was he able to determine that they were indeed in motion about a common center of gravity.

Herschel's other major contribution to astronomy was his discovery of infrared radiation in 1800. After breaking down the sun's light with a prism, he placed thermometers near several parts of it. He noticed that the temperature rose where the red rays fell, but increased even more in regions beyond, where no light was visible. From this he deduced, correctly, that energy was arriving from the sun at wavelengths too long to be perceptible to the human eye.

During this period, Herschel continued to manufacture telescopes, both for his own use and for sale. Interest in astronomy was evidently widespread, for by 1795 he had turned out several hundred instruments, many of which he sold for quite substantial prices to members of the royal family, aristocratic dilettantes, and other amateur observers. His largest instrument—indeed, the largest in the world, and one of the scientific marvels of the time— boasted a 40-foot tube and a mirror over 4 feet in diameter. The mirror alone weighed more than a ton.

Dedicated and self-reliant as he was, William Herschel could hardly have achieved all that he did without the continuous aid of his devoted sister Caroline. As William's constant companion, Caroline performed all of his calculations, summarized his observations, and prepared them for publication. She looked after his daily needs, and even helped in the grinding of his mirrors. She interposed herself between the famous discoverer of Uranus and the hordes of curious who wished to meet him and peer through his telescopes.

More important, she was an excellent astronomer in her own right, discovering seven comets and many nebulas. For these achievements, Caroline received recognition from the scientific community and many awards, including election to the recently founded Royal Astronomical Society. Caroline, incidentally,

John Herschel as a young man.

was one of a long line of illustrious woman astronomers. Among the most famous are the Englishwoman Lady Margaret Huggins and the American Annie Cannon, pioneers in the study of stellar spectra, Henrietta Leavitt, whose work on Cepheid variable stars gave us our yardstick for extragalactic distances (Chapter 9), and the contemporary cosmologist Margaret Burbidge.

When William decided to marry at the age of 50, Caroline was understandably heartbroken, for she saw her position being threatened by a strange woman, but she eventually reconciled herself to the new state of affairs. In 1792 William's wife, Mary, gave birth to a son whom the couple named John Frederick William. John eventually attended Cambridge University, where he studied mathematics. After graduation, he turned first to law, then to chemistry, physics, and astronomy. In 1816, he began reobserving the double stars that had been studied by his father. During this study, John became interested in the problem of stellar parallax. His work on this subject contributed much to the later success of others.

Using his father's powerful telescopes, he also studied the nebulas that had so interested William. "These curious objects," he wrote, "I shall now take into my charge—none else can see them." John Herschel is probably best remembered for his work at the Cape of Good Hope. There he observed and charted the stars of the southern sky, just as his father had done in the northern hemisphere. Between 1834 and 1838, he mapped 69,000 stars and compiled extensive catalogs of nebulas and binary stars.

In 1839, after his return to England, he became interested in the new medium of photography. He discovered the use of sodium hydrosulfate, or hypo, to fix photographs, and coined the terms negative and positive to describe the two aspects of a photographic image. John Herschel pioneered in the development of astrophotography, which he used to help determine stellar distances and magnitudes. Ironically, it was this technique—so indispensable to the modern astronomer—which was ultimately to diminish the importance of direct visual observation, and bring to a close the era of which his father had been the greatest figure.

John Herschel in 1867, at age 75.

There are several reasons why reflectors have supplanted refractors for large modern telescopes. To begin with, there is the reason that led Newton to build his reflecting telescope in the first place. When lenses refract light, they do not bend all colors by the same amount. Short wavelengths are refracted more than long ones. Thus rays of red light does not come to a focus at the same point as do rays of blue light. This effect, known as chromatic aberration, undermines the sharpness of the image that can be produced with a lens. A mirror, by contrast, reflects all colors in exactly the same way. Reflecting telescopes are therefore free of chromatic aberration.

The objective lens of a refractor must be made of extremely clear glass that is free of optical imperfections, and both sides of the lens must be ground and polished with great precision. A mirror, on the other hand, has only one surface that must be carefully ground, and the optical properties of the glass are unimportant, since no light passes through it. For the same reason, a telescope mirror can be supported beneath its entire area. A lens, supported only around its rim, tends to sag under its own weight (which may be thousands of pounds for a 40 inch objective like that of the Yerkes refractor—the largest ever built). This sagging, imperceptible to the naked eye, can be enough to spoil the extremely precise shape of the lens, and impair the optical perfection of the instrument.

Reflectors can thus be made much larger than refractors; no one has attempted to build a refractor with an objective larger than 40 inches since 1897, when the Yerkes telescope was put into service, and there seems no reason why anyone would want to. Even for smaller sizes, though, reflectors are generally cheaper and easier to construct. They are therefore generally the choice of amateurs, many of whom choose to grind their own telescope mirrors. Refractors are the equal or superior of reflectors for certain types of observation, but in our day the reflector has the overall edge. Many refractors found in observatories today are nineteenth century instruments—often very beautiful examples of fine craftsmanship.

Use and Limitations of the Telescope

Telescopes are the subject of several popular misconceptions. It is widely believed that 1) the telescope is the most important astronomical instrument, because 2) it makes distant objects larger; consequently 3) astronomers spend most of their time looking through powerful telescopes. All of these are at best half truths.

It is true that the telescope, for the first 250 years of modern astronomy—from the time Galileo first turned his primitive telescope on the heavens in 1609, to the 1870s—was the chief tool of the astronomer. Then two nineteenth century inventions, photography and spectroscopy,

Figure 3-5

The images of a pair of stars, 1″ of arc apart, in telescopes of various sizes. The larger the objective of the telescope, the greater its resolving power—its ability to form distinct images of close objects. Stars are so distant that they are essentially point sources of light; their images appear as discs because of diffraction.

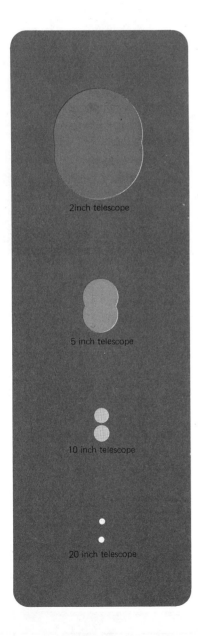

2inch telescope

5 inch telescope

10 inch telescope

20 inch telescope

somewhat changed the situation. While the telescope is as important as ever, it now shares the honors with the camera and the spectrograph.

Most astronomers today spend little of their time actually looking through telescopes. Many astronomical observations now take the form of photographs. Photographs provide inexpensive and permanent records; they can be studied, measured, and compared with other photographs taken hours, days, or even decades later. The use of long exposure times also makes it possible to capture on film objects far too faint for visual observation.

To make such long exposures, it is necessary that the object being photographed remain precisely centered in the telescope's field of view. If the object moves, the picture will be blurred, just as an ordinary snapshot will be if the subject moves during the exposure. It is therefore necessary that the telescope follow the object as it moves across the heavens with the earth's rotation. All professional telescopes, and the better amateur ones as well, are equipped with special mountings and mechanical clock drives for this purpose.

Whether it is used with or without a camera, magnification is not a telescope's most important function. Though some magnification is useful or necessary, the chief superiority of the telescope is its ability to gather more light than the unaided eye, and capture finer detail, in the object observed. Both of these depend largely on the diameter of the telescope's objective lens or mirror.

The ability to record fine detail is called **resolving power,** or resolution. Good resolving power has two chief enemies. One is **diffraction**—the bending or spreading of waves around the edges of objects. We are familiar with this effect from our experience with sound waves. If it were not for diffraction, we would never be able to hear any sound that originated around a corner from us. We do not ordinarily notice diffraction of light waves because the amount of the bending is very tiny for waves of such short wavelength. Still, it is measurable (in fact, it was detected experimentally as early as 1665), and has an effect on the clarity of telescopic images. A point of light, such as a star, for example, produces an enlarged, diffuse spot of light, surrounded by rings. This effect tends to obliterate delicate detail, such as the terrain of a planet, or the components of a closely spaced double star.

Diffraction becomes less bothersome with large telescopes, but size is no help in overcoming another hindrance to good observation: the turbulence of the air. Our atmosphere itself refracts starlight slightly, as if it were an enormous lens. This in itself would not affect the sharpness of telescopic images if the air were still. But the air is seldom still. You may have noticed how hot air rising from a chimney causes objects seen through it to shimmer and blur. The same sort of atmospheric turbulence, in less pronounced form, causes the stars to "twinkle" when seen with the naked eye on many nights. Turbulence also causes the image of

PORTFOLIO

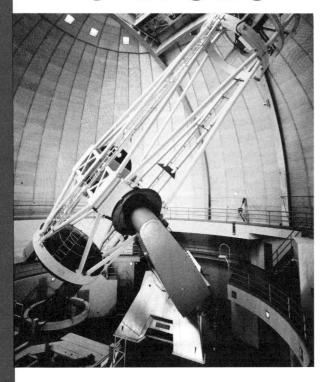

At top left we have the 120-inch reflecting telescope at Lick Observatory. Its mirror is hidden under a protective cap. At top right, the 120-inch mirror is being admired, as it well might be; its shape is accurate to better than one ten-thousandth of a centimeter. The figure at right shows two of the many ways of making the focal plane of a reflecting telescope's mirror accessible to an observer, or a photographic plate. In the largest telescopes the observer can actually sit at the focus of primary mirror without blocking out too much light. This is seen in the photographs on the opposite page. At top left is the 200-inch telescope at Mt. Palomar. The slit in the dome is open. The dome rotates to enable the telescope to be pointed at different objects. The photograph at top right shows an observer riding up, in an elevator, to the observing cage above him. The focus of the 200-inch mirror falls within the cage. Seated within the cage, as in the picture at bottom left, the observer guides the telescope during the long exposures needed to photograph faint celestial objects. Even the best of mechanical guidance systems require small corrections to be made by an alert observer. At bottom right is a picture of the observer's cage of the Lick instrument; it is smaller, oval shaped, and set slightly off-center.

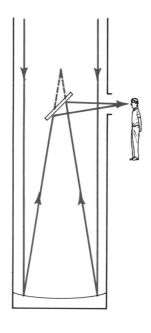

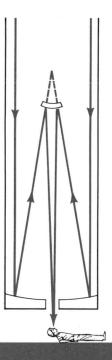

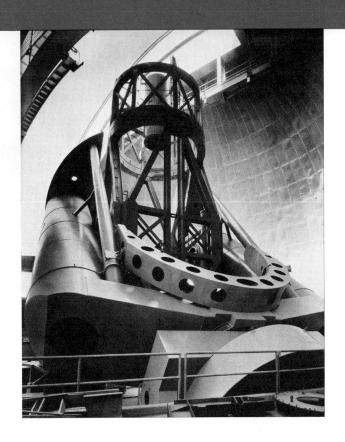

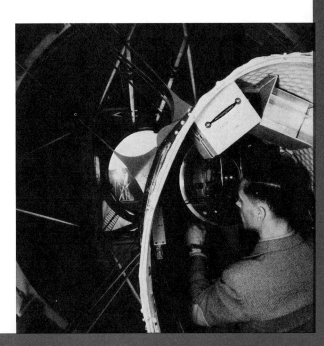

On this page we have two views of the 36-inch refractor at the Lick Observatory. Refracting telescopes usually have very long focal lengths in order to reduce the various defects in the image arising from the optical properties of glass. The length of this telescope can be estimated in comparision with the man seen in the lower picture. Since the telescope is so long, the eyepiece will be at widely differing heights depending on direction in which the telescope is pointing. Observation is made convenient by using a floor that moves up and down. The floor is much lower in the photograph below than in the one at right. The graduated circles enable the telescope to be pointed in the rough direction of a celestial object. Once this is done, it is observed in the wide field of the small finder telescope. The large telescope is then brought to bear on it. This procedure is necessary because the large telescope has a very small field of view.

The refractor of the Harvard College Observatory, at left, looks like what most people think a telescope ought to look. The varying height of the eyepiece is handled by using a movable, adjustable observing seat, seen to the left of the telescope. The 72-inch reflecting telescope of the Domininion Astronomical Observatory, below, has a much more modern appearance. The secondary mirror, seen near the top of the telescope, directs the light to the spectrograph seen at the bottom. The appearance of the 48-inch Schmidt telescope (below left) at Mt. Palomar contrasts greatly with the two other telescopes on this page. A Schmidt telescope, or camera, uses both refraction and reflection to achieve good image quality over a large field. It is used for photographing large regions of the sky.

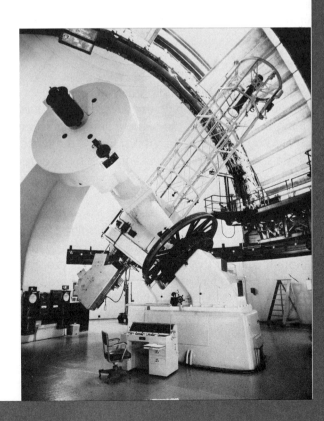

a star seen or photographed through a telescope to blur slightly. Except on a night when the air is exceptionally still and clear, the great 200 inch Mt. Palomar telescope cannot do much better in resolving fine detail than a telescope in the 20 to 40 inch range. Nor will it help to increase the magnification by using a more powerful eyepiece; that will only make matters worse, by magnifying the blur.

The turbulence of the atmosphere limits the ability of telescopes to resolve fine detail

There is nevertheless a crucially important reason for building large telescopes. Most of the objects astronomers study are not too small to be seen. An astronomical object will be visible if it is bright enough, regardless of its size. The stars, for example, seem mere points of light even in the most powerful telescopes; we cannot measure their sizes directly, but we can see their light. Most astronomical objects, however, are too faint to be seen. With the naked eye we can see about 6000 stars and three neighboring galaxies. Yet we know that there are billions of stars in our galaxy, and billions of galaxies. Many of these objects are extremely luminous; we cannot see them merely because they are so distant that very little of their light reaches us. The maximum diameter of the human iris is about .7 cm. The Mt. Palomar telescope, with its 508 cm mirror, captures about 500,000 times as much light as the unaided eye. That is why it enables us to observe or photograph so many objects whose existence we would otherwise not even be aware of. Even a telescope of 6 to 8 inches in diameter—the kind owned by many amateurs—brings into view some half a million stars not visible to the naked eye.

Since most astronomical objects are very faint, the ability of a telescope to gather more light than the naked eye is generally more important than its magnification.

The Spectrograph

When an astronomer photographs a distant star or galaxy, he is using the camera as an accessory to the telescope. The telescope can be said to be doing the real work; the camera is merely helping to record the image created by the telescope. At first glance the same would seem to be true of the spectrograph. Like a camera, a spectrograph is a small instrument (no larger than a trunk, in most instances) attached to a much larger and more costly one. Telescopes can be used without spectrographs, but a spectrograph is practically useless, for astronomical purposes, without a telescope. Yet when a spectrograph is fitted to a telescope, it is in a sense the telescope that becomes the accessory. Its function is simply to obtain enough light for the spectrograph to analyze; the spectrograph does the crucial work. In fact, it is the spectrograph, rather than the telescope, that has given us most of our insight into the nature of the stars.

To learn more about the stars it is necessary to study their light: to determine exactly how much light is emitted at different wavelengths by a particular star. But the light from astronomical objects generally consists of many different wavelengths, and they do not arrive neatly sorted out for us to study. We need some means of separating such a jumble into individual wavelengths before we can learn much about the object which sent them.

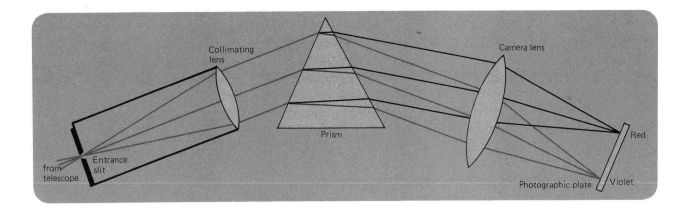

Figure 3-6
A schematic diagram of a prism spectrograph. Light from an object, such as a star, is passed through a narrow vertical slit, so that the spectrum produced consists of a series of images of the slit in light of various wavelengths. That is why spectra seem to consist of lines rather than circular images.

Light from most astronomical objects consists of many different wavelengths. A spectrograph is used to separate them.

This can be done in a very crude way with filters. A certain filter, for example, may pass only red light with wavelengths between 699 and 700 nm. Even assuming that enough different filters, with narrow enough band widths, could be made (and such filters are inefficient and costly), separating wavelengths in this way would be impossibly slow and cumbersome. As an analogy, suppose that you had 1000 apples that you wished to sort by size. You could start by rolling them across a screen with openings 6 cm in diameter. All the apples that dropped through would have diameters of 6 cm or less. You could then repeat the process with a 7-cm grid to separate out the next larger group, and so on. But what if you had to separate the apples into 100 different sizes, rather than five or six? Obviously it would be very convenient to have a device which sorted all the apples simultaneously and lined them up in order of size, regardless of how many different sizes might be present. This is essentially what a **spectrograph** does with light of different wavelengths.

We mentioned earlier that the refraction of light depends on its wavelength, with short wavelengths being bent more than long ones. Thus when light passes through a prism, the various wavelengths that compose it are refracted through different angles. The light that leaves has been spread out into the rainbow array of different colors that we call a spectrum: red, orange, yellow, green, blue, and violet. This separation at different wavelengths of light is known as **dispersion;** it was first studied by Isaac Newton.

Many spectrographs make use of a series of prisms to separate light into its component wavelengths. Another way of increasing the dispersion of a spectrograph is to use a diffraction grating instead of a prism. Diffraction gratings commonly consist of a polished metal reflecting surface, on which many closely spaced parallel lines have been incised. A high fidelity phonograph record can sometimes function as a diffraction grating. You may have noticed the rainbow shimmer of such a record when the light is diffracted from the groove edges. Since the amount of diffraction, like refraction, varies with wavelength, different wavelengths come off the surface at different angles. The more closely spaced the

Light and Its Uses   63

grooves of the grating, the wider the dispersion of the various wavelengths. There may be as many as 10,000 or more lines per cm on a good grating, scratched onto the metal surface with a microscopically fine diamond point. The technology necessary to make such fine gratings is only a few decades old. Though diffraction gratings are still rather expensive to produce, their greater dispersion and lower light loss make them generally more desirable than prisms for modern spectrographs.

ELECTROMAGNETIC
RADIATION

Light, unlike sound waves, rope waves, and water waves, does not need a material medium to travel through. In fact, it travels fastest through space in which there is no matter at all. If a wave is a moving disturbance, what sort of disturbance can there be in "empty" space? Can there be a disturbance in nothing?

The answer to these questions is that space, even when it is empty of matter, is not "nothing." It has certain properties: electrical, magnetic, gravitational, and others as well. These properties can be analyzed, and described by mathematical equations. One of the first men to attempt such a theoretical analysis was the Scottish physicist James Clerk Maxwell (1831–1879). His work led him to predict the existence of **electromagnetic waves:** disturbances of the electrical and magnetic properties of space itself.

Maxwell had not been thinking of light when he began his investigations. But the equations that predicted the existence of electromagnetic waves also enabled Maxwell to determine their velocity. If his theory was correct, they should travel at about 3×10^8 cm/sec. At this time the velocity of light had been measured quite accurately, and was known to be about 3×10^8 cm/sec! Maxwell was sure this could not be a coincidence. Light must be an electromagnetic wave.

Light is an electromagnetic wave — a regular moving disturbance of the electrical and magnetic properties of space.

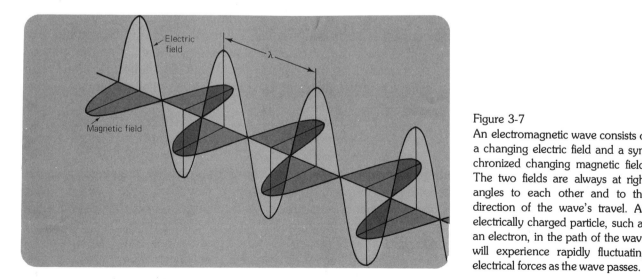

Figure 3-7
An electromagnetic wave consists of a changing electric field and a synchronized changing magnetic field. The two fields are always at right angles to each other and to the direction of the wave's travel. An electrically charged particle, such as an electron, in the path of the wave will experience rapidly fluctuating electrical forces as the wave passes.

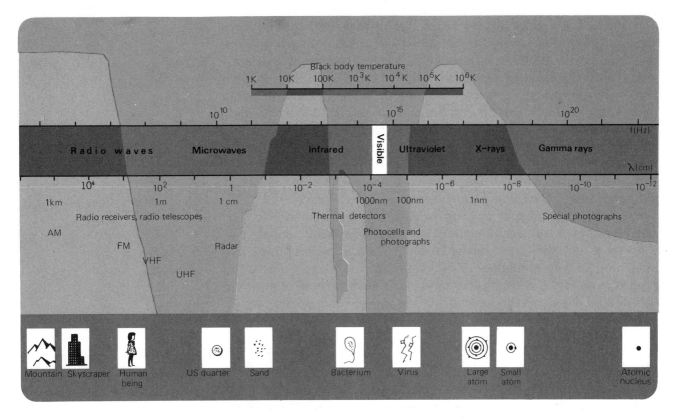

Figure 3-8
Electromagnetic waves of different wavelength have different names and uses. Visible light, with wavelengths of 400–700 nm, occupies only a very small part of the electromagnetic spectrum. The frequency, and corresponding wavelength (along with a scale of sizes for comparison) of various sorts of electromagnetic radiation are shown in the Figure, along with their names, uses, and means of detection. At the top is a scale showing the temperature necessary for a black body radiator to have its radiation peak at that particular wavelength (Wien's law). The atmosphere is opaque to electromagnetic radiation of most wavelengths (colored tint). Only in the visible and radio "windows" is the atmosphere transparent enough to allow observation from the earth's surface.

The Electromagnetic Spectrum

The wavelengths and corresponding frequencies of visible light were known in Maxwell's day. As we saw earlier, light at the extreme red end of the visible spectrum has a wavelength of slightly more than 700 nm while violet light at the opposite end has a wavelength of slightly less than 400 nm. But could electromagnetic waves exist with wavelengths longer or shorter than these? Nothing in Maxwell's hypothesis suggested that electromagnetic waves need be confined to the narrow range of wavelengths characteristic of visible light. Waves of all wavelengths and frequencies should exist—providing only that each frequency was associated with a unique wavelength, given by the formula $c = f \lambda$.

In fact, there was already some evidence for the existence of "light" that was invisible to the human eye. In 1800 William Herschel had found that heat could be produced by rays lying beyond the red end of the visible spectrum. Such rays were eventually named infrared ("below red"). A year later, a chemist found that certain chemical changes could apparently be promoted by rays lying beyond the violet end of the visible spectrum, and these rays were called ultraviolet ("beyond violet"). But Maxwell predicted, not just such close neighbors of visible light as ultraviolet and infrared, but electromagnetic waves with wavelengths differing from those of ordinary light by factors of thousands, or even millions.

Maxwell did not live to see his prediction confirmed. But in 1888 radio waves were discovered—electromagnetic waves with wavelengths on the order of a million times longer than those of visible light. Soon after, in 1895, X-rays were produced in the laboratory—electromagnetic waves with wavelengths on the order of a thousandth those of visible light. Today we know that beyond the X-ray region lie the gamma rays, with wavelengths shorter still by a factor of a thousand or more. In fact, so far as we know, there seems to be no upper or lower limit for the wavelength of electromagnetic radiation. The electromagnetic spectrum is bounded, for astronomers, largely by the practical difficulty of detecting extremely long and extremely short wavelengths.

The wavelengths of visible light represent only a very tiny part of the electromagnetic spectrum.

The accompanying chart gives the names and uses for electromagnetic waves of various wavelengths. It is important to realize, though, that the names are merely convenient labels for various parts of the entire spectrum; they do not represent different **kinds** of radiation. Just as the highest note and the lowest note on a piano keyboard are both sound waves, so radio waves and gamma rays are both electromagnetic waves. In both cases the only difference is the wavelength of the wave.

One of the many things we can learn from light or radio waves is whether their source is moving towards or away from us, and how fast: that is, we can determine the radial velocity of the source with respect to ourselves. This is possible because of a phenomenon known as the **Doppler effect,** an apparent change in wavelength of radiation from a source due to its relative motion toward or away from the observer.

The Doppler effect is a general wave phenomenon, characteristic of all types of waves, not just light. Most of us are familiar with it as a peculiarity of sound waves. You have probably noticed that a train whistle or ambulance siren seems higher in pitch when the vehicle is approaching,

THE DOPPLER EFFECT

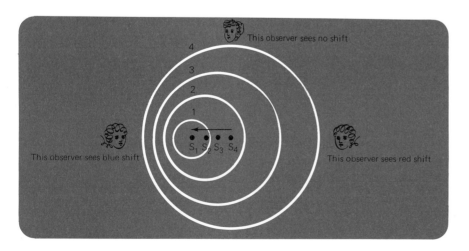

This observer sees no shift

This observer sees blue shift

This observer sees red shift

Figure 3-9
The Doppler effect. The radiation emitted by a moving source seems shorter in wavelength ahead of the source. Thus an observer in that direction will see the light shifted toward the violet end of the spectrum. In the opposite direction, the wavelengths will seem longer, and the light will appear shifted toward the red end of the spectrum. An observer roughly at right angles to the direction of motion, so that the source is neither approaching nor receding, will see no shift in wavelength.

and lower when it is going away. The higher pitch means the frequency of the sound wave is greater, while the lower pitch indicates a lower frequency. This does not mean that the source of the sound wave itself has changed. The siren produces the same number of vibrations each second, regardless of how fast it is moving or in what direction. Because the source is moving, however, we can think of it as chasing after the sound waves ahead of it, and running away from those behind it. The successive crests and troughs will therefore be more closely spaced in the direction of its motion, and more widely spaced in the opposite direction (Figure 3–9). This means that the wavelength of the sound is shorter ahead of the source and longer behind it. To put it another way, a listener ahead of the sound source will receive more crests and troughs each second than a listener behind it, and thus hear a sound of higher frequency. Once again we can see that a higher frequency must correspond to shorter wavelength.

This effect is equally apparent if we regard the wave source as fixed and consider ourselves in motion towards or away from it. As an analogy, think of the crests of a wave as buses running at regular intervals along a road. You stand at a bus stop timing their arrival, and find that they come every 6 minutes. The frequency of this "wave" of buses is therefore 10/hr. If you then start walking in the direction from which the buses are coming, you will find that you meet them more often—say every 5 min. The frequency of the buses has increased from 10/hr to 12/hr.

Because it is possible to measure wavelengths and frequencies of electromagnetic radiation with extraordinary precision, it is possible to detect Doppler changes resulting from relatively slow motions. Cars and baseballs, for example, have very low velocities compared to that of radar waves, which of course travel with the speed of light. A speeding car, or a pitcher's fastball, travel at only about a ten-millionth the velocity of light. Yet many police speed traps, and the "speed guns" used to measure the velocities of baseball pitches make use of Doppler radar. A beam of radar waves is bounced off the moving object. The returning beam will have a slightly different frequency because of the Doppler effect, and the amount of the change reveals the velocity of the moving object.

The Doppler effect also gives astronomers a way of measuring radial velocity. The principle is the same as that used in the speed gun, except

Figure 3-10
Two spectra of the star Arcturus, taken 6 months apart, when the earth was moving in opposite directions in its orbit about the sun. The lines represent particular wavelengths of light not present in the spectrum (this is explained in the following chapter). The change in the wavelength of these lines in the two photos is a Doppler shift produced by the earth's changing radial velocity with respect to the star— about 50 km/sec, in this case. This corresponds to a change in wavelength of 50/300,000, or one part in 6000.

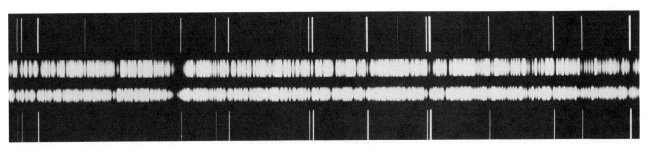

Using the Doppler Effect

The faster the wave source and observer are approaching or receding from each other, the larger will be the Doppler shift that results. This relationship can be expressed in a very simple form. Suppose we let v stand for the velocity of the wave source towards or away from us and V stand for the velocity of the wave itself. The ratio v/V will be equal to the fractional change in wavelength, and also the fractional change in frequency. If we call the unshifted frequency f and the unshifted wavelength λ, and denote the change in frequency and change in wavelength by Δf and $\Delta \lambda$, respectively (Δ is the Greek letter delta), we can write

$$\frac{v}{V} = -\frac{\Delta f}{f} = \frac{\Delta \lambda}{\lambda}$$

The minus sign indicates that, if we consider velocities away from us as positive, the frequency decreases and the wavelength increases.

To see how this formula can be applied, consider the case of a fire engine speeding toward you at 45 mi/hr, or about 20 m/sec. Suppose that its siren is sounding a note whose frequency is 440 Hz—the note we know as A above middle C on the piano, which orchestras use as their reference when tuning up. The velocity of sound at sea level is about 750 mi/hr, or 335 m/sec. Therefore the ratio v_{engine}/V_{sound} is equal to 20/335, or .06. This means that the wavelength of the sound you hear will be 6 percent shorter, and the frequency 6 percent higher. (This is a half tone higher on the musical scale; if you are gifted with good pitch, you will hear the note of the siren as B flat.)

Now let us apply the same formula to finding the radial velocity of a star. A star's velocity is typically about 30 km/sec—a thousand times faster than a baseball, but still only about a ten-thousandth that of light. Suppose we are measuring the radial velocity of a star using visible wavelengths—say in the vicinity of 500 nm. Since v_{star}/c is about .0001, the wavelengths of the star's light will be shifted by about .0001 × 500 nm, or about .05 nm. Such spectral shifts can be measured fairly easily— provided there is a suitable point of reference in the stellar spectrum for us to measure. Of course, in determining the radial velocities of stars in this way, it is necessary to allow for the earth's orbital velocity about the sun. This may add to or diminish the Doppler velocity we detect, depending on the season of the year and the orientation of the earth's orbit with respect to the star.

The Doppler effect shortens the wavelength of radiation from a source that is approaching us (violet shift). It lengthens the wavelength if the source is receding from us (red shift).

that we do not have to bounce any radiation off the stars, since they emit light of their own. (This is fortunate, for otherwise we would have to wait years for the results while our beam made its round trip.) We merely measure the change in wavelength of the light from the star to find its velocity with respect to us (see Box). A shift toward the red end of the visible spectrum, indicative of a lengthening of wavelength, means that the star is receding; a violet shift tells us the star is approaching.

We can only measure such **Doppler shifts,** however, if there is some fixed point in the star's spectrum to use as a reference. Without such a landmark the task would be virtually impossible. If a star's spectrum consisted solely of a continuous unbroken array of colors, representing an infinite number of different wavelengths, all the wavelengths would be changed ever so slightly by the Doppler shift. But the appearance of the spectrum as a whole would be entirely unchanged!

You can understand the problem better if you imagine reviewing a line of soldiers a mile long. They all look alike, and are wearing the same uniform. Halfway down the line you decide to take a break for lunch; later you return to the same spot, but how can you be sure you are really beginning where you left off? What if all the soldiers have taken one step to the right while your back was turned, or three steps to the left? There is obviously no way you can find out: the appearance of the line is unaffected. Suppose, however, that soldier no. 472 was wearing a red carnation in his buttonhole. You would then have a point of reference by which to orient yourself, and would have no trouble determining whether the line had shifted. We will see in the next chapter that stellar spectra, fortunately, have just such landmarks, known as spectral lines, that make the measurement of even very tiny Doppler shifts quite feasible.

SUMMARY

Light is an electromagnetic wave—a moving electrical and magnetic disturbance. It travels fastest in a vacuum, where its velocity (c) is about 3×10^8 m/sec; in various material media, such as glass or water, its velocity is lower. The product of the frequency of a wave (the number of vibrations completed each second, f) and its wavelength (the distance between successive points of maximum disturbance, λ) is always equal to the wave's velocity. Thus in a vacuum, light obeys the equation $c = f\lambda$. This means that if the frequency is known, we can find the corresponding wavelength, and vice versa.

The sensation we call color is produced by light of different wavelengths. The visible spectrum comprises the range of wavelengths to which the human eye is sensitive. The shortest visible wavelength is that of violet light (about 400 nm); the longest is that of red light (about 700 nm). In between we find the other colors of the rainbow: blue, green, yellow, and orange. These wavelengths represent only a tiny fraction of the known electromagnetic spectrum, however.

Infrared, microwave, and radio waves have wavelengths too long for the human eye to respond to. Ultraviolet, X-rays, and gamma rays have wavelengths too short for the human eye to respond to. Observations at all of these wavelengths play an important role in modern astronomy.

Light can be reflected by smooth surfaces such as mirrors. It can also be bent in passing between media in which its velocity is different, such as air and glass. This bending is called refraction. Refraction is the phenomenon that enables lenses to gather light and focus it to form an image. The same function can also be performed by a concave mirror. A telescope uses a magnifying eyepiece to enlarge the image produced by a lens (in refracting telescopes) or mirror (in reflecting telescopes). Because reflectors are cheaper to construct and easier to perfect optically, most modern astronomical telescopes are of this type.

Telescopes are useful chiefly because of their ability to capture fine detail in the subject being observed, and to gather enough light so that very faint objects, such as distant galaxies, can be studied. Both these functions depend on the size of the telescope's objective lens or mirror. The ability to record fine detail, or resolving power, of a large telescope is limited chiefly by the turbulence of the atmosphere, which degrades the quality of the image that can be obtained. The light-gathering power of a telescope can be enhanced by the use of long photographic exposures. To make such exposures possible, telescopes are equipped with mountings and clock drives that keep the instrument pointed at the object being photographed as the earth turns and the stars move across the sky.

The spectrograph is a device for separating light of various wavelengths. It makes use of the principle of dispersion — the refraction of light of different wavelengths by differing amounts. Because the spectrograph can reveal so much about the composition and physical conditions of the stars, it is the workhorse of modern stellar astronomy. One use of the spectrograph is in the determination of radial velocities. This can be done through the measurement of Doppler shifts: changes in the observed wavelengths of light or other electromagnetic radiation when the observer and the source of the radiation are in motion towards or away from each other. If a star, for example, is receding from earth, the wavelengths of all its radiation will appear longer to an observer on earth. If the star is approaching earth, the wavelengths of its radiation will be lengthened. Through measurement of the amount of such Doppler shifts, it is possible to find the radial velocities of stars, galaxies, and even clouds of interstellar gas.

EXERCISES

1. Why do only the brightest stars appear to have colors to the naked eye?
2. What is a wave? Give your own examples of wave motion.
3. How is the velocity of a periodic wave related to its frequency and wavelength? Is it possible to have two light waves in a vacuum with different wavelengths but the same frequency?
4. Suppose trucks in a convoy are spaced at intervals of 500 m, and the convoy is travelling past an observer at 75 km/hr. What is the frequency (in trucks/hr) that he observes?

5. What property of a light wave accounts for our sensation of color? Can you think of an explanation for the existence of colors (such as white and brown) which are not found in the visible spectrum?

6. What is the wavelength (in a vacuum) of light whose frequency is 5×10^{14} Hz? What color would you expect this light to be?

7. What is resolving power? What limits the resolving power of a small telescope? What limits the resolving power of a large telescope, such as the 200 inch Mt. Palomar instrument?

8. Why do astronomers need telescopes with large objective lenses or mirrors? What are the respective advantages of reflecting and refracting telescopes?

9. How does a spectrograph work?

10. Light from the sun takes about 12½ min to reach Mars. How far is Mars from the sun?

11. Light from a star emitted at a wavelength of 400 nm is detected on earth at a wavelength of 400.2 nm. At what rate is the distance between the star and the observer changing? Is it increasing or decreasing?

12. If you had a spectrograph capable of recording stellar spectra, how would you go about proving that the earth moves around the sun? How would you determine the orbital velocity?

13. Although the earth travels with an orbital velocity of nearly 29 km/sec with respect to the sun, this velocity does not produce a Doppler shift in the solar spectrum. Why not?

invisible astronomy

Largest fully steerable radio telescope, 100 metres in diameter, is located southwest of Bonn. Weight of steel construction is 3,200 tons.

We have said that electromagnetic waves are messengers from space. This is just as true of radio waves, X-rays, and other radiation outside the visible range as it is of ordinary light. We can learn nothing from these invisible wavelengths, however, unless 1) they can reach us, and 2) we have some means of detecting them when they do.

It is hard to see, at first, why any radiation should have trouble reaching us from space. After all, leaving aside for the moment the tenuous gas and dust between the stars, there is nothing between ourselves and the astronomical universe but a few miles of thin air. We are accustomed to thinking of our atmosphere as transparent, and this is largely true for visible light. Starlight reaches us easily, with relatively little loss to absorption. The same is not true, however, for radiation in other parts of the electromagnetic spectrum. The air is almost completely opaque to most ultraviolet wavelengths, and virtually all X-rays and gamma rays. Thus messages from the stars may travel for centuries over trillions of kilometers, only to be obliterated in the last .0001 sec of their journey by some 30 km of air.

Figure 3–10 shows the transparency of the atmosphere to various wavelengths. For infrared wavelengths, which we know as heat waves, the chief enemy is water in the air—not the clouds, which consist of tiny droplets of liquid, but water vapor, which is transparent to visible light but absorbs heavily in the infrared. Since water vapor is scarce in the upper atmosphere, it is only the lower atmospheric layers that create a problem. Once we get a few miles above the earth's surface, the "seeing" in the infrared is clear. Thus infrared astronomy is often carried out from mountain tops in areas where the humidity is very low, such as Chile or Hawaii, or with infrared telescopes sent aloft in balloons or aircraft. Ultraviolet and X-ray wavelengths, however, are stopped by even the thin gases of the upper atmosphere. To observe these wavelengths it is necessary to reach altitudes of 100 km or more. Such observations have been made by rocket-borne instruments.

Whether we observe at wavelengths that penetrate the atmosphere, or send our instruments above the atmosphere, there remains the problem of detecting invisible radiation — of making it, in some way, visible. The range of wavelengths that we can see corresponds pretty closely to the visible "window" in the atmosphere. It corresponds even more closely to the wavelengths emitted most strongly by the sun, as we shall see in the following chapter. This is not surprising; our eyes have evolved so as to use the most readily available electromagnetic radiation. For all observations outside the visible range some sort of technological device is needed to replace the human eye as the recorder of information.

With the exception of the near infrared and near ultraviolet, electromagnetic wavelengths outside the visible spectrum were not even known to exist until the end of the nineteenth century. The discovery of radio radiation from the heavens was made accidentally in 1931 by a young engineer named Karl Jansky. Jansky was studying sources of radio interference that hampered transatlantic radiotelephone reception. His antenna picked up a continuous background hiss, which varied with time in a peculiar manner. Eventually Jansky found that the mysterious signal was strongest when his antenna was pointing toward the constellation Sagittarius, where the center of our galaxy lies.

With the discovery that our galaxy was the source of radio waves, a new window was opened on the universe, and an important new branch of astronomy came into existence. After World War II, which produced a great surge of research on radio and radar technology, radio telescopes started springing up like giant inverted mushrooms all over the world. The largest in use today is in Arecibo, Puerto Rico. It was made by bulldozing out a bowl-shaped valley 305 meters in diameter, and lining the interior with metal panels. The radio antenna is suspended at the focus of the bowl by steel cables. The Arecibo instrument is not steerable; the part of the sky toward which it points is determined chiefly by the earth's own

Engineer checks two ultraviolet spectrometers aboard the Orbiting Solar Observatory. The OSO-I is the eighth of a series of satellites.

INTERLUDE

(Overleaf left) Observatory designed by Boeing company (artist's impression). (Center) Output from radio telescopes in a form suitable for analysis. (Right) 1000-foot radio telescope at Arecibo, Puerto Rico. This telescope is being used for radio studies of the planets.

daily rotation, though moving the antenna allows some additional flexibility. Large fully or partly steerable radiotelescopes with diameters of 60 to 100 meters have been built in England, Australia, Germany, and the U.S.

In essence, a radio antenna is nothing but a metal wire that conducts electricity easily. This means that it has many loosely bound electrons, which can move freely through it in response to electrical forces. As we have seen, an electromagnetic wave creates a rhythmic series of such forces, alternating rapidly in direction, as it passes. The electrons in the antenna are pulled first one way and then the other, many times per second. The result is a tiny alternating electrical current in the antenna, which can be magnified by an amplifier and then measured and recorded in a number of ways.

Radio telescopes have certain advantages compared with their optical counterparts. For one

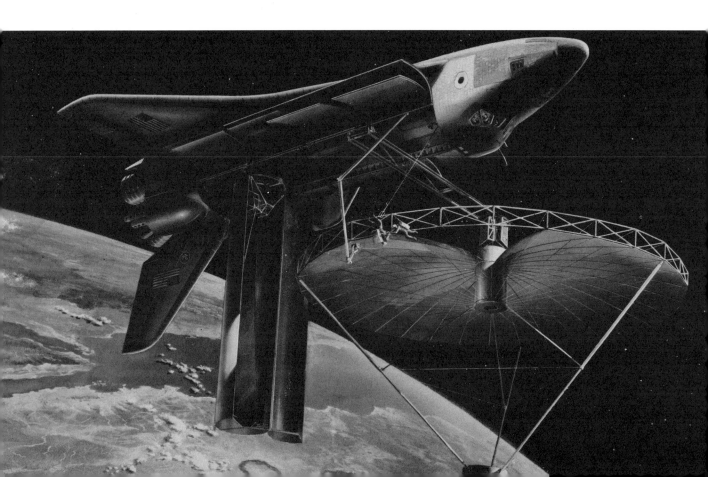

thing, radio waves easily pass through clouds of small particles, such as dust or water droplets. Electromagnetic waves cannot "see" objects much smaller than their own wavelength—the wave passes around them as if they were not there at all. Thus radio astronomers can see through the clouds of tiny particles known as interstellar dust, which obscure many regions of our galaxy from the observations of optical astronomers. Similarly, most radio waves are not obstructed by the ordinary clouds and smog of our atmosphere; a rainy day (provided there is no lightning) is just as suitable for observing as a clear one. Nor is there any difference between day and night. The radio sky is as dark in the daytime as in the nighttime. In fact, it is darker, day or night, than the clearest night sky for optical astronomers. At visible wavelengths, the sky, black though it looks to the naked eye, is never completely dark. There is always a very low level of background light from various sources, including interplanetary dust illuminated by the sun, atmospheric dust illuminated by man-made lights, and the phosphorescence of mole-

cules in the atmosphere's upper layers. This dim glow is enough to limit our ability to observe extremely faint objects, such as distant galaxies—their light is lost against the background.

Still another advantage of radio telescopes is that they can be constructed in larger sizes than their optical counterparts. As we have seen, the mirrors of large optical telescopes must be ground and polished to a high degree of perfection. Not so the reflecting "dishes" of radio instruments. Since radio waves cannot see anything shorter than their wavelength, and since radio astronomers generally work at wavelengths of at least several centimeters, and often of several meters, the surface of a radio dish can be quite rough and still reflect radio waves just as well. In fact, many dishes are made of wire mesh rather than solid metal—a great saving in expense and weight. The radio waves cannot tell the difference, as long as the spaces in the mesh are smaller than the wavelength being used.

Along with these advantages, however, go certain technical problems. Radio wavelengths on the order of several meters have long been fairly easy to work with. Unfortunately, the resolving power that can be obtained at these wavelengths is very poor. It is hard to pinpoint the precise location in the sky which is the source of the radio signal, or make out details of its shape and structure. Shorter wavelengths increase the resolving power, but require a smoother dish—thus several important radio telescopes, including the Arecibo instrument and the pioneering one at Jodrell Bank, England, have had to be upgraded for use at centimeter wavelengths, with smooth metal panels substituted for their original mesh. Moreover, at these wavelengths—and even more at millimeter wavelengths—electronic problems arise. It is hard to amplify the radio signal while keeping the level of extraneous ''noise'' to a minimum. Modern solid state electronics and maser amplifiers have only recently opened up these very short wavelengths for radio observations. At the opposite end of the radio spectrum, at wavelengths of 10 meters or more, the difficulties are of a different nature. At about 10 meters, the radio window ends, and the atmosphere becomes opaque once more. Radio waves with wavelengths greater than 10 meters or so are reflected back into space by the ionosphere—the electrically charged upper layer of the atmosphere —and never reach the earth's surface.

Despite these limitations, radio astronomy has been of immense value to our investigations of the universe. As it turns out, many of the most interesting objects in the cosmos are sources of radio emissions: the nucleus of our own and many other galaxies, supernova remnants (the remains of stars that have blown themselves to bits), the clouds of gas and dust in which stars are born, molecules in interstellar space, and QSO's.

Another useful technique, closely related to (and sometimes included with) radio astronomy is radar astronomy. Whereas the radio astronomer listens to the radio waves being emitted by distant objects, the radar astronomer bounces his own radio signals off objects and listens for the returning echo. From the time it takes these reflected radio waves to return he can determine the dis-

tance of the object, and from their strength he can often determine many things about its size and structure. This is in fact the way aircraft radar is used, and bats hunt in darkness by employing the same technique, though with very high frequency sound waves rather than radio waves. Of course, radar astronomy is feasible only for bodies in the solar system.

If the trend in radio astronomy has been toward observation at shorter and shorter wavelengths, the development of infrared astronomy has been in the opposite direction: toward the detection of ever longer wavelengths. It is not hard to make photographic film sensitive to infrared.

Since infrared radiation is intermediate in wavelength between radio and visible, the mirrors of infrared telescopes must be polished to a corresponding degree of precision. They cannot be quite so rough as radio dishes, but they need not be as smooth as the mirrors of optical reflectors. Thus the construction of instruments with fairly large apertures has not been a problem. A more serious difficulty has been the elimination of "heat pollution." We shall see in the following chapter that all objects radiate electromagnetic energy at wavelengths that depend on their temperature. Objects at room temperatures radiate chiefly in the infrared. In order for an infrared telescope to be effective, all its components must be extremely cold, to keep this thermal radiation to a minimum. In fact, such instruments are usually cooled to a few degrees K by liquid helium. Even then, infrared rays, because of their relatively long wavelengths, tend to diffract around the edges of the telescope aperture, and thus sneak in from the environment. Nevertheless, infrared astronomy has made valuable contributions in recent years, especially in the study of young stars and star formation.

On the other side of the visible spectrum, the visible window remains transparent to wavelengths of the near ultraviolet. Since ordinary photographic film responds easily to these wavelengths, there is no special problem in their detection. But glass absorbs ultraviolet wavelengths, which is

Movable radio telescopes used for interferometry at the Owens Valley Radio Observatory, California.

why you cannot get a suntan through a window, though you can feel the heat of the sun's infrared rays, which penetrate glass easily. Consequently, a reflecting telescope, rather than a refractor, must be used for observations in this region.

The range of ultraviolet wavelengths that the atmosphere will admit, however, is a narrow one. The visible spectrum ends at about 400 nm. By 350 nm, atmospheric absorption is already significant, and by 290 nm it is complete. For wavelengths in the far ultraviolet, observations must be carried out above the atmosphere, from rockets or satellites. Thus the portion of the spectrum with wavelengths shorter than about 300 nm has become known as the "rocket ultraviolet." In the second Color Portfolio, there are several pictures of the sun obtained by the Skylab astronauts at the far ultraviolet wavelength of 30.4 nm.

At even shorter wavelengths, such as that of the X-ray region of the spectrum (which begins at about 1 nm), the problem of the atmosphere's opacity is compounded by a new problem: how to focus the radiation to form an image? X-rays have the ability to penetrate most ordinary matter, as we all know from the doctor's office. Because of their short wavelengths, they can pass between the atoms of most substances, and are thus almost impossible to reflect. In recent years, "glancing-incidence" X-ray telescopes, which make use of the fact that X-rays can be reflected from certain surfaces if they strike at very small angles (that is, almost parallel to the reflecting surface) have been

developed, though not perfected. One of them has been used aboard Skylab to produce X-ray images of the sun, such as that shown in the Color Portfolio. It seems likely that, if all the technological problems involved in X-ray observation can be solved, X-ray astronomy may become one of the most fruitful branches of modern astronomy. The same obstacles that have hindered the development of X-ray astronomy stand in the way of progress in gamma ray astronomy, but they are even more acute. Gamma rays have such short wavelengths and are so energetic that it is almost impossible to focus them. The best that can be done is to determine the direction from which they come. In this way, a number of gamma ray sources have been discovered, many of which emit irregular bursts of radiation. The nature of the sources, however, is not yet known.

So far we have been surveying areas of the electromagnetic spectrum outside the visible band. But in one frontier area of astronomical research, investigators are attempting to detect waves that are not electromagnetic in all: gravitational waves. They are in fact extremely hard to detect, and the attempt involves formidable technological problems. Some years ago, one investigator reported the discovery of such waves emanating from the center of our galaxy. Attempts to duplicate these observations, however, have failed so far, and it may be that the detection of gravitational waves, and their use for astronomical research, still belongs to the future.

CHAPTER FOUR

RADIATION, ATOMS, & SPECTRA

All waves have two general properties in addition to those described in the last chapter. One is that they are carriers of energy. We see that energy at work when a piece of meat is cooked in a few minutes in a microwave oven; or when we find our skin tanned, or even burned, after a day at the beach; or when we see a photograph produced by light falling on film for only 1/100 of a second. The fact that electromagnetic radiation carries energy is such a basic truth of our universe that it is almost impossible to imagine what the cosmos would be like were this not so. The stars, for example, age and die because they gradually lose energy in the form of electromagnetic radiation. Without this loss we would not see them, for it is the energy of light waves that enables them to interact with the light-sensitive pigments in our eyes.

In addition to energy, all waves are carriers of information. Most of what we know about the world around us comes to us in the form of sound waves and light waves. We even use invisible electromagnetic waves to transmit complex information—words, numbers, pictures—over long distances. It is not surprising, then, that almost all our knowledge of the astronomical universe depends on our ability to decipher the information carried by light and other electromagnetic radiation.

In the last chapter we saw how the light that reaches us from a star can tell us the star's radial velocity. This is but one of many things we can learn from the analysis of stellar radiation. Among the others, two of the most important are the temperature of the star and the luminosity per unit area of its surface. This information can often be used, in turn, to find the star's size (if its distance is known), or its distance (if its size is known).

To understand how temperature and luminosity can be deduced from the study of a star's radiation, it is necessary to understand how electromagnetic energy is emitted from and absorbed by matter.

Absorption and Radiation

Electromagnetic radiation can interact with matter in three principal ways. It can be **transmitted**—that is, it can pass through largely unaffected, as visible light passes through window glass. It can be **reflected,** as light is reflected from a polished metal surface or a mirror. Light that is neither transmitted nor reflected by a material body must be **absorbed.** It is rare, however, that any one of these happens to the complete exclusion of the others. Some combination of the three is more likely. A pane of glass will transmit most of the visible light that falls on it, but we know from experience that it will reflect some as well, for we often see the sun glaring from the windows of buildings. It is not as obvious that a small percentage of the light is also absorbed, but this effect can be measured with sensitive instruments.

While many liquids and gases are relatively transparent to light in the visible range, most ordinary solids are quite opaque: that is, they transmit little or none of the light falling on them. The wavelengths of light that an opaque object reflects determine its color. If an object reflects light of all wavelengths, we see it as white. If it absorbs most of the red, yellow, and green wavelengths, reflecting only the blue to our eyes, we see it as blue. A body that absorbs all of the light reaching it appears black to us.

PERFECT RADIATORS   The most obvious fact about the stars is that they radiate electromagnetic energy. What determines the amount of radiation, and the particular wavelengths at which it is emitted? To understand the stars, it is important to know the properties of a radiating body.

Scientists often find it useful to begin their analysis of a problem with an extreme, oversimplified case. They can then apply their findings, with suitable modifications, to the more complicated situations encountered in the real world. An example of such a useful oversimplification is the notion of a **perfect radiator:** an ideal body that radiates electromagnetic energy with the maximum possible efficiency. There is no such thing as a perfect radiator in nature. Stars, however, are generally close enough so that the study of perfect radiators is extremely valuable to astronomers.

It turns out that the same properties that make a body a good radiator also make it a good absorber of radiation. Thus the imaginary perfect radiator is also a **perfect absorber:** one that absorbs all the electromagnetic radiation, of every wavelength, that falls on it. Such a body will appear black at room temperature, since it reflects no light back to our eyes. A perfect radiator is thus often referred to as a "black body." The name is somewhat misleading, though, for it is the *radiation* from black bodies that is of interest to astronomers, and when a black body begins to radiate in the visible range it may be very bright indeed.

Consider what happens as a black body absorbs radiant energy from its surroundings. The energy cannot vanish—it can only be changed to another form. In fact, it is transformed into heat energy, and the body will start to get hotter. If the body itself did not radiate away energy, this process would continue indefinitely. But as the body's temperature rises it will start to emit light, like the coils of an electric broiler when they get hot. (In fact, all bodies radiate in this way even at room temperature. At such relatively cool temperatures, however, most of this **thermal radiation** is in the infrared part of the spectrum. Since these wavelengths lie outside the visible range, we see cool objects solely by the light they reflect.) The hotter the body becomes, the more electromagnetic radiation it emits. Eventually an equilibrium temperature is reached, at which the body is radiating energy away as fast as it is absorbing it from its surroundings.

A black body, by definition, absorbs all wavelengths with equal efficiency, but it does not radiate equally at all wavelengths. At room temperature (about 300 K) most of this radiation is in the infrared. At temperatures over 1000 K, a significant amount is in the red region of the visible spectrum (though much is still in the infrared). Such an object looks "red-hot" to us. But if you have ever seen a welder working with an acetylene torch, you know that at still higher temperatures objects glow white-hot. Both the amount of radiation emitted by a black body and the wavelengths at which it is emitted depend on its temperature.

RADIATION CURVES During the nineteenth century the properties of black body radiation were studied extensively. When the amount of radiation emitted at each wavelength by a black body is carefully measured, the result is a graph like the one shown in Figure 4–1. We can see from the graph that some radiation is emitted at all wavelengths. But there is evidently a single wavelength at which the intensity of radiation reaches a maximum.

If a series of such curves are plotted for black bodies at different temperatures, all of the curves are found to have the same shape. But the height of the curve and the wavelength of greatest radiation both change with temperature. As the temperature rises, more radiation is given off at all wavelengths, and the radiation peak occurs at a shorter wavelength.

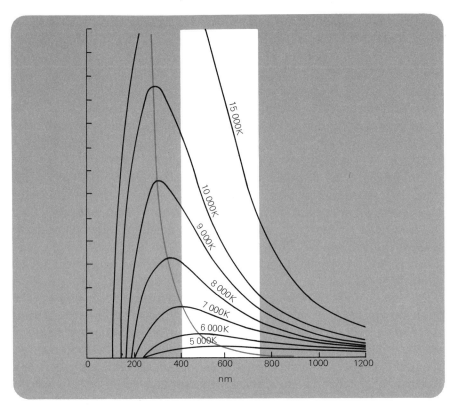

Figure 4-1
These curves represent the amount of electromagnetic energy radiated from each square cm of a black body, plotted against wavelength, for bodies at several different temperatures. All the curves have the same shape (not apparent because of the nature of the scales used in this Figure). The higher the temperature of the body, the more radiation it emits at *all* wavelengths. The wavelength at which the most radiation is emitted—the peak of the curve—is inversely proportional to the body's temperature (Wien's law). Notice how a small increase in temperature results in a very large increase in the total amount of radiation emitted.

WIEN'S LAW These facts were embodied in two laws discovered during the nineteenth century. The first, called **Wien's law,** states that the wavelength at which maximum radiation is emitted is inversely proportional to the temperature of the radiation source. The hotter a body, in other words, the further towards the blue end of the spectrum is its radiation peak. If we measure the temperature on the absolute, or Kelvin scale, and the wavelength in nanometers, the formula is approximately

$$\lambda \ \text{max} = \frac{3 \times 10^6}{T}$$

If we apply this formula to our sun, with a surface temperature of about 6000 K, we can see that the radiation peak will fall at about 500 nm, in the yellow region of the spectrum.

 We can see from this why a radiating body, as its temperature rises, glows first red, and finally blue. But why does it appear white in between? White light, as we have seen, is a mixture of all wavelengths. When a body is hot enough for its radiation peak to lie in the middle of the spectrum—somewhere in the green region—it will also be emitting significant amounts of radiation at wavelengths somewhat longer and shorter—that is, in the red and blue regions. The result will be a combination of all visible wavelengths, which we will perceive as white. Thus we do not ordinarily see any objects that are "green-hot."

The hotter the star, the shorter the wavelength at which it emits most of its energy.

This also explains why we can see bodies whose radiation peaks lie outside the visible range. A black body at 10,000 K, for example, is so hot that most of its radiation is emitted well into the ultraviolet. But a fair amount is also emitted in the violet and blue region, while at the longer visible wavelengths the radiation intensity drops off very sharply. We will therefore see such a body as violet or blue.

THE STEFAN-BOLTZMANN LAW The second law derived from the study of black bodies is called the **Stefan-Boltzmann law.** It states that the total energy per unit area emitted each second by a perfect radiator, summed up over all wavelengths, is proportional to the fourth power of its temperature. Mathematically, we can write this as

$$E \propto T^4$$

To see how this law can be applied, consider a typical yellow star such as the sun, with a surface temperature of about 6000 K. Since stars are fairly good approximations of black bodies, we can compare the sun's output of radiation with that of a black body of the same size at room temperature, 300 K. Since the sun is 20 times hotter, it must radiate 20^4, or 160,000 times as much energy each second.

Now let us compare the sun with other typical stars. There are many red stars with surface temperatures about half that of the sun — 3000 K. The total energy output of such a star from each unit of its surface area is $(1/2)^4$, or $1/16$ that of the sun. A blue star with a surface temperature of 30,000 K, on the other hand, is 5 times hotter than the sun, and thus each square meter of its surface area is 5^4, or 625 times more luminous than an equal area of the sun's surface.

The total amount of energy radiated from each square cm of a black body is proportional to the fourth power of its temperature.

USING THE RADIATION LAWS The two laws presented above allow us to determine quite a bit about stars by following a simple chain of deduction:

1) By observing the color of a star, we can determine how hot it is (Wien's law).

2) From the temperature of the star, we can determine how luminous it is *per unit area of its surface* (Stefan-Boltzmann law).

3) If we know the size of the star, we can multiply the luminosity per unit area by the star's total surface area to find its total luminosity. And, as we saw in the previous chapter, once the luminosity of a star is known, we can find its distance by measuring its apparent brightness in the sky and applying the inverse square law.

4) Alternatively, we can use the logic in step (3) in reverse: if the distance of the star is known, and we can always measure its apparent brightness, we can use the inverse square law to find its true luminosity. If we know its true local luminosity, and also its luminosity per unit from step (2), we can calculate the size of the star.

Using the Radiation Laws

To see how the radiation laws can be applied, let us consider a few examples. Say we are studying two stars, one red and the other blue, that we know to be the same size. Clearly the blue star must be the hotter of the two. Suppose we find that the maximum radiation from the red star is emitted at a wavelength of 700 nm, at the extreme red end of the visible spectrum, while the wavelength at which the blue star emits the greatest amount of energy is 350 nm, in the near ultraviolet. Wien's law tells us that the temperatures of the two stars are inversely proportional to the wavelengths of their radiation peaks. Thus the blue star must be exactly twice as hot as the red one. According to the Stefan-Boltzmann law, therefore, each square meter of its surface must be 2^4, or 16 times more luminous than the same area of the red star.

Since the stars have identical areas, the blue star must be 16 times more luminous than the red star. This fact, together with the apparent brightnesses of the stars and the inverse square law, enables us to find the ratio of their distances. Let us say that the two stars appear equally bright in our telescopes. If the blue one is 16 times more luminous than the red one, it must be four times further away.

On the other hand, suppose that we know nothing about the sizes of our red and blue stars, but we do know their respective distances. From the distances of the stars and their observed brightnesses, we have determined that they are of equal luminosity. Since the blue star is radiating so much more energy from each square meter of its surface, however, it can only have the same luminosity as the red star if it is considerably smaller. In fact, we can easily calculate how much smaller. Since the area of a star's surface is proportional to the square of its radius, the red star must have a radius 4 times that of the blue star to have 16 times the surface area. With 16 times the area, but only $^1/_{16}$ as much energy being radiated from each square meter of its surface, the red star will emit the same total amount of energy as the blue.

Finally, and most simply, we can see from this discussion that if two stars are the same color, they have the same surface temperature (Wien's law), and therefore the same luminosity from each square meter of surface (Stefan-Boltzmann law). Thus if one of the stars is known to be more luminous, it must be larger. Conversely, if one of the stars is known to be smaller, its overall true luminosity must be lower.

In order to have the same luminosity, stars of different surface temperature must have very different sizes.

ATOMS AND SPECTRA

Despite the great usefulness of these radiation laws, one problem continued to frustrate scientists towards the end of the nineteenth century. The radiation curves found by experiment did not conform to those predicted by theory. Obviously something was wrong with the theory of black body radiation. One of the physicists who studied the problem was Max Planck. In 1899 Planck worked out a mathematical expression that described the observed radiation curves very well. In order to get this formula, however, he had to make one assumption that seemed extremely strange to scientists at the time.

The Quantum Hypothesis

Ever since the wave theory of light had been accepted, more than 50 years earlier, people had thought of light as a continuous, unbroken flow, like water from a faucet. Just as you could turn a faucet on and off to get any amount of water you wanted—a gallon, or an ounce, or $\frac{1}{7}$ oz, or .00031 oz, so light could be emitted or absorbed in any quantity whatsoever. Planck's revolutionary notion was that light could only be radiated or absorbed by matter in tiny individual packages of energy called **quanta.** Just as you can receive or spend 1 cent, or $56.95 but never $\frac{1}{2}$ cent, so a single quantum of light can be emitted or absorbed, or a billion, but never $\frac{1}{2}$ a quantum, or 1000.1. Any radiative process must always involve a ***whole number of quanta.***

Electromagnetic energy can only be absorbed or emitted in tiny discrete packets called quanta, or photons.

The amount of energy in each of these packages, Planck found, depended on the frequency of the light. The higher the frequency, the greater the energy of each quantum. A single quantum of blue light carried more energy than a quantum of red light. A quantum of ultraviolet light was more energetic, still, and so on. The precise relationship turned out to be mathematically very simple: the energy of the quantum was directly proportional to the frequency of the light. Putting this in the form of an equation, we can write

$$E = h\,f$$

where E is the energy of a single quantum of light with a frequency f, and h is a very small number known as **Planck's constant.** If we measure energy in ergs and frequency in hertz, the value of Planck's constant is 6.6×10^{-27} erg-sec.

To understand what this means, suppose we find the amount of energy in a quantum of red light with a wavelength of about 656 nm. We can find the frequency of this light by using the formula $c = f\lambda$.

$$3 \times 10^8 \text{ m/sec} = (656 \times 10^{-9}\text{m}) \times f$$
$$f = 4.6 \times 10^{14} \text{ sec}^{-1}$$

Radiation, Atoms, and Spectra 85

We then multiply this frequency by Planck's constant to find the energy of a quantum:

$$E = (6.6 \times 10^{-27} \text{ erg-sec}) \times (4.6 \times 10^{14} \text{ sec}^{-1}) = 3.0 \times 10^{-12} \text{ erg}$$

Each quantum of orange light, therefore, has an energy of about 3 trillionths of an erg. How much energy is that? An erg is itself a very tiny unit of energy. It is roughly the energy of a falling leaf, or the amount of energy a fly must expend to step up from your desk onto a piece of paper lying flat on the desk's surface. Each time you climb one step of an ordinary flight of stairs, you use about a billion ergs. It is easy to see, then, that one quantum of visible light represents a minute amount of energy. Even a single firefly must radiate billions of quanta each time its light flashes.

The shorter the wavelength of electromagnetic radiation, the greater the energy carried by each photon.

If light can only be emitted and absorbed in these tiny packages of energy, it would seem logical to wonder if it also travelled through space in these packages. In 1905 Albert Einstein proposed exactly this interpretation. The quanta of light came to be called **photons.** As we mentioned earlier, it is no more correct to say that light is "really" a stream of photons than to say that light is "really" a wave. Both descriptions contain aspects of the truth, and each helps explain the behavior of light in situations where the other fails.

Line Spectra

Planck's theory enabled him to explain the exact shape of the black body radiation curve. We can see from the graphs of Figure 4–1 that a black body, in theory, radiates some energy at all wavelengths, with no gaps. If we break up the light from a black body with a spectrograph, the result is a **continuous spectrum:** all the colors of the rainbow, from red to violet, blending imperceptibly one into the next. The number of different wavelengths in such a continuous spectrum is infinite.

Continuous spectra, approximating those of ideal black bodies, are produced by solid substances, liquids, and dense gases. But the continuous spectrum was not the only kind known in Planck's day. Thin gases, when made incandescent by high temperature or the passage of an electric current (as in our familiar neon signs) produced a spectrum consisting solely of a handful of bright lines separated by dark space. Such **bright line spectra** meant that only certain wavelengths were being radiated. It was also known that different elements produced different patterns of lines. Each element could be identified solely on the basis of its spectral "fingerprint" — the unique set of wavelengths that it radiated. By the middle of the century new elements, previously unknown to chemists, had even been discovered in this way.

Spectra of still a third type were also known: **dark line spectra.** Early in the century several investigators had observed dark lines crossing

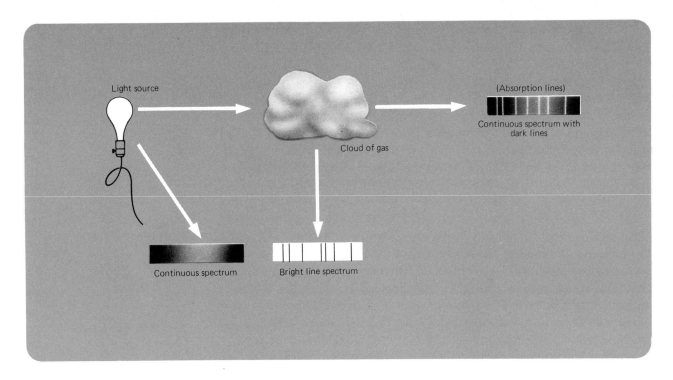

Figure 4-2
The three types of spectra. Objects that radiate like black bodies, such as hot solids, or stars, emit continuous spectra, with some radiation at all wavelengths. A thin, incandescent gas emits a spectrum of discrete bright lines, whose pattern is unique for each element. If a continuous spectrum is passed through such a gas, however, it will absorb exactly the same wavelengths, leaving a pattern of dark lines superimposed on the continuous spectrum.

Each element can either radiate or absorb its own unique set of wavelengths, producing bright line or dark line spectra.

the continuous spectrum of the sun. Eventually hundreds of such lines were discovered and mapped in the solar spectrum. Evidently they represented wavelengths of light not present in sunlight.

Around the middle of the century scientists began to pull these separate facts together into a more unified picture. It was found that dark lines were produced when light containing all wavelengths — that is, a continuous spectrum — was passed through the vapor of an element. The vapor, in other words, *subtracted* certain wavelengths of light, leaving narrow dark gaps. But the wavelengths absorbed by the vapor of a chemical element invariably corresponded to wavelengths that the vapor itself would emit when incandescent. Each element, in other words, both absorbed and radiated the same unique pattern of wavelengths.

The astronomical value of these discoveries was immense. Three things of great importance could immediately be deduced. First, since the sun produced a continuous spectrum, it must be quite dense — a solid, liquid, or (most likely) a gas at fairly high density. Second, the dark **absorption lines** in the solar spectrum showed that the sun was surrounded by an atmosphere of thinner gas. The composition of this gas could be determined from the exact positions of the dark lines. In fact, the element helium was discovered in the sun before it was known on earth because its absorption lines were found in the solar spectrum. Finally, the spectra of the stars generally resembled that of the sun, suggesting that they too were suns like our own. It was even possible to determine the chemical composition of stars trillions of miles from earth.

Why did matter in different states produce these three kinds of spectra? The line spectra of the various elements were particularly tantalizing, for they seemed so orderly. Hydrogen, for example, produced a bright line spectrum in which the lines were spaced in an extremely regular pattern. A mathematical formula was even found that not only accounted perfectly for the known lines, but predicted new ones in the infrared and ultraviolet. These lines were eventually found, at exactly the predicted wavelengths. Yet no one knew why the formula worked. It was the quantum hypothesis of Planck and Einstein that eventually provided the key.

The Atom

Most electromagnetic radiation is produced by atoms. Atoms are called the fundamental units of matter because they are the smallest possible subdivisions of the chemical elements. Elements are the basic substances that remain unchanged in chemical reactions. When charcoal, which is mostly carbon, burns in air, which contains oxygen, the carbon and oxygen combine to form carbon dioxide gas. But the carbon and oxygen remain unaltered; we can break up the carbon dioxide, and we will get the original carbon and oxygen atoms back. If we were to break up an atom of carbon, however—something that can be done using the high-energy accelerators of the nuclear physics lab—we would no longer have the element carbon, but some other element or elements.

There are 92 naturally occurring elements, which means that there are 92 different kinds of atoms. Over a dozen others have recently been artificially made by nuclear reactions, but these are generally unstable and short-lived. Though atoms can be transformed in the laboratory, or in the almost inconceivably hot interior of a star, in ordinary chemical processes they are essentially unaltered.

In the early years of this century it was discovered that each atom has a small, dense, compact center or nucleus, which accounts for most of its mass. Around the nucleus is a large volume of nearly empty space, occupied by much lighter particles called electrons. The nucleus carries a positive electrical charge, while the electrons are negatively charged. Each atom, in its normal state, has its own characteristic number of electrons—just enough to balance the positive charge on the nucleus, which varies from element to element. Thus the atom is ordinarily electrically neutral.

ENERGY LEVELS   The way the electrons are distributed in space about the nucleus determines the energy state of the atom. As a result, we speak of different electron configurations as representing different **energy levels.** It must be understood, however, that when we say that an electron is in a certain energy level, it is not the electron that possesses all the energy. The energy belongs to the atom as a whole—just as, when an arrow is placed in a bow and the bowstring pulled taut, the energy

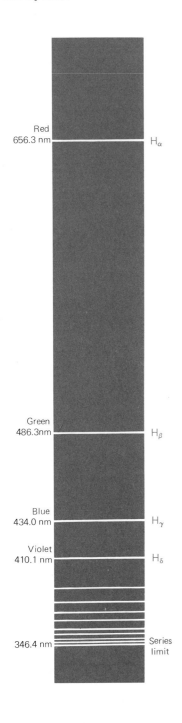

Red
656.3 nm H_α

Green
486.3nm H_β

Blue
434.0 nm H_γ

Violet
410.1 nm H_δ

346.4 nm   Series limit

stored belongs neither to the arrow, nor to the bow, nor to the string, but to the entire system. In the following discussion the terms "electron energy level" and "atomic energy level" are used interchangeably.

When an atom emits electromagnetic radiation, it gives up some of its energy. When an atom absorbs electromagnetic radiation, it gains some energy. But how much energy? This is where Planck's formula is crucial. Consider, for example, the red spectral line of wavelength 656.3 nm in the spectrum of hydrogen. To emit that bright line, trillions of atoms must all be radiating photons of red light of wavelength 656.3 nm. But we know from Planck's theory that each of these photons has exactly the same amount of energy—an energy given by the formula $E = hf$. In fact, we have already found the energy of these photons in an earlier example (p. 86): about 3×10^{-12} ergs.

If we assume that each atom radiates a single photon, then trillions of atoms are each losing exactly this amount of energy—no more, no less —every second. Furthermore, when hydrogen gas absorbs radiation to produce a dark spectral line at that wavelength, trillions of atoms must be absorbing photons with exactly that energy—no more, and no less. The same is true for each of the other spectral lines of hydrogen, or of any other atom.

Each spectral line has a unique wavelength.
Light of each wavelength consists of photons with a unique energy.
Each photon represents the gain or loss of that much energy by an individual atom.

Origin of Spectral Lines

In 1913 the young Danish physicist Niels Bohr drew a bold, and correct, conclusion from these facts. If atoms could only gain or lose certain amounts of energy, and no others, it must mean that the energy of the atom itself is quantized. That is, an atom can only exist in certain energy states, and no others. When it drops to a lower energy state, it radiates a photon. When it absorbs a photon, it jumps to a higher energy state. Each wavelength in the atom's bright or dark line spectrum represents such a jump. The photon energy for that particular wavelength is simply the energy difference between the two states.

We can think of an atom as somewhat analogous to a tall office building with no staircases, only elevators. You can take an elevator from the 20th floor to the 21st, or down to the 10th, or up again to the 99th, but there is no way you can stop between floors; you must always go from one floor to another. Now suppose somebody installs a toll system in the elevators. To move up a floor you must put a dime into a slot in the elevator. If you wish to go up two floors, you must put in 20 cents, and so on. You must always, however, have the exact change required. Fifty

Inside the Atom

The idea that all matter is made up of fundamental, indivisible particles goes back at least to the fifth century BC. It was first championed by two Greek philosophers, Democritus and Leucippus; indeed, our word "atom" comes from the Greek word for indivisible. But their theory was supported by little evidence—it was presented solely on philosophical grounds. That it resembles our modern ideas about matter must be attributed either to profound insight or to luck.

In the nineteenth century, advances in the study of chemistry and electricity led to a revival of the atomic theory, and by the end of the century the existence of atoms could no longer be disputed. Before long, however, it was found that atoms are not simple, indivisible objects after all—they can be taken apart. Atoms are made up of three kinds of particles: protons, neutrons, and electrons.

* **Electrons** were discovered by J. J. Thomson in 1897. They have a negative electric charge and a very small mass: 9.1×10^{28} g. They are stable—that is, they appear to have infinite lifetimes.

* **Protons** were discovered by Ernest Rutherford in 1911. They carry a positive electric charge equal in magnitude to that of the electron. The proton is 1836 times as massive as the electron. Like the electron, it is a stable particle.

* **Neutrons** are neutral particles, with no electric charge. In most other ways they resemble protons. The mass of the neutron is just slightly greater than that of the proton: 1838 electron masses. Outside an atom, a neutron will decay into a proton and an electron in about 16 minutes. Neutrons were discovered by James Chadwick in 1932.

In atoms, the protons and neutrons are concentrated in a small, dense central **nucleus.** The nucleus is surrounded by shells of electrons, bound to it by electrical attraction—the electrons are negatively charged, and the nucleus, because of the protons that it contains, is positively charged. The diameter of the atom is essentially that of the electron shells. A typical atomic diameter is about 10^{-8} cm. The nucleus is some 10,000 times smaller: about 10^{-12} cm. Thus most of the atom is nearly empty space.

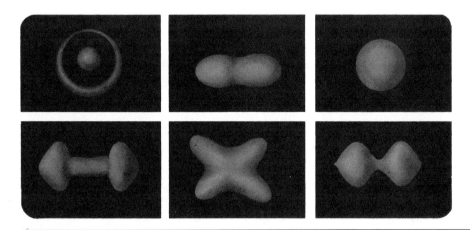

Figure 4-B
Some typical electron cloud distributions, corresponding to various energy states.

The protons and neutrons in nuclei are tightly bound together. The energy required to break up a nucleus is about a million times that required to remove an electron from an atom. How is it that the nucleus can hold together so strongly, when it contains positively charged protons which repel each other electrically? The answer is that the **nucleons**—protons and neutrons—are bound together by an even more powerful force. This **strong force** is not fully understood. One of its strange properties is its very limited range. Beyond about 10^{-13} cm it has virtually no effect, but within this distance its power is very great.

There are many things we do not yet know about the atomic nucleus. We do know, however, that it can remain in various energy states. We know that some nuclei are stable, but others are unstable, and change by emitting electrons or **alpha particles.** (These are the nuclei of helium atoms, consisting of two protons and two neutrons.) This process is known as radioactive decay. We also know that under certain circumstances, two nuclei can fuse to form a heavier nucleus, generally releasing a great deal of energy. As we shall see in Chapter 6, fusion is the source of energy for most stars, as well as for the hydrogen bomb.

When the components of the atom were first becoming known, around the beginning of the present century, it was suggested that the atom might resemble a miniature solar system, with electrons in orbit about the central nucleus much as the planets circle the sun. Such an atom could theoretically exist in a continuous range of energy states. According to classical physics, however, it would not exist at all for very long. It should radiate all its energy away in a tiny fraction of a second and collapse. In order to explain the fact that atoms did not collapse, but seemed able to exist only in certain specific energy states, this model was modified by Niels Bohr in 1913. In Bohr's original theory, the electron was thought of as able to occupy only certain permitted stable orbits about the nucleus.

This model provided the basis for the development of modern ideas, but it too had to be modified. Our present conception, part of the general framework of quantum mechanics, has been shaped by a deep theoretical insight known as the **uncertainty principle.** The uncertainty principle establishes that it is impossible to observe a system without disrupting it by a certain irreducible minimum amount. This does not matter in everyday life, when we are dealing with large objects such as baseballs or planets. Where electrons are concerned, however, it is crucial. Any attempt to observe an electron in orbit would itself change the orbit. Thus it is impossible to assign meaningful orbits to electrons.

Instead, quantum mechanics deals with the possible states of energy and angular momentum of the atom. Each state corresponds to a certain probability distribution for the electron. In other words, we can calculate, not where an electron is, but where it is *most likely* to be for each energy state. The higher the energy state, the farther from the nucleus the electrons are likely to be found. Quantum mechanics is a highly mathematical theory which involves subtle and unfamiliar ways of thinking. Its justification is that it gives us unmatched power in describing and predicting phenomena on the atomic level.

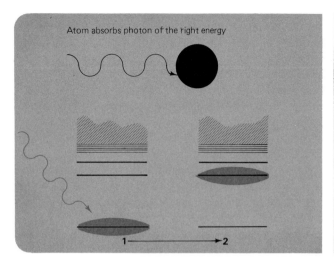

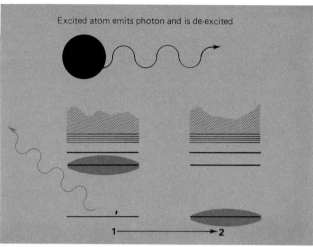

cents is required to ascend five floors; but 49 cents or 52 cents will get you nowhere. In this toll system, overpayment is as bad as underpayment.

Unlike most toll systems, however, this one costs you money only when you move up. When you move down, you get money back at the same rate. If you take the elevator from the 25th floor down to the 20th you get 50 cents; if you choose to descend to the 15th, the elevator lets you have a dollar. In this analogy the floors correspond to the different allowable energy levels of the atom. The money corresponds to the radiation the atom must absorb to move up, or emit when it moves down. Notice that when an atom changes energy levels, it must do so by absorbing or emitting a photon with *exactly the necessary amount of energy.*

The only flaw in this analogy is that, unlike the floors of most buildings, the energy levels of an atom are not equally spaced. Thus the amount of energy needed to go from level 1 to level 2 is not the same as the amount of energy needed to go from level 2 to level 3. For the hydrogen atom, the levels become more closely spaced as we move up the energy scale. These levels are depicted schematically in Figure 4–5. The precise spacing of these energy levels is characteristic of each element; that is why we can identify elements by their spectra.

Of course, just as you can take the elevator more than one floor at a time, atoms can jump several energy levels at once. An atom, for example, can jump from the first energy level directly to the 4th — provided it absorbs a photon of exactly the necessary energy, corresponding to the energy difference between these two levels. Similarly, an atom dropping from the 4th level down to the first can drop all the way at once, losing a lot of energy. This means that it will emit a photon with high energy: that is, one with a very short wavelength. The corresponding spectral line, for hydrogen, will lie in the ultraviolet portion of the spectrum.

Figure 4-4
An electron can jump to a higher energy level in two ways. It may gain energy during a collision between two atoms, or absorb a photon whose energy is exactly equal to the difference between the two energy

An atom can change its energy state by absorbing or radiating a photon of exactly the necessary energy.

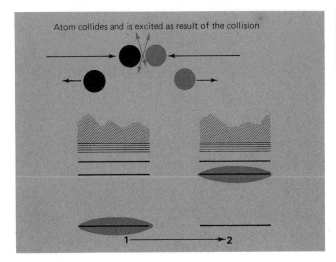

Atom collides and is excited as result of the collision

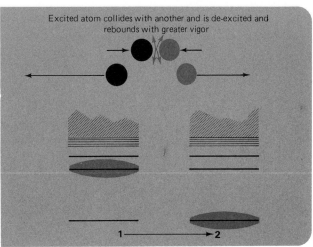

Excited atom collides with another and is de-excited and rebounds with greater vigor

levels. Similarly, an electron can drop to a lower energy level as a result of an atomic collision (surrendering some of its energy to increase the kinetic energy of the atoms) or by emitting a photon with the proper energy.

Each spectral line is the product of a particular atomic energy transition.

THE HYDROGEN ATOM   The hydrogen atom, which ordinarily has only a single electron, was the first to have its energy levels analyzed by Bohr's methods. The visible lines of the hydrogen spectrum all belong to a single family known as the Balmer series. As emission lines, these correspond to jumps from higher energy levels down to level 2. The longest wavelength of the series, designated α (Greek letter alpha), is emitted when an electron falls from level 3 to level 2. The resulting photon has a wavelength of 656.3 nm, and lies in the red portion of the spectrum.

The next line in the series is produced when an electron drops into the second level from level 4. Since this is a greater drop in energy, a more energetic photon is emitted, with a wavelength of 486.1 nm (green). When an electron makes the transition from level 5 to level 2, the result is a blue photon of wavelength 434.1 nm, and so on. The series limit, at 365.3 nm, represents the radiation from electrons dropping to the second level from extremely high levels — so high that they are almost out of range of the nucleus' attractive force, and are on the verge of being lost from the atom.

These same Balmer lines also can appear as absorption lines. An electron in the second level can absorb a photon of wavelength 656.3 nm, which carries just enough energy to boost it to the third level. A photon of wavelength 486.1 nm has the right energy to kick the electron up two levels, from level 2 to level 4, when it is absorbed, and so on.

Electrons dropping into or out of other levels are responsible for other families of lines. Transitions to or from the lowest energy level, or ground state, produce the spectral lines of the Lyman series. We can see from the accompanying diagram that the difference between the first level and the second is by far the largest. Thus all the lines of the Lyman series involve large changes of energy; their wavelengths are short, and lie in the ultraviolet part of the spectrum. Electrons dropping down to, or being boosted up out of, the third level account for the Paschen series.

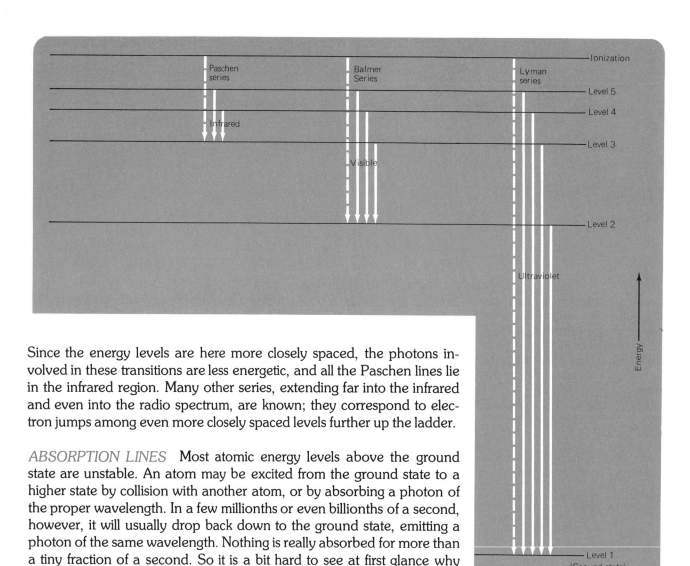

Since the energy levels are here more closely spaced, the photons involved in these transitions are less energetic, and all the Paschen lines lie in the infrared region. Many other series, extending far into the infrared and even into the radio spectrum, are known; they correspond to electron jumps among even more closely spaced levels further up the ladder.

ABSORPTION LINES   Most atomic energy levels above the ground state are unstable. An atom may be excited from the ground state to a higher state by collision with another atom, or by absorbing a photon of the proper wavelength. In a few millionths or even billionths of a second, however, it will usually drop back down to the ground state, emitting a photon of the same wavelength. Nothing is really absorbed for more than a tiny fraction of a second. So it is a bit hard to see at first glance why absorption lines should ever be produced.

The explanation is simply a matter of geometry. When an atom emits a photon, it can do so in any direction. Thus 1000 atoms may each absorb one photon from the deeper layers of a star, and in a billionth of a second reradiate them, but those 1000 photons will be radiated in 1000 random directions. Perhaps only one of them will be emitted in such a direction as to reach our spectrograph on earth. Thus even the center of an absorption line is never completely dark—there are always some reradiated photons coming our way. Their number, however, is small compared with the number that have been absorbed and reradiated in other directions. It is also small compared to the number of photons of other wavelengths which reach us directly, without being absorbed along the way. Thus an absorption wavelength is dark compared to adjacent wavelengths where no absorption happens to be taking place.

Figure 4-5
The energy levels of the hydrogen atom, showing how three series of spectral lines are formed. Notice how large is the jump from the first level (the ground state) to the second level, compared to all the other transitions.

At stellar temperatures, many atoms are found in various states of ionization — that is, missing one or more electrons.

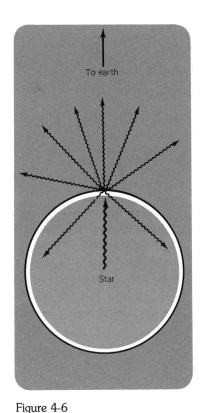

Figure 4-6
The creation of absorption lines takes place when certain wavelengths of radiation from a star's interior are absorbed and reradiated by atoms in the thin stellar atmosphere. Since the reradiation takes place equally in all directions, only a very small fraction of the original radiation is emitted towards us. At each wavelength at which absorption is taking place, therefore, there is a dark line in the star's spectrum.

IONIZATION   Sometimes an atomic electron gains so much energy that it can escape from the attractive force of the nucleus. When one or more electrons escape in this way, the atom is said to be **ionized.** The energy necessary to strip an electron from an atom, or **ionization energy,** can be supplied in several ways — by a very energetic photon, for example, or by a violent collision with another atom. Such collisions take place most frequently when the temperature and density are high, for in hot gases the atoms travel with enormous velocities, and high density means that the atoms are crowded closely together. At the temperatures and densities found in most stellar atmospheres, many atoms are ionized at any given moment.

An ionized atom, having lost one or more negatively charged electrons, carries a net positive charge. Because of this, the remaining electrons are generally held more tightly. All of its energy levels are consequently somewhat different from those of the un-ionized atom, and its spectral lines reveal this. Most stellar spectra contain lines of many elements in various stages of ionization.

In hot stars, where there has been much ionization, there are many free electrons, and ions may recapture them. To be recaptured, the electron must lose some energy, radiated away in the form of a photon. The exact amount of energy that must be given up will depend upon two factors: how much energy of motion the free electron possessed at the moment of recapture, and the energy level into which it then drops. A great deal of energy may be lost, in the form of a photon of short wavelength, allowing the electron to drop directly into a low energy level. If only a little energy is lost, however, in the form of a photon of longer wavelength, the electron will drop into one of the higher energy levels. From there it may drop further in a series of downward jumps, like a ball bouncing down a staircase. Each time it falls to a lower energy level, of course, the corresponding photon will be radiated. This process of recapture and fall from high energy levels is an important source of radiation from the clouds of interstellar gas surrounding many hot young stars.

THE CONTINUOUS SPECTRUM   If most electron transitions result in the emission of isolated spectral lines with sharply defined wavelengths, why do we see a continuous, or black-body, spectrum from the sun and other stars? The answer is that bright line spectra, as we mentioned earlier, are produced by very thin gases. At these low densities, collisions between atoms are relatively infrequent, and the atoms do not interact extensively with each other. At much higher densities, however, such as those found in solids, or the gases of the sun, atoms, ions, and free electrons are crowded together, and their interactions are constant. Some of these interactions are *not* quantized — that is, they may involve the gain or loss of *any* amount of energy, and result in the emission of a photon of *any* wavelength. Interactions of this sort are responsible for the continuous spectrum of stars, which consist of a vast number of different wavelengths blending together.

So far we have discussed spectral lines in a very abstract way. We have talked about them as though they were all uniformly strong—that is, equally bright or equally dark. We have spoken of their precise wavelengths, as if they were purely one-dimensional lines like those in geometry books, without any measurable widths. Figure 4–8 are actual stellar spectra as they appear when photographed on black and white film. Looking at this image, it is evident that real spectral lines vary in both intensity and width. A stellar spectrum, therefore, provides us with at least three kinds of information: the locations of the lines, their relative intensities, and their widths.

It is hard to judge line intensity and width from a photograph using only the naked eye, but they can be measured very accurately with instruments. Even more detail can be obtained if we use a photosensitive cell instead of a photographic film to record the precise amount of light at each wavelength across a star's spectrum. The record produced by such a device is shown in Figure 4–7. Here the horizontal axis indicates

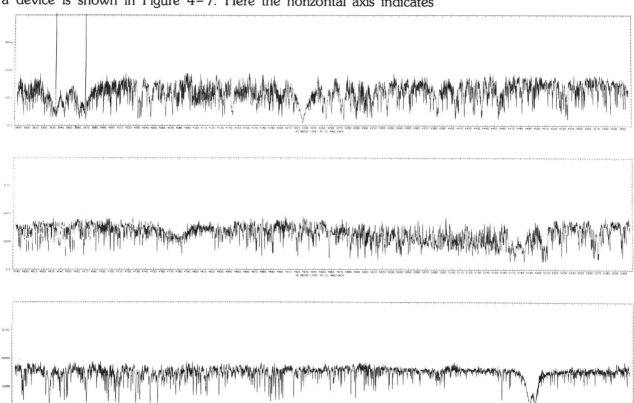

wavelength, just as in a photographic spectrum. The height indicates the number of photons received per second at each wavelength. Displayed in this form, each spectral emission line might more properly be called a peak, and each absorption line a valley. The height (or depth) of each line indicates its strength—that is, the amount of emission (or absorption) taking place at that wavelength. The width of each line indicates the range of wavelengths, above and below the line's theoretical wavelength, at which a significant amount of radiation (or absorption) is occurring. Finally, each line has a characteristic contour, or shape.

All of these characteristics—the positions of spectral lines, the relative intensity of different lines, their widths and shapes—convey information about their source. Properly interpreted, the lines of a stellar spectrum can tell us such things as the star's temperature, its chemical composition, the pressure of the gases in its atmosphere and their motions, the strength of the star, surface gravity and magnetic field, and even how fast the star is rotating.

Temperature

It would seem logical to assume that the more atoms of a particular element there are in a stellar atmosphere, the more radiation they should absorb at their characteristic wavelengths. The strength of the stellar absorption lines, in other words, should tell us the abundance of various elements. This is true, but in actuality, the strength of various lines is strongly influenced by the temperature as well. Both the composition of a star and its temperature can be learned from the study of stellar absorption lines, but the effects of the two must first be carefully disentangled.

Figure 4-7
The spectrum of a type M Main Sequence star, recorded by a solid-state electronic device containing 1024 tiny photosensitive elements. The three panels at left cover the range from 389 to 600 nm—from the extreme violet edge of the visible range to the borderline between orange and red. The panel at right

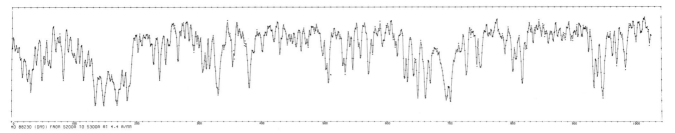

HD 88230 (OMO) FROM 5200A TO 5300A AT 4.4 A/nm

is a magnified detail of the region from 520 to 530 nm. An enormous number of different spectral lines can be distinguished. The two sharp peaks at upper left, near 393 and 397 nm, are emission lines of ionized calcium. The double valley near 589 nm represents a pair of sodium absorption lines.

To see why temperature should play such a key role, consider the hydrogen line at 656.3 nm. This line is produced as an emission line when an atom of hydrogen drops from energy level 3 to level 2. As an absorption line, it appears when an atom in level 2 absorbs a photon and jumps to level 3. The strength of that particular spectral absorption line is determined not merely by the number of hydrogen atoms in the stellar atmosphere, but also by how many of those atoms are in level 2 at any one time.

The number of atoms in any one energy state depends in part on the kind of atom and the particular energy state. Some energy states are easier to reach than others, and some will last longer than others. These factors are known, and can be allowed for in calculation. In the case of hydrogen, for example, atoms in the second energy state normally fall back spontaneously to the ground state in about a hundred-millionth of a second. We would expect, then, that there would never be very many atoms in the second energy level. But in a hot gas like that of the sun's atmosphere, with atoms rushing around at tremendous velocities, collisions keep boosting atoms up to the second level. So even though they drop back so quickly, there are always quite a few of them in the second level of any one instant.

The number of atoms in any given energy state is determined in large part by temperature.

The chief variable factor that determines the population of a given energy state is therefore the temperature. But the relationship between temperature and the strength of these lines is not simple. At temperatures much lower than that of the sun, fewer atoms are boosted up to the second energy level, so absorption by atoms in that level is weaker. In stars very much hotter than the sun, on the other hand, there are also fewer atoms in the second energy level, because most of the atoms are ionized. The absorption lines of hydrogen, therefore, are relatively strong in stars of moderate temperature (about 10,000 K), but weak in both very hot and very cool stars, even though the atmospheres of these stars may be just as rich in hydrogen.

The fact that temperature plays such a crucial role in determining the number of atoms in each energy and ionization state is very useful to the astronomer. By comparing the intensities of certain selected lines, he can find out how the atoms of a particular element are distributed among their various possible energy and ionization states. This information gives him the temperature of the star. In practice, this method is easier and more reliable than trying to determine the wavelength of the star's radiation peak (Wien's law).

Figure 4-8
Examples of the principle types of stellar spectra. The Balmer lines of hydrogen are strongest in the spectrum of class A star; notice how they grow weaker in stars that are hotter or cooler. In O and B stars, the lines of helium are visible, but they are never stronger than the hydrogen lines. Class F and G stars have spectra that are crowded with lines of metals. Molecular bands make their appearance in the spectra of K and M stars; the are particularly prominent in the M5 spectrum.

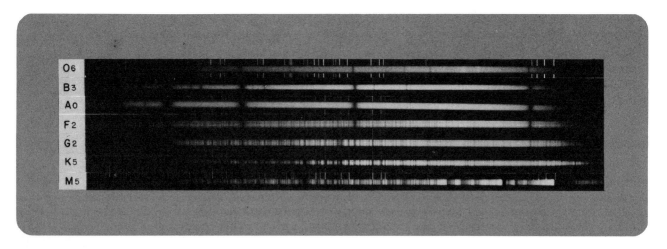

Figure 4-9
The strengths of certain types of spectral lines as a function of temperature. Hydrogen and helium lines are seen in high temperature stars, metalic lines in stars of moderate temperature, and molecular bands in cooler stars.

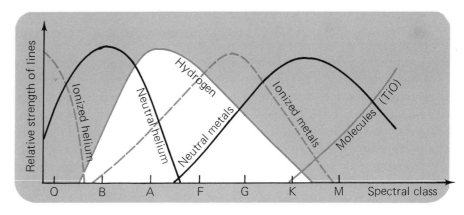

Stellar spectra are classified, in order of decreasing temperature, into types O, B, A, F, G, K, and M.

THE SPECTRAL SEQUENCE   Early in the present century, a spectral classification system was devised. Stellar spectra were sorted out into more than a dozen categories, each designated with a different letter of the alphabet, on the basis of the strength of certain lines, particularly those of hydrogen. Later, when the relationship of temperature to line intensity was better understood, these categories were rearranged and consolidated to form a sequence based on temperature. In order of decreasing temperature, the modern categories are O, B A F G, K, M. (Generations of astronomy students have memorized this sequence with the aid of the sentence. "*O*h, *b*e *a* *f*ine *g*irl, *k*iss *m*e.") The temperatures corresponding to these classes, and the most prominent features of their spectra, are shown in Figure 4–8. For still finer discriminations, each letter class is broken down into 10 numerical subclasses. The sun, for example, is a G2 star; its surface temperature is about 6000 K. From Wien's law, we know that each temperature will correspond to a different color. Class O and B stars, with radiation peaks well into the ultraviolet, will appear blue. Class A and F stars are generally white, class G stars yellow, class K stars orange, and class M stars red.

Table 4–1 The Spectral Sequence

Spectral class	Color	Surface Temperature (K)	Characteristics	Example
O	Blue	above 25,000	Few absorption lines—highly ionized atoms, especially helium; hydrogen lines weak.	10 Lacertae
B	Blue	11,000–25,000	Neutral helium, ionized oxygen, silicon, and magnesium; hydrogen lines stronger.	Spica Rigel
A	Blue	7500–11,000	Strong hydrogen lines, lines of ionized metals; some neutral metals.	Vega Sirius
F	Blue–White	6000–7500	Hydrogen lines weaker than in class A stars; lines of singly ionized and neutral metals.	Canopus Procyon
G	White–Yellow	5000–6000	Lines of ionized calcium prominent; many ionized and neutral metals; hydrogen lines weak; conspicuous bands of CH radical.	Sun Capella
K	Orange–Red	3500–5000	Neutral metals predominate; CH bands.	Arcturus Aldebaran
M	Red	below 3500	Neutral metals; molecular bands, especially titanium oxide.	Betelgeuse Antares

Composition

When the temperature of a stellar atmosphere is known, its effects on all the spectral lines can be taken into account. It then becomes possible to determine the abundances of various elements in the star. (It is important to remember, though, that we learn all this from the absorption lines produced in the star's outer layers. We cannot determine the composition of stellar interiors directly, and must rely on theoretical models.) From the study of stellar spectra, we know that most stars are about 50 to 80 percent hydrogen, by mass. Most of the rest is helium. In some stars all the other elements together account for no more than a tiny fraction of a percent; in others, the figure may be as high as three or four percent. Among the elements that are relatively common in stars are oxygen, carbon, nitrogen, neon, argon, magnesium, silicon, iron, and sulfur.

It is also possible to detect the presence of atoms chemically bonded together, or **molecules,** in some of the cooler stars. Most stars are hot enough to break down all molecules into their component atoms. But in cool stars, especially those of spectral type M (about 3000 K), some of the more stable molecules can survive and imprint their characteristic absorption lines on the star's spectrum. The energy levels of molecules, however, are much more complex and numerous than those of atoms. As a result, molecules produce groups of many closely spaced lines called **molecular bands.**

PRESSURE   Though temperature and composition are the most important factors in determining line intensities, pressure also plays a subtle role. When atoms are ionized, for example, the rate at which they recombine depends in part on the pressure. The greater the pressure, the more encounters between ions and electrons, and the more often electrons will be recaptured. Thus the pressure will influence the ratio of ionized to un-ionized atoms. In this and other ways it will affect the star's spectrum. It is important to be able to detect these pressure effects, for knowing the pressure in a stellar atmosphere enables us to determine the strength of the star's gravity, which in turn gives us information about its size and mass.

DOPPLER BROADENING   So far we have been discussing the strength or intensity of spectral lines. The width and shape of the lines is also important. There are many factors that affect line width. One of the most revealing is the Doppler effect. The atoms that emit radiation are always in motion. Usually these motions are random. At any given moment some atoms will be moving toward us, some away from us, and some at right angles to our line of sight. Because of the Doppler effect, the radiation from atoms moving toward us will have a slightly shorter wavelength. Radiation from atoms moving away from us will have a

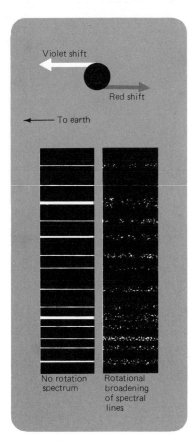

Figure 4-10
If a star is rotating, part of its surface will be moving toward us and part away at any given moment. Light from the approaching side is violet-shifted, light from the receding side is red-shifted. Light from the entire star contributes to the creation of the spectral lines, which are thus spread out over a range of wavelengths, or broadened.

slightly longer wavelength. The radiation we receive, therefore, will not be confined to a single, sharply defined wavelength, but will be spread out over a range of wavelengths. The spectral line, in other words, will be broadened. Since atoms travel faster as the temperature increases, this type of broadening is most prominent in hot stars.

In addition to these random motions of individual atoms, there may be large numbers of atoms all moving towards or away from us. Some stars, for example, have expanding atmospheres. The lines in the spectra of such stars will all be Doppler shifted toward the violet end of the spectrum since the contributing atoms all have some component of motion toward us. Other stars have turbulent atmospheres, with great masses of gas moving upwards or downwards. This will produce broadening of the spectral lines, just as the motion of individual atoms does.

Doppler broadening can also be used to determine a star's rotation. If a star is rotating (provided its axis of rotation is not pointing directly toward us), the atoms at one edge of the star will be approaching us, while those at the opposite edge will be receding from us. Some of the star's radiation will thus be shifted toward the red end of the spectrum, and some toward the violet end. Except in the case of the sun, however, we cannot separate the light reaching us from various parts of a star's surface. The red-shifted, violet-shifted, and unshifted wavelengths all arrive mixed together indiscriminately. The result is a broadening of the lines in the stellar spectrum. This phenomenon gives to spectral lines a characteristic contour, so that rotational broadening can usually be distinguished from that due to other factors.

MAGNETIC FIELD The presence of a magnetic field will change most of the energy levels of an atom slightly. Generally each level will be replaced by two or more levels differing slightly in energy. The result is that single spectral lines may be split into two or more closely spaced lines, each representing a jump to or from these multiple energy levels. The strength of a stellar magnetic field can be determined by the degree of splitting that takes place. It is also possible, by closer examination of the individual lines, to determine the direction of the field.

SUMMARY

Stars radiate electromagnetic energy, and to understand them it is important to know the properties of a radiating body. Since a body that radiates with perfect efficiency also absorbs radiation with perfect efficiency, scientists study ideally perfect absorbers. A body that absorbs all the radiation, of whatever wavelength, that falls on it will appear black at room temperature. Thus the hypothetical perfect absorber is sometimes called a black body, though such a body will appear very luminous when it is hot enough to radiate in the visible range. Stars are fairly good approximations of black bodies.

Black bodies radiate some energy at all wavelengths, but the amount of radiation at each wavelength depends on the temperature of the body. Wien's law states that the wavelength at which such a body radiates the greatest amount of energy is inversely proportional to its temperature. If the temperature is then expressed in degrees Kelvin (absolute) and the wavelength in nm, the relationship is roughly $\lambda = 3000/T$. The Stefan-Boltzmann law states that the total energy radiated each second from each square meter of surface of a perfect radiator is proportional to the 4th power of its temperature: $E \propto T^4$.

From these laws many stellar properties can be determined. The wavelength at which the radiation peak occurs tells us the surface temperature of a star (Wien's law), which in turn tells us the luminosity of each square meter of its surface (Stefan-Boltzmann law). Knowing this, we can find the size of the star (if its total luminosity is known), or the luminosity of the star (if its size is known). From the star's luminosity, we can determine its distance by measuring its apparent brightness and applying the inverse square law.

Solids, liquids, and dense gases emit continuous spectra, approximating that of a black body. Two other kinds of spectra were discovered in the nineteenth century. The rarified vapor of each element, when incandescent, emits a bright line spectrum—its own unique pattern of discrete wavelengths, with little or no radiation in between. But when a continuous spectrum is passed through the vapor of an element, the vapor absorbs radiation of precisely the same wavelengths that it emits when incandescent, so that the continuous spectrum is crossed by numerous thin, dark lines. This is known as an absorption, or dark line spectrum.

In explaining the characteristics of black body radiation, Max Planck had to make a revolutionary assumption about light: that it could only be radiated and absorbed in small discrete packets of energy which he called quanta. The quanta of light were later named photons. The energy of each photon is directly proportional to the frequency of the light.

The quantum hypothesis enables us to understand how line spectra are produced. Most radiation is emitted by atoms, the tiny building blocks of matter. Each chemical element is composed of a different kind of atom. The energy levels of an atom depend on the way its electrons are distributed in space about the central nucleus. Only certain energy levels are possible for each type of atom. An atom can jump from one permitted energy level to another by emitting or absorbing a photon, if the energy of the photon is exactly equal to the energy difference between the two levels.

Atoms may jump to higher energy levels by absorbing particular wavelengths, or drop to lower energy levels by emitting particular wavelengths. The former process produces dark (absorption) lines, the latter bright (emission) lines. In the hydrogen atom, for example, when electrons in the second level absorb photons and are boosted to higher levels, the result is a series of absorption lines known as the Balmer series, several of which lie in the visible spectrum.

Sometimes an atom absorbs enough energy to tear one or more of its electrons completely loose. This is called ionization, and the atom lacking its normal complement of electrons is known as an ion. The energy levels of an ionized atom, and thus its spectrum, are different from those of a normal atom. At the high temperatures found in most stars, many of the atoms are ionized at any given moment. When an ionized atom recaptures a free electron, it will emit a photon. The energy of this photon is not quantized.

Because the matter in the interior of stars is hot and dense, it emits a continuous spectrum. This is produced when atoms are packed so closely that they interact with each other, causing energy changes that are not quantized, and may involve the emission of light of any wavelength. Absorption takes place in the thinner, cooler regions of the star's atmosphere, giving rise to the dark lines in stellar spectra. From the intensities, widths, and shapes of these absorption lines, astronomers can learn much about the conditions in the stellar atmospheres where they are produced. For example, since the lines of each series all represent transitions from a single energy level, their intensities reflect the number of atoms in that particular level, which in turn depends chiefly on the star's temperature. Stars are generally classified on the basis of their surface temperatures into seven principal classes of the spectral series: O, B, A, F, G, K, and M.

Once the temperature has been taken into account, the intensities of the spectral lines of various elements allow us to estimate their abundances within the star. The spectra of some cooler stars also reveal the presence of certain chemical compounds, whose many, closely-spaced absorption lines constitute the so-called molecular bands. Among the other physical data that can be extracted from stellar spectra are the pressure in the stellar atmosphere (which reveals much about the size and mass of the star), the degree of turbulence, the star's rate of rotation, and the strength of its magnetic field.

EXERCISES

1. Why do you think human beings exchange information chiefly through hearing and sight? How are these senses particularly suited for communication of great amounts of information over long distances?
2. What is it that determines the color of an object? How would a coat of many colors appear in a room lit by light of a single wavelength?
3. What happens when light falls on an opaque, absorbent object?
4. What is a "black body?" Does a "black body" really appear black? At what temperatures? Why do astronomers find it useful to study "black bodies?"
5. In what part of the spectrum does most of the radiation of an ordinary household bulb lie? Is it a very efficient light source? (The temperature of the filament is about 3000 K.)
6. Two stars of equal size have temperatures of 4000 K and 12,000 K. Which radiates more energy per second? How much more? What color will each appear? Which radiates more energy in the red part of the spectrum?
7. Why does a black body never appear green, though it may appear red or blue at various temperatures?
8. If a star has a surface temperature of 12,000 K and a luminosity equal to that of the sun, what can you conclude about its size?
9. Two stars are observed to have the same parallax and the same brightness. One is red and the other is blue. Which star is larger?
10. How is it possible to have a beam of red light that is brighter than a beam of blue light, when blue-light photons are more energetic than those of red light?

CHAPTER FIVE

WEIGHING AND MEASURING THE STARS

In the last three chapters we have seen how astronomers can learn a good deal about where the stars are and what they are made of. We would also like to know how large the stars are, and how much matter they contain. In most branches of science, measuring and weighing are among the simpler tasks for the investigator, but this is not the case in astronomy, where the objects of our study are so far away.

Of the two problems, determining the sizes of stars is the easier. A few nearby stars have had their diameters determined directly with a device known as an interferometer (Chapter 9). And as we saw in the last chapter, the radiation laws enable us to estimate the size of any star whose spectral type and distance have been determined. In the following pages we shall encounter other methods by which the diameters of some stars may be measured with even greater accuracy.

Determining the amount of matter in a star is a more complex problem. On earth, we do this by weighing the object we are interested in; but how does one weigh a star? The weight of an object is simply the force which gravity exerts on it. Trillions of kilometers from earth, where our planet's gravitational attraction becomes too tiny to measure, the very concept of weight becomes all but meaningless.

Most stars, however, are part of double or multiple star systems, and move in orbits about each other. It is often possible to study the orbital motion of such stars, and thus determine their mutual gravitational influence. In this way we can find, not the weight of the stars, but a more fundamental and more useful property: their masses. The mass of a star can be thought of as a measure of how much matter the star contains, but as we shall see, it really means something more than this, and more precise. To understand the concept of mass, it is necessary to understand a series of physical laws, deriving chiefly from the work of Isaac Newton, that describe the motion of bodies under the influence of various forces, and the gravitational force that all bodies in the universe exert upon each other.

MOTION, FORCE, AND MASS

To understand the laws that govern the motion of bodies, we must deal with certain fundamental quantities. When we say that an object moves, we mean that it changes its location in space over a period of time. To speak of motion at all, therefore, we must have some way of measuring two basic elements: length or distance in space, and duration in time. From these we derive our notion of **speed**: the *rate* at which a body moves, or changes its position. If a plane covers the 320 km between New York and Washington in 40 min (2400 sec), its average speed is $320/2400 = .13$ km/sec. If a ball rolls 150 cm in 5 sec, its speed is 30 cm/sec.

In everyday language we use the terms speed and velocity interchangeably. If a pitcher has a strong fastball, we say he gets "good velocity" on the ball. But in physics, the term **velocity** includes not only the distance travelled each second, but also the *direction.* To specify a velocity, we must give both the speed and the direction: 980 cm/sec downward, 25 m/sec southwest, 55 km/sec towards the galactic center, and so on. Velocity is thus a vector quantity: one that has both magnitude and direction. When either is altered, the velocity is no longer the same. If you are driving west at 70 km/hr, you change your velocity not only whenever you slow down or speed up slightly, but also each time you turn the wheel to the right or left.

Inertia

An object at rest resists being set into motion, and the heavier the object, the greater its resistance. Once in motion objects also resist our attempts to stop or change their motion in any way. Again, the heavier the object, the greater its resistance to change. A massive revolving door provides a good illustration: it is hard to set into motion, but once in motion it is equally hard to stop. A pinwheel, by contrast, is much easier to start or to stop.

The first scientist to undertake a systematic study of objects in motion was Galileo Galilei (1564–1642). Though others had no doubt noticed the resistance of objects to any change in their motion, Galileo was the first to recognize the fundamental importance of this property. He called it **inertia,** which means laziness, and formulated the law of inertia. The first part of this law states that bodies at rest tend to remain at rest. At the time Galileo did his work this was not especially surprising, for rest was thought to be the natural state of all matter. It was generally believed that forces were required both to set bodies into motion and to keep them in motion, and common sense seemed to support these conclusions.

Inertia is the tendency of a body in motion to remain in motion, and a body at rest to remain at rest.

Galileo, however, realized that motion was as "natural" a state as rest. From his experiments, he saw that if friction is reduced, a moving object will move further before coming to rest. A block of wood slides further on smooth ice, for example, than on a wooden floor. He concluded, correctly, that in the absence of a resisting force such as friction, a body would continue in its initial state of motion forever. Once a body is set in motion, no additional force is needed to keep it in motion. On the contrary, an outside force (such as friction) is needed to slow or stop it.

Newton's First Law

Galileo's discoveries were incorporated into the work of Isaac Newton some 50 years later. The chief results of Newton's analysis of motion are commonly summarized in three laws. The first law states that

> A body at rest will remain at rest, and a body in motion will remain in motion with constant speed in a straight line, unless disturbed by an outside force.

In other words, the velocity of a body remains unchanged unless some outside force acts upon it.

Two things are notable in this statement. First, as Galileo showed, the property of inertia extends to objects in motion as well as at rest. Matter in motion resists **any** change in that motion. An excellent example of this is the experience we have all had of falling forward when the vehicle in which we are riding comes to a sudden stop. We are in motion with the vehicle, and tend to continue in motion until we are restrained by our seat belt, or by bumping into the seat in front of us.

But Newton's formulation represents an improvement on Galileo's in one respect. Galileo considered the circular motion of the moon and planets to be an undisturbed state of motion, but Newton did not. He realized that uniform motion in a straight line was the only kind of motion possible without the action of some external force. Just as we fall forward when a car stops abruptly, and are pressed back into our seats when it starts abruptly, so we feel ourselves thrust sideways when the car rounds a sharp curve. Here too we are resisting a change in our state of motion: a change, not in speed, but in direction. The car turns, but our bodies try to continue in a straight line. We are not really being thrown sideways; rather, the car is turning out from under us. Only the forces imposed on us by the car itself, as we bump against the door, or pull against our seat belt, enable us to move with the car around the curve rather than flying off on a tangent. Indeed, if we were riding on top, we probably would fly off. So too in the astronomical realm. The moon and planets, to pursue their circular paths, must be experiencing some force operating from outside themselves. We shall see the conclusions Newton drew from this crucial point when we discuss his theory of gravitation later in this chapter.

ACCELERATION   Any change in velocity, whether it takes the form of a change in **speed** of motion, or **direction** of motion, or both, involves **acceleration.** In ordinary language, we use this term to mean a speeding up. In physics, however, even deceleration (slowing down) is included under the general term acceleration, for slowing down represents a change in velocity. So does turning a corner; if you are moving south at 80 km/hr and turn west, you have undergone acceleration, even if your speed doesn't change at all. Acceleration can be defined as the **rate of change of velocity.** It tells how fast the velocity of a body is changing.

Acceleration is the rate of change of velocity; it includes both changes in speed and changes in direction.

For simplicity let us consider a body starting from rest and accelerating at 2 cm per second per second ($2 \frac{cm/sec}{sec}$, usually written 2 cm/sec^2). In the first second the velocity will increase from zero to 2 cm/sec; in the next second, from 2 cm/sec to 4 cm/sec; at the end of the third second it will be 6 cm/sec, and so on. The acceleration does not depend on the initial velocity. If the object were traveling at 100 cm/sec, rather than being at rest, its velocity at the end of the first second would be 102 cm/sec; after the next second it would be 104 cm/sec, and so on. The same acceleration is needed to increase velocity from 1 cm/sec to 2 cm/sec as from 100,001 cm/sec to 100,002 cm/sec.

In the above example we have assumed that the acceleration was in the same direction as the original velocity, so that the velocity increased. But acceleration, like velocity, is a vector quantity; its direction is as important as its magnitude. If the acceleration had been in the opposite direction, the velocity of the object would have decreased. At the end of the first second it would have been cut to 98 cm/sec; in the next second it would have dropped to 96 cm/sec, and so on. We impose such a negative acceleration whenever we step on the brake in a car. Acceleration can also take place in any direction in between these two extreme cases. In the special case where the acceleration is always perpendicular to the motion, only the direction of the velocity changes, the magnitude remaining constant. This situation has special significance for orbiting bodies.

Newton's Second Law

Newton's first law describes the motion of bodies when no external force acts upon them: they remain at rest, or move with constant velocity. Another way of saying this is that they undergo no acceleration. The clear implication of this statement is that when a force **does** act on a body, an acceleration results. This is confirmed by the second law, which describes how the acceleration produced by a force acting on a body depends both on the strength of the force and the mass of the body.

In the preceding discussion of inertia, it was pointed out that the **degree** to which an object resists any change in its state of motion is related to its weight. A heavy object is harder to set into motion, harder to stop, harder to turn. Intuitively, we are well aware of this property. If a ball comes rolling in our direction, we do not hesitate to put out our foot to stop it. If a car comes rolling toward us at the same speed, however, we quickly get out of the way. Newton found that if the same force were applied to various objects, the acceleration each acquired was inversely proportional to its weight. The heavier the object, in other words, the smaller its acceleration. Newton used this relationship to define a very important property called **mass**. The mass of a body is a measure of its inertia—its resistance to any change in its velocity.

If Newton found that inertia is directly related to weight, then mass would seem to be identical to weight. Why then bother to introduce the concept of mass at all? The answer is that mass and weight, though they seem interchangeable in everyday experience, are fundamentally different quantities. Mass is actually the more basic and useful of the two. Weight is a measure of the gravitational **force** which the earth exerts on a particular object. Thus weight can change with location. A block of butter that weighs one pound at sea level will weigh slightly less on top of Mount Ranier, where the strength of the earth's gravitational attraction is less. On the moon it would weigh considerably less—only about $1/6$ lb—because the moon's gravity is only $1/6$ that of earth. And in empty space, far from any planet, it will "weigh" nothing at all. But as long as no matter is added to or subtracted from the block, its resistance to acceleration, and thus its mass, will be the same in all three places.

Mass is a measure of a body's resistence to any change in its state of motion. In everyday experience, we associate it with the amount of matter in the body.

For a constant force, therefore, the acceleration acquired by a body is inversely proportional to its mass: $a \propto 1/m$. Once we fix our unit of mass, we can then compare any mass by measuring its acceleration in response to the same force. In the system of units employed in this book, the unit of mass is the gram (g). This is defined as the mass of one cubic centimeter of water at $4°$ C (the temperature at which water is densest). On the earth's surface, 1 g of mass weighs about .035 ounces.

Newton also found that when different forces were applied to the same mass, the acceleration produced was directly proportional to the force. Doubling the force doubled the acceleration; tripling the force tripled the acceleration, and so forth. We can state this in compact form by writing $a \propto F$. Combining these two relationships, we get Newton's second law of motion: $a \propto F/m$, or

$$F \propto ma$$

This expression states that the acceleration of an object is directly proportional to, and in the same direction as, the force acting on it, and is inversely proportional to its mass. If we give the mass in grams, and the acceleration in cm/sec², then we can define our unit of force as that force which will give a mass of 1 g an acceleration of 1 cm/sec². This force is called one dyne, and it is a very small unit. The weight of an ordinary aspirin tablet — that is, the force exerted on it by the earth's gravitational attraction — is several hundred dynes.

Newton's Third Law

When you kick a solid object such as a football, you are obviously exerting a force on it. This force imparts an acceleration to the ball, in accordance with the second law. At the same time, however, the ball exerts a force on the foot kicking it, that is why few people would willingly kick a ball barefoot. Similarly, when you press your finger against a wall, you can feel the wall pressing back. In fact, you cannot press against anything unless it *does* press back — a fact you can verify by trying to lean against the air in your room. Newton summarized these insights in his third law of motion:

> Whenever an object exerts a force on a second object (action), the second object exerts an equal but opposite force back on the first object (reaction).

This law is familiarly stated as, *for every action, there is an equal and opposite reaction.* When you throw a baseball, you are exerting a force on it to set it into motion. The ball, therefore, exerts an equal and opposite force on you. You are usually not aware of this, because the force is transmitted through your feet to the earth (that is why it is important to plant your feet firmly when you throw). Owing to its enormous mass, however, the earth experiences only an infinitesimal acceleration as a result of this force. If everyone on earth were to gather in one spot and

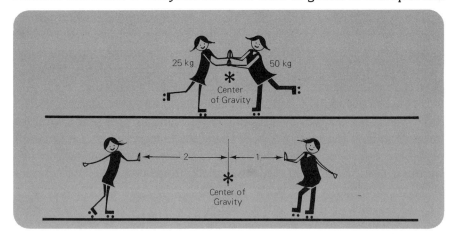

Figure 5-1
The heavier and the lighter girl push on each other with the same force (Newton's third law). The lighter girl is accelerated to a higher velocity. Since the forces are internal, the center of mass is not altered by the girls' movements. The lighter girl is always proportionately farther away from the center of mass.

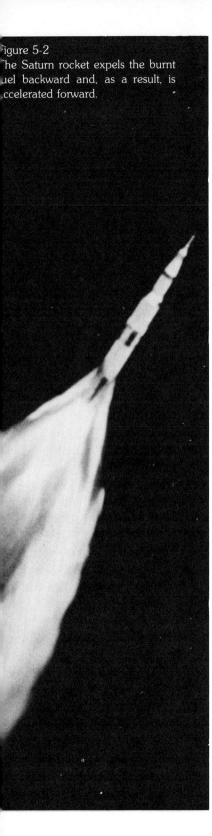

throw baseballs simultaneously the effect on the earth's motion might just be measurable with modern equipment. But if an astronaut taking a space walk were to throw a ball, the reaction would propel him backwards with a significant acceleration. In fact, this is the principle on which jets and rockets work; expelling hot gases backward propels the rocket forward.

CIRCULAR MOTION The laws of motion apply to objects traveling on curved paths — including such closed curves as circles and ellipses. In fact, the laws of motion have had their greatest successes in dealing with the motions of satellites, planets, and stars, which move in circular or elliptical orbits.

To begin with, suppose we consider objects moving in circular paths right here on earth. A ball whirled about at the end of a rope is such an object. We can see that the ball is not moving with constant velocity. Even though its speed may be constant, its direction changes continuously. According to the first law, therefore, a force must be operating on the ball. The second law tells us that this force is producing an acceleration. What sort of acceleration will change a body's direction of motion, but not its speed? We saw earlier that this is characteristic of an acceleration which always remains perpendicular to the body's velocity.

If the acceleration is always perpendicular to the ball's direction of movement, the force producing it must be also. In fact, they are both directed inward toward the center of the circular orbit. Such a force, and its resultant acceleration, are known as **centripetal** (center-seeking). What is responsible for the centripetal force? Imagine what would happen if the rope were suddenly cut. The ball would continue to move in a straight line in the direction it was going at that instant. That is, with no external forces on it, the ball would continue in uniform straight-line motion, in accordance with the first law. The force, then, must be the result of the tension in the rope pulling inward on the ball. The third law tells us that the ball is pulling equally hard outward on the rope (Figure 5–3).

It is significant to note that if there were no centripetal acceleration a mass initially located at point A would move to point P in a given interval of time. Because of the centripetal acceleration, however, it moves along the curve to point B. The mass has, in effect, "fallen" a distance PB toward the center of the circle. This falling inward is a characteristic of all orbital motion. The centripetal acceleration keeps the mass on its circular path. From an analysis of this situation, it is possible to demonstrate that the centripetal acceleration (a_c) can be expressed in terms of the velocity of the body (v) and the radius of the orbit (r) according to the formula $a_c = v^2/r$. That is, the acceleration toward the center is proportional to the square of the body's velocity, and inversely proportional to the radius of the orbit. For instance, if the velocity of the body in orbit doubles while the radius remains unchanged, the inward force that must

The Conservation of Momentum

When Newton originally published his second law, he did not state it in precisely the form we have given above. Instead, he wrote that the "change in motion" of a body was proportional to the force acting upon it. By "motion," Newton did not mean simply velocity, but the velocity of the body **multiplied by its mass,** or **mv.** Thus a 2 kg mass moving with a velocity of 10 m/sec would represent twice the amount of "motion" as a 1 kg mass moving with the same velocity. Today, to avoid confusion, we use a different term for what Newton called "motion": **momentum.** Since the definition of momentum includes a vector quantity (velocity), it is itself a vector—that is, it has direction as well as magnitude. A 1 kg mass moving north at 10 m/sec does not have the same momentum as a 1 kg mass moving south at 10 m/sec, but rather an exactly opposite momentum. If you added the momenta of the two, the total would be zero.

It is not hard to show that Newton's version of the second law is equivalent to the one given earlier, $F \propto ma.$ In Newton's formulation, the force acting on a body is proportional to the rate of change of its momentum, **mv.** Assuming that the mass remains constant, the rate of change of **mv** is the same as **m** times the rate of change of **v,** its velocity. But the rate of change of velocity is, by definition, acceleration. Thus the force acting on a body is proportional to its mass multiplied by the acceleration that it undergoes—the same expression that we originally presented.

The usefulness of the concept of momentum lies in the fact that it is **conserved**—that is, in any system of bodies, there is always a constant amount of momentum, and this amount can be altered only by the operation of some force from outside the system. The conservation of momentum can be discovered experimentally, by studying the collisions of various bodies in situations where external forces such as friction or air resistance are negligible. It is also possible, however, to derive the law of conservation of momentum from Newton's second and third laws.

We can illustrate this with a very simple example. Suppose that two bodies collide, and thus for a very brief time exert forces upon each other. According to Newton's third law (action and reaction), the forces on the two bodies must be equal and opposite. From the original version of Newton's second law, therefore, we can conclude that their **rates of change of momentum** must also be equal and opposite. And since the time during which the forces act is evidently the same for both bodies, the resulting changes in the momenta of the two bodies during their interaction must be equal and opposite. Thus the total mo-

Figure 5-A
Peggy Fleming begins her spin with outstretched arms. As she pulls her arms and free leg close to her body, she spins much faster.

mentum of the system is the same after the collision as before. We can infer from this that no internal force can alter the momentum of the system, and the same can be shown to be true for systems of more than two bodies as well.

So far we have been discussing only linear momentum—the momentum of bodies moving in straight lines. But the momentum of rotating or revolving bodies—**angular momentum**—is also conserved. The definition of angular momentum is slightly more complex than that of linear momentum, for it involves not only the velocity of the body and its mass, but also the distribution of that mass relative to the center of rotation. Consider, for example, a very simple case: a ball whirling about at the end of a rope. It seems logical to expect that doubling the ball's velocity will double its angular momentum, and this is true. We might also expect that doubling the mass of the ball will double the angular momentum, and this too is correct. Less obvious, however, is the fact that, if the ball's velocity and mass remain constant, doubling the length of the rope will also double the angular momentum.

In general, the angular momentum of a body that can be regarded as essentially a point mass, such as a ball on the end of a rope or a planet in orbit about the sun, is equal to *mvr,* where *m* is the body's mass, *v* it s velocity, and *r* its distance from the center of rotation. This quantity is conserved—that is, within any system of bodies it remains unchanged unless some external force acts on the system. Thus a planet, in order for its angular momentum to remain constant, must travel faster when it approaches closer to the sun. This fact is already familiar to us from Kepler's second law, and in fact Kepler's law can be derived from the principle of conservation of angular momentum.

In the case of a spinning body, such as the earth turning on its axis, different parts of the body lie at different distances from the center of rotation. Since *r* is not constant, the actual value of the total angular momentum is more difficult to calculate. Nevertheless, it is obvious that if the body becomes smaller without any change in its mass, it will spin faster. A familiar example is that of a figure skater, who pulls her arms in closer to her body in order to increase her rate of spin. In astronomy, we often encounter similar situations—rotating bodies that change their radius over a period of time. We shall see in Chapter 7, when we deal with the birth and death of stars, in Chapter 9, when we discuss the formation of galaxies, and in Chapter 14, when we consider the origin of the solar system, that the conservation of angular momentum plays a key role in all these processes.

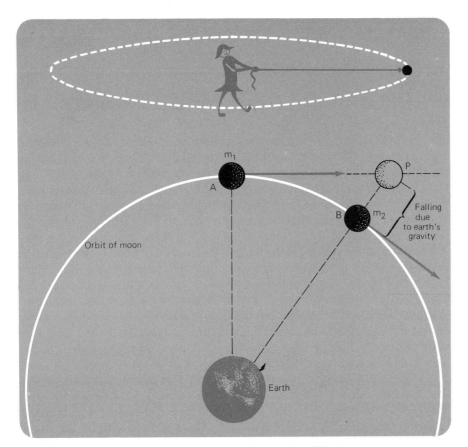

Figure 5-3
In the absence of a force, the moon would travel in a straight line from A to P. But the earth's gravitation causes the moon to "fall" toward the earth, so it actually goes to B.

be exerted to keep the body in its circular path will be four times greater. If the radius of the orbit is doubled without changing the velocity of the body, the inward force need be only half as great to hold it in its orbit. If you imagine swinging a ball on a rope in this way, you will see that these conclusions are consistent with experience. The faster the ball is spun or the smaller the circle in which it moves, the harder you will have to pull on the rope.

To keep a body moving in a circular path, a force directed inward toward the center of the circle is needed. This is called a centripetal force.

Gravitation

For a ball whirling at the end of a rope, the source of the centripetal force is clear. But for a planet orbiting the sun, there is no easily visible force at work. This was not considered a problem in the early days of the heliocentric theory, for until Newton's time, no one suspected that a force was necessary to account for orbital motion. Newton, however, not only showed that such a force was needed, but was able to analyze its nature and describe it mathematically.

The force of gravity, which pulls objects towards the earth, was obviously familiar in Newton's time. Galileo before him had measured the acceleration it produces: 980 cm/sec² at sea level, in modern units. It was Galileo who first stated that, contrary to the belief of the times, all bodies, regardless of their masses, fall towards the earth with this same acceleration. The only reason we do not usually observe this to be true is that air resistance slows some objects more than others. In a vacuum a feather falls at exactly the same speed as a marble. The full significance of this fact, however, could not be appreciated until Newton's analysis of gravitation.

Newton wondered what type of force could cause the moon to remain in orbit around the earth. He realized that, as we have shown, orbital motion can be described as involving a constant "falling" in toward the center of the orbit. The moon, therefore, is always falling toward the earth. Could this fall be, in fact, a result of the earth's gravitational attraction—the same force that caused an apple to fall from the branch of a tree? And could the same force keep the planets in their paths around the sun? Newton boldly guessed that this might be the case. But how could he confirm this hypothesis?

To determine the nature of this force, Newton made several assumptions. First, he assumed that the three laws of motion which he discovered here on earth were applicable to planetary motion—that they were, in fact, universally applicable. Second, he assumed that a mutual gravitational attraction exists between each object in the universe and every other object. Finally, he supposed that this attraction provides the centripetal force that holds the moon in its orbit around the earth, and the planets in their orbits around the sun.

In order to analyze mathematically the force that produces the planetary orbits, Newton had to know their characteristics. These had already been determined by Johannes Kepler in the early years of the seventeenth century, and described in his three laws of planetary motion. First, Kepler had found that the orbits were **ellipses** with the sun located at one focus.

Ellipses are closed curves with rather simple mathematical properties which identify them as close relatives of the circle. (In fact, a circle can be regarded as merely a special kind of ellipse.) An ellipse encloses two points, analogous to the center of a circle, called **foci.** The sum of the distances from each focus to any point on the circumference of the curve is always constant. Because of this property, it is easy to draw an ellipse. Place a loop of string around two tacks pushed through a sheet of paper into a drawing board. The string loop must be slack. Now, place a pencil point inside the loop. Keeping the loop taut, trace a closed curve around the tacks. The result will be an ellipse, with the tacks as foci (Figure 5–4a). The diameter of an ellipse varies from a maximum value in one direction (the major axis) to a minimum value in a direction 90 degrees away (the minor axis). The size of an ellipse is generally specified by its semimajor axis—that is, half of the major axis—and its shape by its **eccentricity**—a measure of how elongated it is.

The orbits of the planets are ellipses with the sun at one focus.

Kepler's second law, the law of areas, states that a radius drawn from the sun to an orbiting planet will sweep out equal areas in equal times, no matter where the planet is in its orbit (Figure 5–4b). From this fact, Newton was able to deduce mathematically that the force responsible for the planetary orbits must be directed towards one point, and that the sun lies at that point. The force in question, in other words, must be a *central* force.

Kepler's third law, known as the law of periods, states that the square of the period of revolution, **P,** of any planet is proportional to the cube of its orbit's semimajor axis, **a.** In mathematical form, we write

$$P^2 \propto a^3 \quad \text{or} \quad P \propto a\sqrt{a}$$

From this fact, Newton was able to find the magnitude of the centripetal force exerted on a planet to keep it in its orbit. He showed that it must be inversely proportional to the square of the planet's distance from the sun. Gravitation, in other words, operates according to an inverse square law, just like the intensity of light.

To check this point, Newton considered the case of the moon in its orbit around the earth. Newton reasoned that since the moon is about 60 times further from the earth's center than is the surface of the earth, the gravitational acceleration experienced by the moon should be $(\frac{1}{60})^2$, or $\frac{1}{3600}$ of it value on earth. Dividing the acceleration of gravity on the surface of the earth, 980 cm/sec², by 3600 gives a value of about .272 cm/sec². Newton then calculated the centripetal acceleration that the moon must actually be experiencing to keep it in its orbit. The result was .274 cm/sec². This remarkably good agreement was a brilliant confirmation of Newton's hypothesis that the moon's orbit is shaped by gravitational force operating in accordance with an inverse square law.

Newton had found how the gravitational force between two bodies varies with their distance. It seemed obvious that the force must also depend in some way on the masses of the bodies involved. After all, wasn't it the huge mass of the earth that accounted for its powerful gravitational pull, that could influence objects as far away as the moon? To determine how mass was related to gravitational force, Newton drew on two already established results: his own second law of motion, and Galileo's discovery that the acceleration of gravity is constant for all masses at the earth's surface.

Newton's analysis can be recapitulated as follows. Consider a body of mass **m** falling freely towards the earth. Its acceleration a_g is 980 cm/sec². The acceleration is produced by the force of the earth's gravity on the body, and this force is what we experience as the body's weight, **w.** Since by Newton's second law, the force on a body must equal the product of its mass and the acceleration it undergoes, we can write

$$w = m\, a_g$$

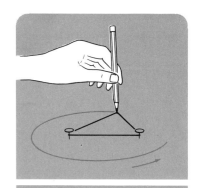

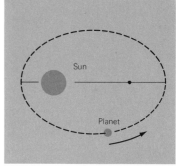

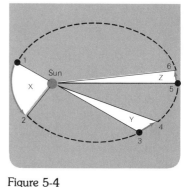

Figure 5-4
The orbit of a planet is an ellipse (you can draw one with a pencil, drawing pins, and a loop of string) with the sun at one focus. When a planet is far from the sun, it travels slowly; it takes as long for the planet to move between 5 and 6 as it does for it to move between 1 and 2. The areas of spaces X, Y, and Z are all equal.

But since a_g is constant, the gravitational force **w** experienced by a body must always be proportional to its mass.

We can look at this in another way. We know that a 500 kg person falls with the same acceleration as a 250 kg person. But the more massive person has twice the inertia. It must take twice the force to give him this acceleration. Gravity, therefore, must be pulling him towards the earth twice as hard as it does the 250 kg person—that is, he weighs twice as much. Again, we must conclude that the gravitational force experienced by a body is proportional to its mass.

In these examples we have not considered the earth's mass, for it is of course constant. Newton extended his reasoning to include the role of **both** masses in determining the gravitational force between two bodies. To do this, he drew on his third law of motion—the law of action and reaction. The third law tells us that if the earth pulls down on a mass—you, for example—the earth must experience an equal but opposite force. If the earth attracts you with a force of 150 lbs (your weight), then a force of 150 lbs is also tugging up at the earth. Thus the forces on you and the earth are equal and opposite. In fact, just as you fall toward the earth with an acceleration of 980 cm/sec² in response to this mutual gravitational attraction, the earth must experience an upward acceleration in response to the same force. The earth, in other words, falls up toward you. But because the earth is enormously more massive than you are, its inertia is proportionately greater. Therefore the acceleration it experiences is infinitesimal, and you do not see the earth rushing up to meet you when you fall (except figuratively).

Newton concluded that the force of attraction between you and the earth is proportional **both to your mass and the earth's mass.** We can demonstrate this relationship in a very simple way. Suppose that we place two blocks, each of mass **m,** in space, and call their mutual force of attraction **F** (Figure 5–5a). Now suppose that we add more blocks of the same mass, so that the body on the left now has a total mass of **2m,** and that on the right **3m.** Each of the 1-**m** blocks on the left attracts each of the 1-**m** blocks on the right, and vice versa. Since the distance hasn't changed, the force of each block on every other block will be **F.** It is evident from the diagram (Figure 5–5b) that the total force on each large block will now be 6**F.** Thus we can see that the attractive force is proportional to the product (3×2) of the two masses.

If we combine this result with the inverse square law, we have a complete account of the factors that influence the gravitational force between two bodies: it is proportional to the product of their masses, and inversely proportional to the distance separating them:

$$F \propto \frac{m_1 m_2}{r^2}$$

When the constant of proportionality is determined, this expression can be used to find the mass of the earth. We simply weigh an object of

Figure 5-5
Every gram of one object exerts the same force on each of the grams of the other object. Here the force exerted on each object is 3×2; or 6.

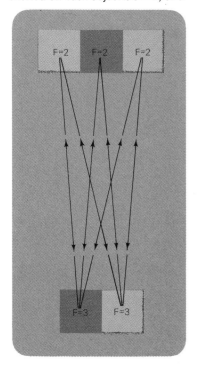

known mass, giving us values for **F** (the weight) and **m**. For **r** we use the distance to the center of the earth (its radius), since Newton showed that the gravitational force of a uniform spherical body is the same as if all its mass were concentrated at its center. We can then solve for m_2, the earth's mass. (The result is about 6×10^{24} kg.)

It is clear from the gravitation formula that *every* body in the universe exerts a force on *every* other body, and that *none* of these forces ever vanishes, no matter how far apart the bodies. The force between two bodies merely becomes insignificantly small when the distance between them becomes very great.

Mutual Orbits

When orbits are discussed, the cases commonly used are those of the earth in orbit about the sun, or an artificial satellite about the earth. In both instances, the mass of the smaller body is so much less than that of the larger that we can neglect its influence on the latter. In such situations we can regard the more massive body as essentially immobile, and the less massive one in orbit around it. In reality, however, any two gravitationally bound bodies are in orbit about each other. More precisely, they both orbit about a common **center of mass,** or **barycenter.**

For a rigid system, such as a solid object, the center of mass is that point at which the entire mass can be thought to be concentrated. In a uniform sphere, for instance, the center of mass coincides with its geometrical center. (That is why we could consider the earth's mass as lying at its center when we "weighed" the earth in the previous section.) For a two-body system, such as the earth and moon, there is also a center of mass. It lies on a line connecting their centers. As with two children of unequal mass balanced on a see-saw, the more massive body will always be closer to the center of mass than the less massive one. In fact, the ratio of the two masses will always be inversely proportional to the ratio of their distances from the barycenter:

$$\frac{m_1}{m_2} = \frac{r_2}{r_1}$$

In other words, if one body is twice as massive as the other, it will always be twice as near the barycenter; if it is only a tenth as massive, it will always be ten times farther away, and so on. When two bodies are in orbit about each other, both orbits are ellipses, with the center of mass at one focus.

Although it is convenient to consider the moon as revolving about a stationary earth, in reality the earth too moves about their common center of mass. It has been found that small variations in the apparent motions of the planets are in reality due to the earth's wobble as it rotates around the earth-moon barycenter. Careful measurements of this motion have shown that the barycenter is about 4700 km from the center of the

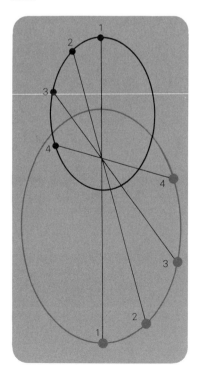

Figure 5-6
Two stars orbit their center of mass, with the more massive star always proportionately closer to the center. The two orbits have the same shape, but the massive star has a smaller orbit.

earth. Thus it actually lies some 1670 km beneath the earth's surface! The moon is about 384,000 km from the earth's center, or about 81 times more distant from the barycenter. Its mass, therefore, is about $1/81$ that of the earth: 7.35×10^{22}kg.

We have been dealing in this example with a two-body system. Actually, though, the notion of a "system" is completely arbitrary. We can regard any collection of particles or bodies as constituting a system. Once we have done so, we must start to distinguish between internal and external forces. The center of mass is uniquely defined for any system, and it can be shown from Newton's laws that its location cannot be affected by any internal forces. Suppose, for example, that two bodies in a system of masses, one five times as massive as the other, are falling together under the force of their mutual gravitational attraction. The forces on each are of course equal. The more massive body, however, has five times the inertia of the less massive body, and will thus acquire only a fifth as much acceleration. The less massive body will approach the center of mass five times more rapidly. Both objects will eventually collide exactly at the center of mass. (Figure 5–7).

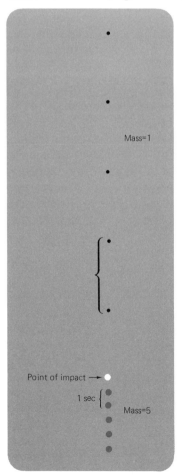

Figure 5-7
Two mutually attracting masses fall together to collide at the center of mass. The moving rocket and the expelled fuel also have a center of mass. Since only internal forces are acting, they move so that their center of mass always remains fixed.

Predicting Planets

In the century and a half after Newton, the science of celestial mechanics—the study of the motions of heavenly bodies—developed rapidly. New mathematical methods were developed to deal with such complex problems as perturbation, the influence of one planet on the orbit of another. The greatest triumph of Newtonian mechanics came in 1846. In that year two men, working independently, startled the world by predicting the location of a previously undiscovered planet, solely from its effect on the path of a planetary neighbor.

The neighboring planet was Uranus, which had been discovered accidentally by William Herschel in 1781. It soon became apparent that Uranus had been seen before—as many as 20 times, in fact—but had always been mistaken for a star. These early sightings, dating back to 1690, should have been a great help in plotting a precise orbit for Uranus. But in 1820, when the French astronomer Alexis Bouvard tried to do this, he found that it was impossible to reconcile all the observations. His solution was to disregard the older observations as probably inaccurate, and base his calculation of the orbit only on recent data. But as the years passed, Uranus was once again found to be straying from its predicted path. The longer the planet was observed, the greater the discrepancy became. Something was wrong.

There seemed only two possible explanations. Either the laws of Newtonian mechanics were in error, or Uranus was being deflected from its calculated orbit by the gravitational influence of some unknown body. Most scientists were naturally reluctant to question Newton's laws, which had proved themselves so well for over a century and a half. The most logical conclusion was the existence of an undiscovered planet outside the orbit of Uranus.

But predicting the existence of a planet was one thing; finding it was quite another. In theory, it was possible to determine the mass and orbit of the hypothetical planet from the perturbations of Uranus' path. In practice the task was enormously difficult. Unknown to each other, however, two men were determined to tackle it. One was the French mathematician Urbain LeVerrier; the other was a young English mathematician named John Couch Adams.

In 1841, while still a 22 year old student at Cambridge University, Adams was already interested in the problem. He was sure that there was no flaw in Newton's laws. As he later wrote,

> Some had even supposed that, at the great distance of **Uranus** from the sun, the law of attraction becomes different from that of the inverse square of the distance. But the law of gravitation was too firmly established for this to be admitted till every other hypothesis had failed, and I felt convinced that in this, as in every previous instance of the kind, the discrepancies which had for a time thrown doubts on the truth of the law, would eventually afford the most striking confirmation of it.

But Adams was not content to wait for others to vindicate his confidence in the law of gravitation; he determined to do it himself. On July 3, 1841, he jotted down the following note:

> Formed a design, in the beginning of this week, of investigating, as soon as possible after taking my degree, the irregularities of the motion of Uranus, wh. are yet unaccounted for; in order to find whether they may be attributed to the action of an undiscovered planet beyond it; and if possible thence to determine the elements of its orbit, &c. approximately wh wd. probably lead to its discovery.

Adams began his laborious calculations in 1843, and by the fall of 1845 he had obtained his first results, which he sent to the Astronomer Royal, Sir George Airy. Adams hoped that a telescopic search for the new planet could begin at once. Airy, however, had several reasons to be suspicious of Adams' prediction. For one thing, Airy did not believe that a planet was responsible for the peculiar behavior of Uranus. He was one of those who thought that Newton's laws required modification. He seems also to have been skeptical about the ability of so young and unproven a mathematician to handle such an intimidating task. Instead of ordering a search for the planet, Airy wrote back with a rather trivial question to Adams about his calculations. Discouraged by Airy's evident lack of interest in his work, Adams did not pursue the matter.

Meanwhile, a race was on, though neither of the participants knew it. LeVerrier presented accounts of his work to the French Academy of Sciences in November 1845 (a month after Adams had written to Airy) and June 1846. When this second paper came to Airy's attention, he saw that LeVerrier's prediction was remarkably close to the one Adams had sent him the previous fall. He then finally requested a search for the hypothetical planet by James Challis at the Cambridge Observatory. Lacking star maps of high quality for the indicated region, however, Challis had to compare observations on different nights, in hope of detecting the planet by its motion against the background of stars. It later turned out that the planet was actually seen on August 4, but Challis neglected to compare his records with those of the previous night, and failed to recognize it.

Later in the same month, LeVerrier presented his third and final report, and soon afterward wrote to Johann Galle at the Berlin Observatory, asking him to look for the planet. The search was begun on the evening of September 23. Excellent new star charts were available at Berlin. With their help, Galle needed just one hour to find the object. The new member of the solar system was named Neptune, and at first LeVerrier was hailed as the man responsible for the discovery. When Adams' work became known, however, he was given a share of the credit.

This remarkable piece of scientific detective work conferred great prestige, not only on Adams and LeVerrier, but also on the science of celestial mechanics. Today we know that luck, as well as imagination and hard work, was a factor in the discovery. Both Adams and LeVerrier had based their calculations on several false assumptions, and the values they obtained for the mass and orbit of Neptune were significantly in error. Had their work been done 20 years earlier or later, Neptune would have been far from its predicted position.

A photograph with the 200-inch telescope, showing the movement of Pluto in a 24-hour period.

The discovery of Neptune led to the search for two other planets in the same way, but the success of Adams and LeVerrier has never been repeated. When peculiarities in the motion of Mercury were noticed, many astronomers for a time believed they must be caused by another planet closer to the sun. A name was reserved for the new body—Vulcan, god of fire—and some astronomers even claimed to have sighted it. But in 1905, Einstein's theory of relativity provided a completely different explanation for the oddities of Mercury's orbit, and little has been heard about Vulcan since.

Neptune itself also exhibited irregularities in its motions that led astronomers to wonder if there might not be yet another unseen member of the solar system out beyond it. Percival Lowell calculated the orbit of such a planet, and initiated a 25 year search for it, using blink microscopes (Chapter 2) and a telescope specially designed for the purpose. In 1930, after Lowell's death, the planet was found, and named Pluto. Many elements in Lowell's calculations, however, were later shown to be in error—so much so that, in the words of a recent writer,

Pluto, at first considered as the second planet, after Neptune, to have been discovered through prediction, is actually the second planet, after Uranus, to have been discovered by accident.

As the bodies fall towards each other, however, the center of mass will remain in exactly the same place. These purely internal forces will not affect it. The center of mass is affected only by forces external to the system. In fact, it obeys Newton's laws just as if it were itself a single physical object. If it is at rest, it will remain at rest unless some external force acts on the system; if it is in motion, it will continue in uniform straight-line motion in the absence of an external force.

Once Newton had developed his laws of motion, conservation of momentum, and gravitation, he was able to use them to derive Kepler's three laws of planetary orbits. Kepler's laws, in other words, were shown to follow from the more general principles of Newtonian physics. Newton was also able to put Kepler's third law into a more general form. For a two-body system, the law becomes

$$(m_1 + m_2) \, P^2 \propto r^3$$

We shall see in the following section that it is this law, along with the previously derived expression giving the relative distances of two orbiting masses from their barycenter, that enables astronomers to determine the masses of binary stars.

BINARY STARS

When you look at the heavens through even a small telescope, what the naked eye saw as a single star often turns out to be two closely spaced stars. The first such **double,** or **binary star** — Mizar, the middle star in the handle of the Big Dipper — was discovered in 1650, and astronomers soon found many others. In fact, double or multiple star systems are apparently the rule, rather than the exception. The majority of stars in our neighborhood belong to such systems.

For a century and a half, astronomers could not be sure whether the components of binary star systems were really near each other in space. It seemed possible that their apparent closeness was merely an optical effect: a chance alignment of two stars, one much more distant than the other, that happen to lie in nearly the same direction seen from earth. Since the first stellar distances were not determined until 1838, there was no evidence to settle the question directly. But as early as 1767, the Rev. John Michell made a statistical study of double stars, and showed that such **optical binaries** should be very rare. The number of known double stars was too great to be accounted for in this way, and Michell concluded that most pairs we see must really be physically associated.

We can observe a similar situation, and construct a similar (though nonmathematical) argument, observing people in the park on a sunny day. Many of the people that we see are not alone, but in couples or small groups. There would not be so many groups solely as a result of chance — people who just happen to be walking in the same direction with the same speed at the same time. We can conclude, therefore, that most of the groupings we observe are not random. There is some underlying reason for people walking together: they are friends, lovers, members of a family, and so on.

People are often reluctant to believe a purely statistical argument without additional evidence. One of those who doubted Michell's reasoning was William Herschel. Herschel conducted an exhaustive search for binary stars, and over a 40 year period found some 800 of them. His goal in this search was the determination of stellar distances. If one component of a double star was really much closer than the other, it should show a proportionately greater parallax.

Herschel failed to find these parallaxes, but in 1804 he discovered something that proved far more important. Comparing observations of the same binary pair made twenty years apart, he found that the two stars appeared to be in motion about each other. The first double star orbits were plotted some years later, and found to be elliptical, like those of the planets. This demonstrated that gravity operated in the remote stellar universe, not just within the confines of the solar system. Today we tend to take this idea for granted, but up until the early nineteenth century no evidence for it was available.

A majority of stars are part of binary or multiple star systems, travelling together through space in mutual orbits.

Visual Binaries

Double stars that can be observed telescopically are known as **visual binaries.** In order to be resolvable by our telescopes, the stars of a binary system must be quite widely separated: usually by dozens, or even hundreds of astronomical units. Kepler's third law tells us that the larger the stars' orbits, the more slowly they travel, and the longer their periods.

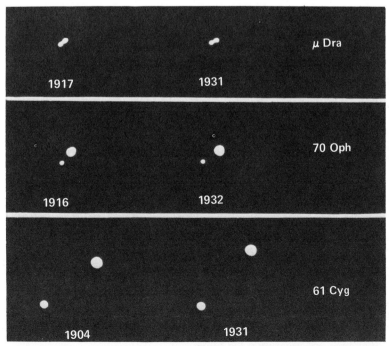

Figure 5-8
These photographs of three binary stars, taken many years apart, show a change of position due to orbital motion. If you assume that the orbits are roughly circular, can you estimate their periods?

The orbital period of the typical visual binary pair is measured in centuries. Often the movement of the stars is so slow that even decades of observation are not sufficient to plot an accurate orbit. In fact, the orbits of only about 100 binary systems have been precisely determined, with another 100 or so known somewhat more roughly.

Even where orbital motion cannot be detected, however, we can infer that two very close stars are part of a binary system if their distances, proper motions, and radial velocities seem identical. Such stars must be moving through space together, and it is a safe assumption that they are gravitationally bound to each other. Nearly 700,000 such systems are known. Some of these are multiple star systems, with as many as six components. An example is the "double double," Epsilon Lyrae, well known to amateur astronomers. A pair of binoculars shows it to be a double star, but seen through a larger telescope, each of the components turns out to be itself double.

Determining Stellar Masses

The only way the mass of a celestial body can be precisely determined is by its gravitational effect on another body nearby. Thus the satellites of planets provide us with a very convenient way of measuring planetary masses. But the average individual star is so isolated in space that its gravitational effect on any other object is negligible. If a star is a binary, however, we have an opportunity to measure the gravitational effect of one component on the other. Studying visual binaries whose orbits have been plotted is thus the best way (and almost the only way) of measuring stellar masses.

If the binary system shows an appreciable proper motion, the observed motion of the two component stars against the sky will generally consist of two intersecting, undulating curves (Figure 5–9a). The center of mass moves with constant velocity, through space, while the undulations are the result of the stars' motion around the center of mass. The most convenient way to establish the orbit of the stars is to ignore the proper motion, regard one of the stars (generally the brighter) as fixed, and record the motion of the other star about (Figure 5–9c). This gives the **apparent relative orbit** of the stars.

The apparent relative orbit does not represent the true orbit of the stars in space. In reality, as we have seen, **both** stars follow elliptical paths about their barycenter. But the apparent relative orbit is nevertheless very useful. For one thing, the apparent relative orbit is also an ellipse, and its period is the same as that of the stars' true orbits about the barycenter. This is important, because the period is related to the masses of the two stars by Kepler's third law:

$$m_1 + m_2 = a^3 / P^2$$

Figure 5-9
When a binary system is initially observed, it looks like the drawing at the bottom. But when we adjust for the proper motion of the binary system, we see that the two stars are actually in orbit around their center of mass.

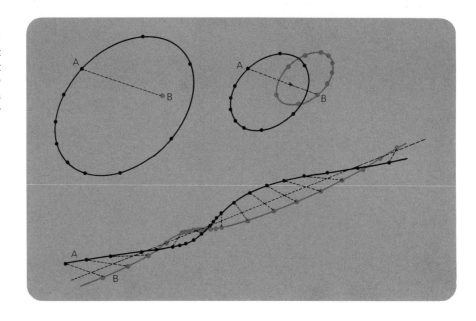

If we can plot the orbits of visual binary stars, and know their distance from us, we can determine their masses.

Moreover, from the apparent relative orbit we can find the other unknown in this equation: the semimajor axis of the true orbit. To do this we need two additional pieces of information. We must know how far away the binary system is, and we must know at what angle we are viewing the plane of its orbit. In general we cannot assume that we are looking at an orbit that is really perpendicular to our line of sight. In most cases the orbit is tilted, and our estimate of its size will be distorted as a result. (This effect, known as foreshortening, is familiar to anyone who has ever tried to draw an object in perspective.) Fortunately, however, there are geometrical methods of determining the angle that the orbital plane makes with our line of sight. The orbit can then be mathematically "untilted" so that its true dimensions can be found. With both the orbital period and the length of the semimajor axis known, we can use Kepler's third law to calculate the sum of the stars' masses.

We would like, of course, to be able to find the masses of the individual stars. Using only Kepler's third law, we cannot do this, for there are two unknown quantities (m_1 and m_2) in a single equation. To solve for both the unknowns, we need at least two equations. We can get our second equation, however, simply by abandoning the fiction (which has been convenient up to this point) that one star is revolving about the other, and locate the center of mass about which they both revolve (Figure 5–9b). We saw earlier that the ratio of the stars' distances from this point is constant, and inversely proportional to the ratio of their masses:

$$\frac{m_1}{m_2} = \frac{r_2}{r_1}$$

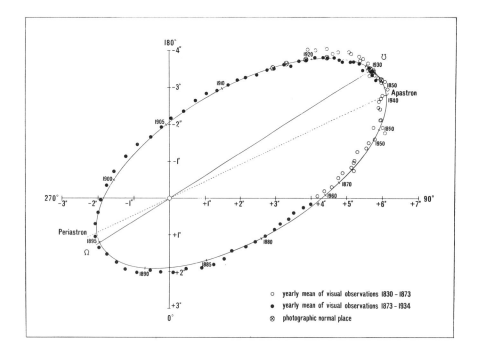

Figure 5-10
An ellipse is fitted to the various points that represent observations of a star's position. Why don't all the points lie on the ellipse?

This will be true at any point in their orbits. Thus we have two equations, one giving the sum of the stars' masses and the other their ratio. We can easily solve these equations to find the actual mass of each star. If the sum of the masses is 8 solar masses, for example, and one of the stars is always three times farther from the center of mass than its companion, the mass of the star with the larger orbit must be 2 solar masses, and that of the star with the smaller orbit 6 solar masses.

ASTROMETRIC BINARIES   Even if one component of a binary star system is too faint to be seen by our telescopes, its gravitational influence on its visible companion may be noticeable. The proper motion of the brighter star will not be a straight line, as we would expect if it were travelling alone through space, but will have a detectable undulation, or "wave." This is the result of its revolution about the center of mass of the two stars. In 1844, for example, it was found that Sirius, the brightest star in the sky, exhibits such a motion, with a period of about 50 years. It was not until 18 years later that the American telescope maker Alvan Clark, testing a new 18½ inch lens, actually observed the small, faint companion star responsible.

A star system which includes an unseen component, whose presence is detected solely by its gravitational influence on the visible star, is called an **astrometric binary.** The period of such a system can often be determined by observation, as in the case of Sirius. If the mass of the visible star can be estimated (means for doing this are discussed in the following chapter), the formulas used in the previous section for the sum

Figure 5-11
The upper photo shows the way we customarily see Sirius A and Sirius B. The lower photo, taken under unusually good observational conditions, shows the tiny dot that is Sirius B, a white dwarf about 10,000 times fainter than A.

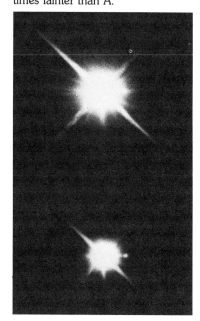

of the stars' masses and the ratio of the masses can be used to find the mass of the invisible star. This has been done for some nearby stars, with very interesting results. In at least one case, and possibly in others, the companion body turns out to have a mass so small that it is probably not a star at all. Presumably, then, it must be a planet! This constitutes the only direct evidence of other planets outside our own solar system, (though we have reason to think they might be relatively common in the universe at large). The reason this method is not very helpful in discovering many planetary systems (if in fact they do exist) is that only a very massive planet, circling a nearby star of unusually low mass, is likely to produce enough of a wiggle in the star's motion to be observable. A less massive planet circling a more massive or more distant star would remain undetected. It is extremely unlikely, for example, that an observer elsewhere in the galaxy possessing astronomical instruments similar to our own could determine that our sun had a family of planets.

Spectroscopic Binaries

Another way of detecting binary stars that cannot be resolved visually is by means of the spectroscope. As two stars revolve in their orbits, they will at times be moving towards us, and at other times away from us. (The only exceptions will be those few systems that we happen to see "full face" — that is, whose orbital planes are exactly perpendicular to our line of sight.) We can see from Figure 5–6 that the stars always remain on opposite sides of the center of mass; thus when one is approaching us, the other will be receding. It is also evident that twice in every revolution both stars will for a brief time be travelling across our line of sight, neither approaching nor receding.

We saw in Chapter 3 that the spectral lines of an approaching star are shifted toward the violet end of the spectrum by the Doppler effect — that is, their observed wavelengths are slightly shorter than they would be if the star were not in motion. Similarly, the spectral lines of a receding star shift towards the red end of the spectrum. When the stars of a binary

Wobbles in the motion of nearby stars may indicate the existence of unseen companions, or even of planets.

Figure 5-12, 5-13
At top, a single-lined spectroscopic binary is recognized by the periodic shift in its spectral wavelengths. At bottom, a double-lined spectroscopic binary has a composite spectrum. When the lines of one star shift toward red, the lines of the other shift toward blue.

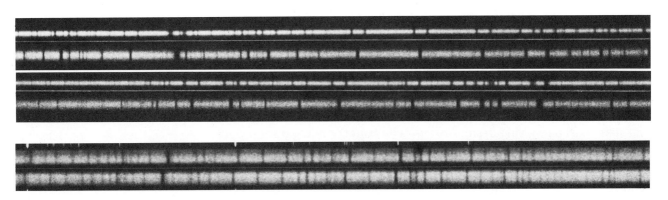

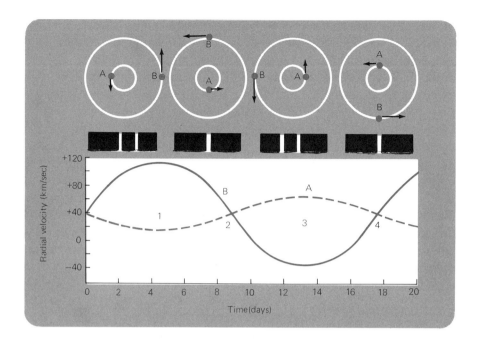

Figure 5-14
Stars A and B are shown in several positions as they orbit around their common center of mass. Underneath are shown their spectral lines. The graph at the bottom shows that the average radial velocity of both stars is about 40 km/sec. Which star would you say is the more massive?

system are moving across our line of sight, their spectral lines should coincide. At other times, however, they should appear double—the lines of the approaching star Doppler-shifted towards the violet, those of its companion shifted toward the red. The amount of these shifts should vary continuously as the stars revolve about their common center of mass.

Star systems which can be identified as double or multiple only by these recurrent Doppler shifts are called **spectroscopic binaries,** and over a thousand of them are known. Some multiple star systems are both visual and spectroscopic. Castor, in Gemini, is an example. Seen through a small telescope, it appears to be two stars. There is also a remote companion, revolving slowly about the pair in a much larger orbit. Each of these three visual components, however, turns out to be double when examined spectroscopically, making Castor a system of at least six stars.

As a rule, the periods of spectroscopic binaries are much shorter than those of visual binaries; they are commonly measured in days rather than centuries. Here again Kepler's third law provides the explanation. The long periods of visual binaries indicate that the stars are moving in large orbits. If this were not so, they would be too close to be resolved as separate stars in our telescopes. Spectroscopic binaries, by contrast, are detected through their Doppler shifts, and these are more likely to be noticeable if the stars are moving very fast. To have high orbital velocities, the stars must be moving in small orbits, with correspondingly short periods. In fact, of all the thousands of known visual binary systems, only three stars have radial velocities great enough to be detected spectroscopically.

From the recurrent Doppler shifts of spectroscopic binaries, the radial velocities of the component stars can be determined. Those for a simple hypothetical case, with circular orbits, are plotted in Figure 5–13. If the orbits are elliptical rather than circular, the curves will have a less symmetrical form, from which the shape and orientation of the stars' orbits can be deduced. It is also possible to find the ratio of the stars' masses from such a curve. From Kepler's third law, we know that the more massive star will always be closer to the barycenter of the system, and will therefore move more slowly than its less massive companion, which has a larger orbit to complete in the same period of time.

We saw in the previous section that if the **ratio** of the masses and the **sum** of the masses of binary components are known, we can find the masses of the individual stars. In the case of spectroscopic binaries, we cannot see the size of the stars' orbits, which we need to determine the sum of the masses. It would seem, though, that the velocity curves should give us both the period of the orbit and the velocities of the individual stars. Knowing how fast the stars go and how long it takes them to complete their revolution, it should be easy to find the sizes of the two orbits.

Unfortunately, however, we do not really know the stellar velocities —only their **radial** components. The radial velocity is only part of the star's true velocity. Exactly how large a part depends on the tilt of the stars' orbital plane. If the plane is nearly perpendicular to our line of sight, the radial component will be only a tiny fraction of each star's true velocity; if the plane is nearly edge on to us, the radial velocity will be nearly equal to the true velocity. In general there is no way of knowing this tilt, and so no way of knowing the true velocities of the stars.

With this incomplete information, we can establish only a lower limit for the masses of spectroscopic binary stars. We know that the true stellar velocities will always be greater than the radial velocities that we measure. If we think the stars are moving more slowly than they really are, we will conclude that they cover less ground than they really do, and so we will underestimate the size of their orbits. If we think they are travelling in smaller orbits than is actually the case, Kepler's third law will give us too low a figure for their masses. The true masses, therefore, will always be greater than the masses calculated in this way.

SPECTRUM BINARIES   We mentioned earlier that if the plane of a binary star orbit is perpendicular to our line of sight, the stars will have no radial velocities. Such systems can never be detected as spectroscopic binaries, since the stars will have no Doppler shifts to measure. The stars may also be too close to be resolved as a visual binary. It is just this fact, however, that sometimes provides a way of discovering the presence of more than one star. If the stars cannot be resolved telescopically, the spectrum they produce will be a composite one, consisting of light from

Spectroscopic binary systems are detected by the periodic Doppler shifts of the lines of the component stars.

both stars. If the two components are of very different spectral types, the composite spectrum may contain features not ordinarily found together —lines typical of a very hot star, for example, along with lines typical of a very cool star. Such evidence tells an astronomer that he is observing a **spectrum binary.**

Eclipsing Binaries

Of particular value to astronomers are those double stars known as **eclipsing binaries.** These are binary systems whose orbits we happen to see edge-on, or nearly so. As a result, each star passes in front of the other at regular intervals, obstructing its light. There are two such **eclipses** in each orbital period⁶one when star A blocks off the light from star B, and one when the positions of the stars are reversed, with star B in the foreground blocking off the light from A. If we record the brightness of such a system on an hourly or daily basis, we find the resulting **light curve** has two noticeable dips, or minima, corresponding to these two eclipses.

The first eclipsing system discovered was Algol, in Perseus. Normally a fairly bright star, it dims quite noticeably every 2.9 days, only to return to normal a few hours later. Western astronomers, curiously, do not seem to have observed this striking behavior until 1669, and it was more than a century later that the precise period and brightness range were recorded. Today more than 1500 eclipsing binaries, with periods ranging from 80 minutes to 27 years, have been found.

A great deal can be learned by studying such systems. For one thing, the missing factor that prevents us from finding the masses of spectroscopic binaries—the tilt of the system's orbital plane—is known for eclipsing pairs. In order for eclipses to take place, we must be viewing the system at an angle of about 90° to its orbital plane. Thus if we can measure

Eclipsing binaries travel in orbits that we see edge-on, so that each component star periodically blocks off the other's light.

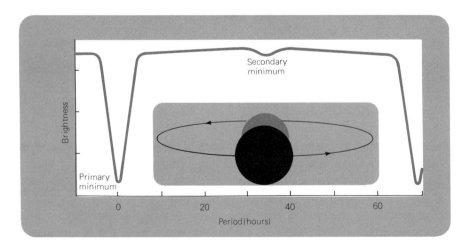

Figure 5-15
Algol is a well known eclipsing binary with a period of 2.8673285 days. During the eclipse the appearance of the star changes dramatically, becoming 3 times fainter in the space of about 3 hours. The secondary minimum which occurs mid-way between the primary minima is due to the eclipse of the brighter star by the fainter.

Figure 5-16

In an eclipsing binary, the shape of the light curve tells us something about the system. Flat-bottomed eclipses (AR Lac, IH Cas) indicate total or annular eclipses. Sharp-bottomed minima (RX Her, Algol) indicate partial eclipses. The position of the secondary minimum between the primary minima is a clue to the shape and orientation of the orbit. Varying light between eclipse (β Lyrae) indicates that the components are distorted non-spherical stars.

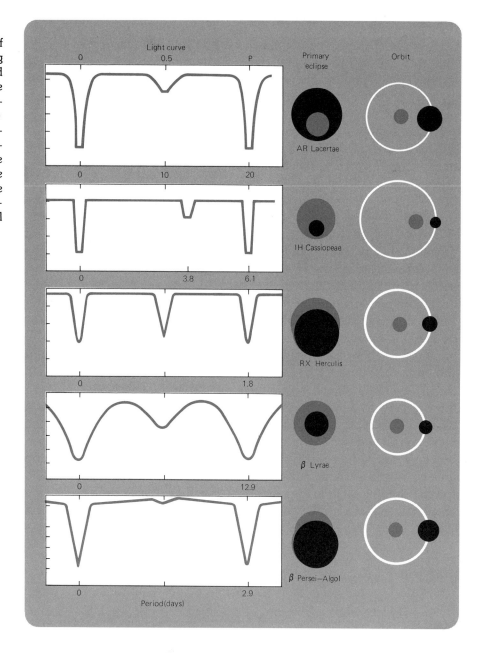

the Doppler shifts of the stars in an eclipsing system, we can determine their true orbital velocities (rather than just a lower limit, as in the case of most spectroscopic binaries). From these we can calculate the true dimensions of the stars' orbits, and finally the actual masses of the two stars.

The appearance of the light curve of an eclipsing binary can tell us the precise type of eclipse that is taking place. If the minima are flat-bottomed, one of the eclipses is total (they will not both be total unless the stars are nearly equal in size, of course). The flat portion of the curve corresponds to the time when the smaller star is entirely in front of or behind the larger one. During this time the light intensity does not change. If the bottoms of the minima are pointed, however, we know that the eclipses must be partial (Figure 5–16), with each star obstructing only a portion of its companion.

The duration and spacing of the eclipses provide additional information about the stars' orbits. If the orbits are circular, both the durations of the two minima and the times between them will be equal, since the stars are always equally far apart and each moves with constant velocity in its orbit. Suppose, though, that the stars move in elliptical orbits that are quite elongated. If the long axis of the ellipse is pointing towards earth, one of the eclipses will take place when the stars are closest together, and are thus moving fastest in their orbits. At the other eclipse the stars will be at their greatest separation, and moving most slowly, according to Kepler's second law (Figure 5–17). The time **between** eclipses will still be equal, but the **durations** of the two eclipses will be quite different. If the short axis of the ellipse is pointed toward us, on the other hand, the **durations** of the eclipses will be identical, but the times **between** them will be unequal. The effects on the light curve of these and other orientations are shown in the figure.

The amount of light lost during each eclipse—that is, the depth of these valleys or minima in the stellar light curve—depends not only on whether the eclipse is total or partial, but also on the temperatures of the two stars. If the orbits are circular, then the area obstructed at each eclipse is the same regardless of which star is in front—it is always the area of the smaller star. But stars of different temperatures, as we saw in the last chapter, differ in the amount of light they radiate from equivalent areas of their surface. A white star with a surface temperature of 10,000 K

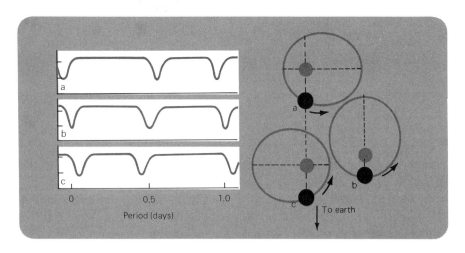

Figure 5-17
The position of the secondary minimun between the primary minima is determined by the orientation of the eclipsing binary orbit with respect to the observer.

radiates, according to the Stefan-Boltzmann law, some 16 times as much energy from each square meter of its surface as an orange star whose surface temperature is 5000 K. Thus when a cool star eclipses a hot companion, the amount of light lost is large. The resulting dip in the light curve, called the **primary minimum,** is very deep. But later in the orbital cycle, when the hot star obstructs light from the cooler one, the amount of light lost is considerably smaller. The result is a much shallower dip in the light curve, known as a **secondary minimum.** The relative depths of these two minima provide a convenient way of calculating the temperatures of the two stellar components.

SIZES OF STARS   When the orbits and velocities of eclipsing stars are known, it is usually fairly easy to find their sizes from the light curve of the system. Between point 1 on the curve, when the small star first begins to pass behind its companion, and point 2, when it disappears completely, it must move a distance exactly equal to its own diameter. Similarly, between the time it reaches point 1 and the time it starts to emerge from behind its companion at point 3, it will have moved a distance exactly equal to the diameter of the larger star. These times can be recorded very precisely, and the relative velocities of the stars can be calculated from our knowledge of their orbits. The two stellar diameters can then be found simply by multiplying the velocity by the appropriate time interval.

We saw in Chapter 4 that the sizes of stars can also be found from the amount of energy they radiate. Since stars are not perfect black bodies, however, that method gives only an approximation. Much greater accuracy can be obtained from eclipsing binary measurements, which thus offer a check on the more generally applicable but less reliable results obtained from the radiation curves. Usually the two methods are found to agree fairly well.

A great deal of information can be deduced from careful interpretation of binary light curves. To mention just one of the more interesting examples, it is possible to detect distortions in the shapes of stars. Stars are not always perfectly spherical. Rapid rotation may cause them to bulge at the equator and be somewhat flattened at the poles. In other instances, the gravitational pull of a massive companion may cause them to become distorted into shapes resembling eggs or footballs. Such stars always have their long axes lined up, pointing towards each other, so that as they revolve about their center of mass we tend to see them "broadside" between eclipses and end-on just before and after eclipses. As a result of this view, the light curves of such systems have a characteristically humped appearance, with a maximum brightness between eclipses. Other peculiarities found in some stellar light curves point to the presence of "hot spots" on the surface of one component, the product of reflected radiation from its nearby companion. In such cases the light peak will usually occur just before or after an eclipse.

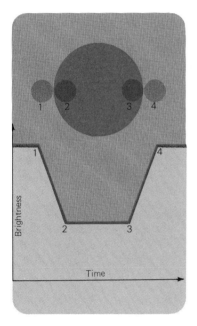

Figure 5-18
The sizes of the two stars of an eclipsing binary system can be calculated from the durations of the various parts of their light curve.

From the light curve of an eclipsing binary system, it is possible to determine the sizes and the temperatures of the component stars.

The discoveries of Kepler and Galileo were elaborated, generalized, and explained by the work of Isaac Newton. Newton's contributions to the study of motion are generally summarized in the form of three laws. Newton's first law asserts that the velocity of a body remains constant unless it is acted upon by some outside force. Acceleration is the rate of change of a body's velocity. It can take the form of a change in speed or a change in direction, or both. Newton's second law states that when a force does act upon a body, the resulting acceleration is directly proportional to, and in the same direction as the force, but inversely proportional to the body's mass. Newton's third law states that whenever an object exerts a force on another object, the second object exerts an equal but opposite force on the first (action and reaction).

Newton realized that in order to maintain their curved paths, rather than flying off into space in a straight line, the planets must be experiencing a constant acceleration inward, toward the sun, which lies at the focus of their orbits. He attributed this acceleration to the force of the sun's gravity. Working from Kepler's laws of planetary orbits and his own laws of motion, Newton was able to devise a theory of universal gravitation. Each particle of matter in the universe attracts every other particle of matter with a force that is proportional to the product of their masses, and inversely proportional to the square of the distance between them: $F \propto m_1 m_2 / r^2$.

When two bodies are bound together by gravitational attraction, they move in elliptical orbits about their common center of mass. The center of mass will thus lie at one focus of both elliptical orbits, and the two bodies will always be on opposite sides of it. The more massive of the two bodies will always be closer to the center of mass.

Binary or multiple star systems are actually more common than solitary stars. Visual binaries, which can be observed telescopically, must be quite far apart. Consequently they have very long periods, and must be observed for many years for their orbits to be plotted. When the orbits of a binary pair can be determined, it is usually possible to find the masses of the two stars. Kepler's third law gives us the sum of the masses in terms of the period (known from observation) and the size of the stars' orbits, which can also be determined if our distance from the system is known. The ratio of the masses can be calculated from the ratio of the stars' distances from their common center of mass. These two relationships provide enough information to solve for the individual mass of each star.

Binary stars that cannot be resolved telescopically can often be detected by their spectra. In all such spectroscopic binary systems, when one of the components is approaching us, the other will be receding. The spectral lines of the approaching star will be shifted toward the violet end of the spectrum by the Doppler effect, while the receding star's spectral lines will show a red shift. For a brief period of time, when both stars are moving across our line of sight, the spectral lines of the two components should coincide; at other times, however, the spectrum of such a pair will show a characteristically double set of lines.

1. A force of 200 dynes produces an acceleration of 25 cm/ sec². What is the mass of the body on which it is acting?

2. A space ship is at rest at point A in empty space. By using its rocket engines, it travels to point B. How is this possible, since a body cannot change its momentum by using only internal forces?

3. When a person pulls a wagon, Newton's third law tells us that the wagon must be pulling back on him with an equal and opposite force. How then do they manage to move at all? (Hint: consider what other forces may be at work in the system.)

4. How is it possible for an astronaut to hurt himself by bumping against the walls of his space capsule, if he has no weight in space?

5. A comet has an orbit which brings it very close to the sun, and then takes it out to a maximum distance of 200 AU from the sun. What is its orbital period?

6. The period of a planet is 8 years. What is its mean distance from the sun?

7. Two stars in a binary system approach to within 15 AU of each other and then recede to a distance of 27 AU. At their closest approach their relative velocity is 90 km/ sec. What is their relative velocity when they are farthest apart? (Hint: use Kepler's second law.)

8. Ordinarily, a spacecraft accelerates by firing its rockets backward and slows down by firing them forward (in the direction in which it is moving). If a satellite in orbit fires a rocket backward, however, it ends up moving more slowly, while if it fires a rocket ahead of it, it ends up moving faster. Can you think of an explanation for this fact? (Hint: consider the effect on the satellite's orbit, and use Kepler's third law.)

9. Suppose the masses of **all** bodies of the solar system doubled, but their orbits remained the same, what could you predict about the periods of the planets in this new system?

10. What facts do we need to know about a satellite in order to calculate the mass of the mother planet?

11. According to Newton's theory, how far does the force of gravity extend?

12. Why does the moon have a surface gravity that is $1/6$ that of earth, though its mass is only $1/81$ that of earth?

13. An imaginary planet has a density the same as that of the earth, but a diameter 3 times as great. What is the surface gravity in terms of the earth's?

14. If the force between two 1-g masses placed 1 cm apart is taken as one unit, what is the force between a 6 g mass and an 8 g mass 8 cm apart?

15. A cannonball and a feather are dropped from a cliff on the moon. Describe exactly how they fall, and explain why.

16. How do we know that gravity operates beyond the surface of the earth? How do we know it operates outside the solar system?

17. How was it possible to predict that the members of most binary systems were really close together in space, not just in the same line of sight from earth, before any stellar distances had been measured?

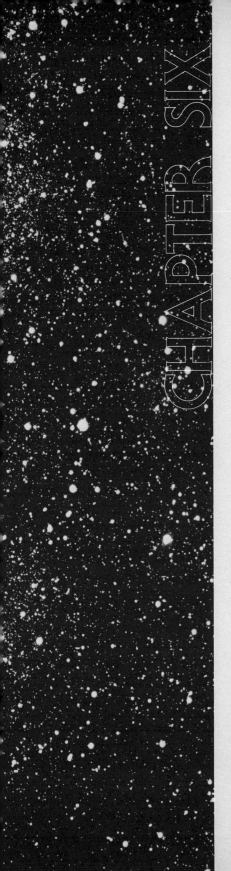

CHAPTER SIX

STELLAR STRUCTURE

There is a general procedure which scientists often follow in their attempts to understand nature. Whenever they examine a set of objects — crystals, or mice, or stars — and observe the same two properties in each one of them, they make a graph plotting one property against the other. This is an easy way of determining whether the two properties are related, and what kind of relationship they might have.

To see the usefulness of this method, consider two varying properties we can observe in Scotch whiskey: the price and the age. By examining the various brands on the market, we can construct a diagram that will show how these two properties are related (Figure 6–1a). The graph readily shows that whiskey aged for a longer time tends to cost more. We can use this graph to predict the probable price of a particular whiskey solely from a knowledge of its age, or the approximate age of a whiskey solely from a knowledge of its price. On the other hand, if we plotted the price against the number of letters in the brand name, we might get a diagram resembling Figure 6–1b. There is obviously no pattern on this graph, indicating that the two properties we are examining are not related in any consistent way.

In the last three chapters we have studied the methods used in order to get facts about stars. We have seen how the masses, luminosities, temperatures, radii and chemical compositions of stars may be deduced from observations. We thus have available to us many disconnected facts about the stars. To formulate an adequate theory of stellar structure and evolution, we must find the hidden connections among these observed properties. This is done by plotting one observed property against another in the way just described.

Property	How Determined	Chapter
Luminosity	Brightness + distance (Inverse square law)	2
Surface Temperature	i. Strength of spectral lines (Spectral sequence)	4
	ii. Color (Wien's law)	4
Radius	i. Luminosity + Surface temperature (Stefan-Boltzmann law)	4
	ii. Eclipsing binaries (light curve)	5
Mass	i. Visual binaries	5
	ii. Spectroscopic eclipsing binaries	5
Composition	Spectroscopic analysis	4

Stellar Luminosity

One of the first things an astronomer needs to know about a star is its luminosity. Unlike lightbulbs, however, stars do not come marked with their luminosities. As we saw in Chapter 2, it is usually necessary to measure the distance of a star before we can tell how luminous it truly is.

Once the luminosities of a large number of stars are known, the next question is whether there are equal numbers of stars of every luminosity, or whether some luminosities are more common than others. Just as a sociologist or census taker wishes to know how many people in a given population have incomes between $5000 and $10,000 per year, how many between $10,000 and $15,000, how many between $15,000 and $20,000, the astronomer wishes to know how many stars there are at each level of luminosity.

Figure 6–2 shows the distribution of luminosities among stars in the neighborhood of our sun. The horizontal axis represents twenty different levels of luminosity, while the vertical axis indicates how many stars there are at each level. The total range of stellar luminosities extends

Dim stars are far more common than luminous ones; the great majority of stars are less luminous than the sun.

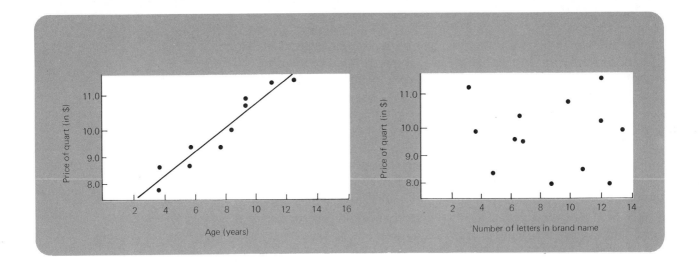

Figure 6-1
When we plot a whiskey's age against its price, we notice that a straight line can be drawn through most of the points, indicating a simple relationship between the two quantities. This is in conformity with our everyday experience that old whiskey costs more. The number of letters in the brand name, not surprisingly, has no relation to the price; no simple line or curve can be drawn through this array of points.

from about one millionth that of the sun to about a million times that of the sun, but the stars are obviously not distributed equally over that range. The diagram clearly shows that there are many more dim stars than luminous ones, for fully ⅔ of the stars near us are only $1/100$ as bright as the sun or less. The superabundance of dim stars does not extend all the way down the scale of luminosities, however. Beyond a certain point the number of dim stars falls off, and there seem to be very few stars less than $1/25,000$ as luminous as the sun. It is hard to tell, though, whether this indicates a real absence of such stars, or whether it merely reflects our difficulty in observing them. It is possible that the galaxy is teeming with very dim stars too inconspicuous for us to notice.

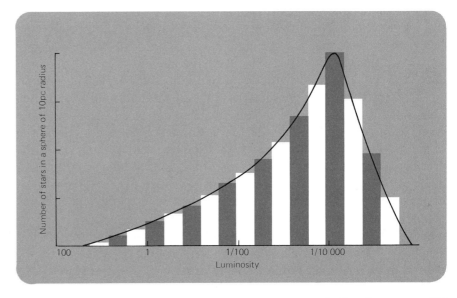

Figure 6-2
A graph of the luminosity function shows the number of stars of various luminosities. We can see that highly luminous stars are very rare. The majority of stars are far less luminous than the sun.

Star Samples

With modern telescopes astronomers can observe or photograph literally millions of stars. It is impossible to study them all. One of the astronomer's important tasks is choosing samples of stars that are truly typical of the stars of our galaxy. This is not as easy as it appears, however. If you were studying the political attitudes of people in America, you would probably realize that interviewing only bankers or only convicted criminals might produce results that were not truly representative. In an attempt to avoid such distortions, you might decide to interview people chosen at random in the subway. But are subway riders a fair cross-section of the general population? You might find that housewives and poor people who ride the subway to work were over-represented, and that suburbanites who drive to work and wealthy people who travel in taxis or limousines were under-represented. Astronomers face the same sorts of problems in selecting a representative sample of stars.

THE BRIGHTEST STARS   Bright stars are the easiest to observe, and so it is always tempting to choose the brightest stars for study whenever possible. But the brightest stars are not a good sample of all the stars.

To see why this is so, it is necessary to consider whether the brightest stars in the sky are bright because they are truly very luminous, or whether they are bright because of their closeness to us. We have already seen that the brightness of a star increases inversely with the square of its distance. If two stars are equally luminous and one is five times as distant as the other, the nearer one will be 25 times as bright. Since distance plays such a powerful role in determining brightness, we might expect that the brightest stars we see are simply the ones closest to us.

Yet this turns out not to be true. There are 254 stellar systems within 11 pc of the sun. Of these nearby stars, only seven appear on a list of the 100 brightest stars. Evidently, then, most of the brightest stars must be bright because they are truly luminous, not because they are close to us. This is doubly surprising, since we have already seen that luminous stars are very rare, and constitute a minority of the stellar population.

The solution to this paradox lies in the **observability** of highly luminous stars. We can measure the observability of a star by the volume of space within which it can be seen. To illustrate the point, consider a star of the sun's luminosity; it will be visible with the naked eye to any observer within 16 pc of the star. If by comparison, we consider a star 100 times as bright, the inverse square law tells us that such a star will appear as bright as the first star from ten times as far away. It will thus be visible to any naked eye observer within 160 pc.

Now compare the volume of space from which each star is visible. The first can be seen from any point within a sphere of radius 16 pc, the second from any point within a sphere of radius 160 pc (Figure 6–3).

Most of the brightest stars in our skies are truly very luminous, rather than merely close to us; the high observability of such stars more than makes up for their scarcity.

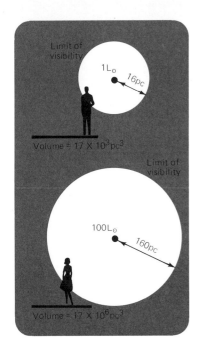

Figure 6-3
A star of solar luminosity can be seen as far away as 16 pc with the naked eye. A star that is 100 times brighter can be seen as far away as 160 pc. The brighter star is therefore visible from anywhere within a volume that is 100 times greater.

But the volume of a sphere is proportional to the cube of its radius. The larger sphere, with a radius ten times that of the smaller, has a volume of 10^3 or 1000 times greater. The more luminous star, in other words, will be visible from a much larger region of space. If we consider an observer located at some random point in space, he is 1000 times as likely to find himself within naked eye sight of the more luminous star than within naked eye sight of the dim star—just as an observer plunked down at random in the United States is about 220 times as likely to find himself in Texas as he is to find himself in Rhode Island.

Looking at Figure 6–2, we can see that stars 100 times more luminous than the sun are about 30 times scarcer than stars of solar luminosity. But such stars are also 1000 times more observable, which more than makes up for their scarcity. That is why these stars are the ones we see as the brightest in our skies. Astronomers must keep this point in mind when they select sample populations of stars for study. Such samples are usually dominated by stars of high luminosity, and it is necessary to compensate for this fact. We do the same thing in our everyday lives. We do not really think, for example, that the population of the country consists principally of rock stars, politicians, and murderers, although they dominate the pages of our newspapers. We realize that these persons merely have the highest observability.

THE NEAREST STARS   If the brightest stars are not a representative cross-section of the stellar population, perhaps the nearest stars can provide a more typical sample.

There are 39 known stars within 5 pc of the sun. Some of them are known to be double or multiple systems, and if the components are counted as separate stars then the figure comes to at least 52. Of these, only three stars—Sirius, Alpha Centauri, and Procyon—appear in the list of the 100 brightest stars, and only eight are visible to the naked eye at all. The majority of the nearest stars have extremely low luminosities, and even their closeness to us does not make them prominent objects. In fact, it is quite possible that many stars within 5 pc have been missed because they are so faint.

Although we are not sure whether we have accounted for all the stars within 5 pc, we can be quite certain that we have missed many dim stars more distant than 5 pc. For evidence of this all we have to do is look at the density of stars in our neighborhood. Within the 5 pc radius considered above, the stellar density is about one star for every ten cubic pc, or .10 stars per pc^3. Now consider the 254 stars within 11 pc of the sun, occupying a volume over ten times as great ($5^3 = 125$, $11^3 = 1331$). The density of stars in this region is somewhat less than .05 stars/pc^3. This suggests that we have probably missed over half the stars.

Another way of looking at the situation is to consider the problem of finding stars of low luminosity. In order to recognize that a faint star is truly of low luminosity, it is necessary to determine that it is a nearby star (rather than a very luminous but distant star, for example). This means measuring the star's parallax. But it would be pointless to try to measure the parallax of each of the hundreds of millions of recorded stars—especially since most will turn out to be too distant to have any measurable parallax anyway. In practice, astronomers will generally select stars with large proper motions for parallax measurement, on the theory that they are probably nearby objects.

Consider, for example, the least luminous star known. Its luminosity is only about $1/500{,}000$ that of the sun. If this star were photographed through a telescope in the course of a sky survey, it would be one faint image on a photographic plate containing thousands of faint images. Our only clue to its closeness would be its relatively large proper motion. But this could only be noticed by comparing two photographs of the same region of the sky, taken several years apart. The task is like that of examining pictures of the entire Red army on review in Peking, in the hope of finding the one soldier who has a tendency to sway at attention. Even with the blink microscope, we would have to be very diligent or very lucky. And if by chance its proper motion happened to be very small, we would never notice it all.

Evidently, then, a selection of stars that depends heavily on those nearest us may underrepresent very faint stars, but this distortion will only increase as we include larger regions of space in our sample. And as we have seen, a selection of the brightest stars hopelessly overrepresents stars of high luminosity. The stars in our immediate neighborhood, therefore, probably constitute as good a sample as we can get for many purposes. As we shall see in Chapter 8, however, they do not necessarily represent all the stars in other parts of the galaxy, or in other galaxies.

Stellar Masses

The second fundamental property of any star is its mass. Stellar masses are a key factor in all our theories of stellar structure and evolution. So important is this information that many astronomers are now working on the determination of the masses of various types of stars. Despite the demand for this data, however, the masses of only about 50 stars have been determined with precision.

MASS, SIZE AND DENSITY   Is the mass of a star related to any other of its observable properties? One place to begin looking for such correlations is with regard to stellar sizes. We might expect more massive stars to be larger than less massive ones, and if we plot the masses of stars against their radii we find that on the whole this is true. Stellar radii do seem to increase with increasing mass, though somewhat more slowly, for about 90 percent of the stars studied. A typical star of 10 solar masses, for example, will have a radius about 5 times that of the sun.

There are, however, exceptions. Two classes of stars in particular do not seem to fit this pattern. One class consists of stars that are unusually small, with radii about $1/100$ that of the sun. Many of these stars are white, indicating that they are very hot. As a result they are known as **white dwarfs.** The second class consists of cool, red stars of exceptionally large size. There are stars of this type which, if they replaced our sun, would swallow up the entire solar system out to the orbit of Jupiter. These stars are called **red giants;** the largest ones are known as **supergiants.**

The differences among these three classes of stars become dramatically clear if we calculate their densities. The **density** of any object is the amount of mass it contains for each unit of its volume. A cubic centimeter of water, for example, contains about one gram of mass under ordinary conditions. We can therefore say that the density of water is 1 g/cm³. The density of pure gold is about 19 times that of water. Air, by contrast, is about $1/1000$ as dense as water under ordinary conditions. If we know the radius of a star, its volume is easy to find; it is $4 \pi r^3/3$, where r is the radius. The mass of the star divided by its volume then gives the density:

$$d = \frac{3\ m}{4\ \pi\ r^3}$$

When this calculation is performed for a large number of stars, we find that stellar densities fall into three very distinct ranges, corresponding to the three families of stars described above. The majority of stars have densities ranging from about 10 times that of water to $1/100$ that of water. When we calculate the densities of the giants and dwarfs, though, we find some astonishing results. Since the volume of a sphere varies with the cube of its radius, a dwarf star with a radius $1/100$ that of the sun has a volume of $(1/100)^3$, or one millionth that of our sun. A giant with a radius 100 times that of the sun has a volume 100^3 or a million times that of the sun. But the masses of these stars do not extend over anything like a comparable range. Stellar masses, in fact, vary relatively little. Most stars fall between $1/10$ the mass of the sun and 50 times the mass of the sun. Thus a red giant may have a mass equal to that of the sun or only slightly greater, spread out over a volume of space that is a million times larger! The density of such a star would be about $1/1,000,000$ that of water. This is a density so low that it would be considered a vacuum here on earth.

But the density of the typical white dwarf is even more remarkable. Imagine, once again, a mass equal, or nearly equal, to that of our sun — but now packed into a volume only $1/1,000,000$ as great. The density of a white dwarf must therefore be on the order of 1,000,000 times that of water! The densest substance known on earth, the metal osmium, is only about $22\frac{1}{2}$ times denser than water. White dwarfs are so dense that a single teaspoon of matter from such a star would weigh 20 tons. It would require a special crane just to lift that spoon.

Main sequence stars have densities roughly comparable to that of water. White dwarfs are a million times denser; red giants are so rarified that their outer layers would pass for a vacuum here on earth.

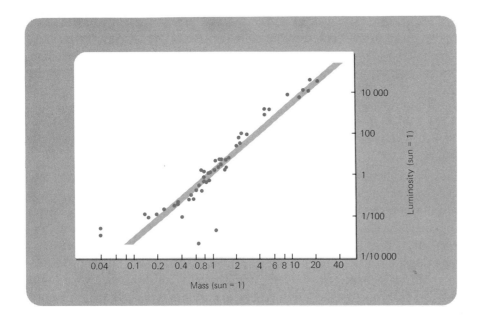

Figure 6-4
The luminosity of a star is closely related to its mass. Only about 10% of all stars fail to fall along the straight line that graphs the mass–luminosity relationship.

MASS AND LUMINOSITY   Having explored the range of stellar luminosities and masses, the next logical step is to see what relationship, if any, can be found between these two properties. In Figure 6–4 masses of a number of stars are plotted on the horizontal axis against their luminosities on the vertical axis. We can see that most stars fall on a narrow band running diagonally across the diagram. It is also evident that a small increase in mass corresponds to a very large increase in luminosity. A star of twice the sun's mass, for example, has a luminosity about 12 times greater. A star of 10 solar masses is roughly 6000 times as luminous as the sun.

The correlation shown on the graph can be put in the form of a mathematical relationship. The luminosity of a star is usually roughly proportional to the $3^{1}/_{2}$ power of its mass. A star that is **n** times as massive as the sun, in other words, will be $n^{3.5}$, or $(n) \times (n) \times (n) \times (\sqrt{n})$, times more luminous. Any attempt to understand how stars work must explain why these two basic properties, mass and luminosity, are linked in precisely this way. One of the reasons we have confidence in our modern theories of stellar structure is that, as we shall see, they actually predict this **mass-luminosity relationship**.

The H-R Diagram

Before a theory of stellar structure can be developed, one more relationship must be studied. This provides the last crucial item of observational evidence—the final piece in the puzzle. It is the relationship between the luminosity of a star and its surface temperature.

In 1913 the American astronomer Henry Norris Russell was investigating the properties of stars in the solar neighborhood whose distances

The luminosities of most stars are proportional to the 3.5 power of their masses.

Lifetimes of Stars

The mass-luminosity relationship can tell us a lot about stellar life expectancies. We know that the stars cannot be eternal. As they shine, they are radiating vast quantities of energy away into space, and this energy has to come from somewhere. Whatever the process, it cannot go on forever; eventually the available store of energy will be used up. The star will no longer be luminous, and that will be the end of its life as a star.

Even without knowing the mechanism by which the energy of stars is produced, we can make certain plausible assumptions about it. Suppose we assume that a star's total store of energy is proportional to its mass. We have no proof of this, but it seems a reasonable guess on the basis of everyday experience. The amount of heat energy that can be obtained from a pile of coal, for example, is proportional to the mass of coal in the pile. The star's energy too must come from some internal process, and a more massive star has more matter to take part in this process. Let us also assume that the luminosity of a star remains more or less constant throughout most of its life.

These simple assumptions, in conjunction with the mass-luminosity relationship, make it possible to draw some very useful conclusions about stellar lifetimes. Suppose, for example, we compare two stars, one with a mass equal to that of the sun, the other ten times as massive. The second star is drawing on a store of energy ten times as large as the first. The mass-luminosity relationship, however, shows that the second star is about $10^{3.5}$, or 3160 times more luminous. This means that it is radiating away its energy at 3160 times the rate of the smaller star. The larger star, with 10 times more fuel to start with, will nevertheless burn out sooner; the smaller star, using up its skimpy supply of fuel much more slowly, will last about 316 times as long.

It is easy to generalize this relationship into a simple mathematical formula. The lifetime of a star will evidently be equal to its fuel supply divided by the rate at which it uses up that fuel. The amount of fuel available we are assuming to be proportional to the star's mass, **m.** From the mass-luminosity law, we see that the rate of consumption is proportional to $m^{3.5}$. We can write,

$$\text{Lifetime} = \frac{\text{Fuel Supply}}{\text{Rate of Use}} \propto \frac{m}{m \times m \times m \times \sqrt{m}} \propto \frac{1}{m \times m \times \sqrt{m}}$$

A star of $\frac{1}{4}$ solar mass, therefore, will last only about $1/(\frac{1}{4} \times \frac{1}{4} \times \frac{1}{2})$, or 32 times as long as the sun.

Superstars, in other words, live fast and die young. This result gives us a clue as to why we find so few stars of high luminosity. These stars have such short life-spans that we are far more likely than not to miss their act. Not so for the fainter stars, the bit players in the drama. They remain around for long periods of time. If we miss any of them it is only because they are so inconspicuous.

had been found by parallax measurement. Since their distances were known, their luminosities could also be easily determined. He plotted the luminosity of each star against its spectral class to produce a diagram, a modern version of which is shown in Figure 6–5. The vertical axis represents luminosity and the horizontal axis is marked to show spectral type, running from O on the left to M on the right. Since spectral type, as we have seen in Chapter 4, is really an indication of temperature, the horizontal scale can also be thought of as a temperature scale, running from about 50,000 K at the extreme left and decreasing to about 2500 K on the right.

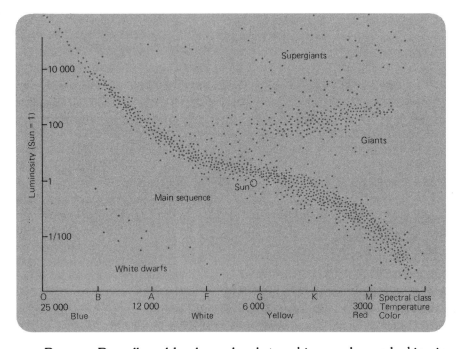

Figure 6-5
In a Hertzsprung–Russell diagram, luminosity is plotted against surface temperature, as determined from the star's spectral type. Ninety percent of all stars lie along the Main Sequence band, which runs diagonally across the diagram.

Because Russell could only use local stars, his sample was lacking in stars of high luminosity, for these are very scarce in our vicinity. There were enough moderately luminous and dim stars, however, for a distinct pattern to emerge on the diagram. Most of the stars lie in a fairly narrow band stretching from the upper left (hot, luminous stars) to the lower right (cool, dim stars). This band is called the **Main Sequence.**

Russell was not the only astronomer thinking along these lines. Two years before Russell's work, the Danish astronomer E. Hertzsprung had been studying star clusters. He had taken all the stars in a particular cluster and plotted the brightness of each against its color index. The color index, a way of quantitatively measuring the color of a star, is closely related to the star's temperature. It was not, in those days, quite so good an indicator of temperature as the spectral type, but involved less time and effort. So both Hertzsprung and Russell, in their different ways, were plotting temperature on the horizontal axis.

An H-R diagram is a graph of temperature against luminosity for a group of stars.

But there is a real difference between the brightness of a star, used by Hertzsprung, and its luminosity, used by Russell. Luminosity is something intrinsic to a star; brightness depends in part on the accidental location of the star with respect to the observer. Hertzsprung was studying clusters of stars, however, where the individual stars are relatively close together. Thus all the members of a particular cluster can be considered about equally distant from us. The relative brightnesses of the stars within each cluster are consequently a good indication of their relative true luminosities. If one star of a cluster appears five times brighter than another, we can assume that it is really about five times more luminous, since the distance of both stars must be nearly identical. Thus Hertzsprung's diagrams can also be regarded as plots of temperature against luminosity.

It would obviously be useful to us if we could fit all of these diagrams together, to form a composite diagram. Similarly, we might combine Hertzsprung's diagrams with that prepared by Russell for nearby stars. The difficulty is that we do not know the actual luminosities of the stars in these diagrams, only the relative luminosities within each cluster.

CLUSTER FITTING   Suppose we assume that the stars making up the main sequence are essentially similar in all the clusters studied. We can then superimpose the diagrams so that the temperature scales along the horizontal axis are lined up over each other (Figure 6–6). We next move the diagrams up or down until the main sequence bands coincide. The result is a composite diagram. The beauty of this method is that it can be done without knowing the distances of the clusters plotted. Better still, it actually enables us to determine those distances. To do this it is only necessary to have one diagram with stars of known luminosity, such as Russell's original diagram of nearby stars, and to assume that these stars are similar to the distant cluster stars. When the diagram of another cluster of unknown distance is superimposed in the manner described above, the true luminosities of all its stars can be read off the vertical scale. Comparing the true luminosities of these stars with their observed brightnesses then makes it easy to calculate their distance, using the inverse square law.

GIANTS AND DWARFS   Figure 6–5 is really a modern composite **Hertzsprung-Russell diagram,** showing the relationship between temperature and luninosity. The most striking feature of the diagram is that stars are not distributed all over it. About 90 percent of all stars lie in the Main Sequence band. This means that most of the highly luminous stars have high surface temperature (upper left corner of the diagram). Stars of low luminosity are cool (lower right corner of the diagram). The sun lies at about the middle of the Main Sequence, and can thus be considered a very typical star.

About 90 percent of all stars lie on the Main Sequence band of the H-R diagram, showing that there is a definite relationship between temperature and luminosity for most stars.

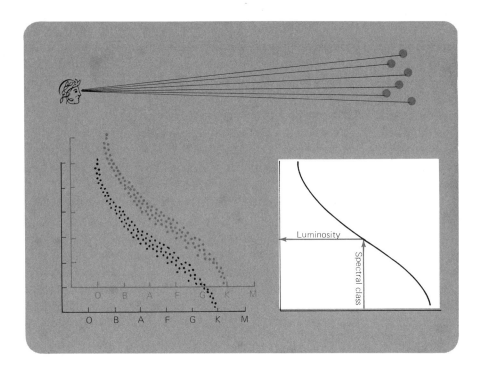

Figure 6-6
All the stars of a cluster can be regarded as about equally distant from us. The Main Sequence of a cluster at an unknown distance can be fitted to the Main Sequence of a cluster whose distance is known. In this way, the luminosities of the stars can be found, and it is possible to determine their distance. The same method can be used, with slightly less accuracy, for individual stars.

Figure 6-8a
Red giant stars such as Aldebaran owe their high luminosities to their large sizes.

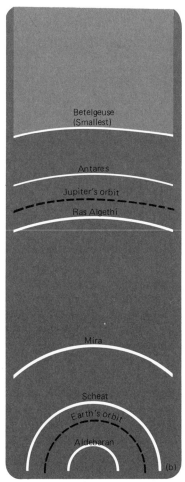

At the lower left there is a group of very hot stars whose luminosities are much lower than Main Sequence stars of the same temperature. We have seen in Chapter 4 that hot stars must radiate large amounts of energy from each unit of their surface area. The only way we can account for such low luminosity in such hot stars is to suppose that they are very small. In fact, this turns out to be the case. These small, hot stars are none other than the white dwarfs. In the neighborhood of our sun, white dwarfs appear to constitute about 10 percent of the stellar population.

In the upper right hand part of the diagram there is a group of cool yellow and red stars, hundreds of times more luminous than Main Sequence stars of the same temperature. To be so cool and yet so luminous, these stars must be very large. They are in fact the red giants. Above the giants, but extending further to the left on the diagram, are the supergiants, hundreds of times more luminous even than the giants. The giants and supergiants are thought to comprise less than one percent of all stars.

RADII ON THE H-R DIAGRAM   In discussing the giants and dwarfs we have seen how the position of a star on the H-R diagram can give an indication of its size. To make this clearer, consider three stars of equal radius but different temperatures. Suppose that star B is twice as hot as star A, while star C is only half as hot. Since B has the highest temperature, it will lie farthest to the left on the diagram. Since the stars are the same size,

Figure 6-7

The diagonals on this H–R diagram are lines of constant radius. The Main Sequence is nearly parallel to these lines, indicating that the sizes of Main Sequence stars do not vary enormously. We can see from this diagram that red giants are much larger than Main Sequence stars, and white dwarfs much smaller.

Figure 6-8b

Supergiants such as Betelgeuse and Antares dwarf even the giants. They have been described as red-hot vacuums.

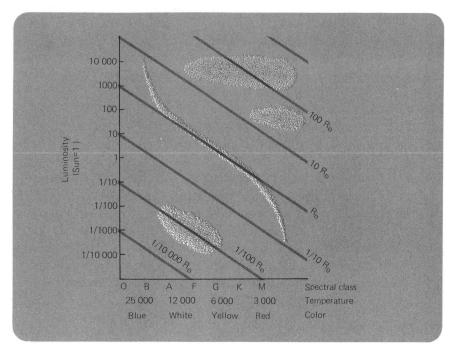

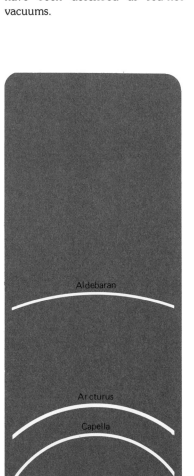

the surface area of B is the same as the surface area of A. Because B is hotter, however, each square meter of B's surface is emitting more radiation than each square meter of A's. Therefore B must be more luminous than A, and will lie above it on the diagram. Using the same arguments, we can see that star C, because it is cooler, will be less luminous than star B. It will therefore lie down below it and further right on the diagram. Notice that stars A, B, and C all lie on the same line.

Stars twice as big as A, B, and C, will all lie on a straight line parallel to the first line but above it. Stars half the size of A, B, and C will lie on a lower line. Several such lines, corresponding to different sizes, are plotted on Figure 6–7. All the stars on each line have the same radius; they differ only in temperature and luminosity.

DISTANCES FROM THE H-R DIAGRAM   An important application of the H-R diagram is its use in finding stellar distances. If we know a star's spectral type, we can immediately find its luminosity by plotting its position on the Main Sequence of the H-R diagram. Knowing the brightness and the luminosity we can then calculate the distance of the star by using the inverse square law. To use this method we must be sure that the star really belongs on the Main Sequence; it will not work for a giant or dwarf. These stars, fortunately, are easily recognizable on the basis of their spectra.

Astronomers like to say that the interior of the sun is better understood than the interior of the earth. Although we can see only the outermost layers of the sun, messages from the interior reach us constantly. There is the sun's gravitational force, for instance, which cannot be stopped by any physical obstacles. There is the electromagnetic radiation from the interior, which eventually makes its way to the surface and is radiated out into space. There are the neutrinos emitted from the solar core, which astronomers are now attempting to detect and study. And there is important evidence from geology and paleontology, which indicates that the sun has not changed very much for a long time.

With all of this information, it is possible to construct theoretical models of the sun which agree very well with our observations. We can then use our understanding of the sun to study other stars. To do this, however, we must make one important assumption. We must assume that the physical laws that operate in our terrestrial laboratories also hold good in the interior of the sun and the other stars, under conditions which are so different from any we have experienced that we can hardly imagine them. So far as we know, that assumption has proven sound.

The Sun

The accompanying table lists some of the observed facts about the sun. From just these few pieces of information, it is possible to construct a model of the sun's internal structure and source of energy. The first question which arises is: What holds the sun together? This is easily answered. Gravitational forces keep the sun together. Gravitational forces that are completely negligible between objects of low mass, such as houses and trees and tractors, become very powerful when we deal with objects as massive as the earth or the sun. It is the mutual gravitational attraction between the particles composing the sun that keeps it together.

Solar Data	
Mass	2×10^{33} g
Radius	7×10^5 km
Density	1.4 g/cm^3
Luminosity	4×10^{33} erg/sec
Temperature (surface)	6000 K
Period of rotation	25–30 days

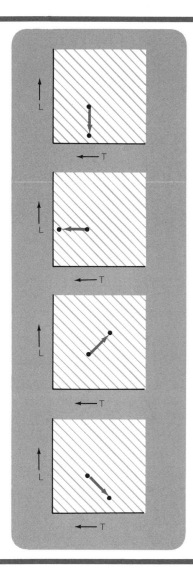

An Exercise with the H-R Diagram

One way to become more familiar with the H-R diagram and what it can tell us is to imagine ways that a star might change, and plot those changes on the diagram. Not all of these changes are physically possible for a real star; some are presented solely as an exercise. Others, however, represent events that do happen to stars in the course of their lifetimes, as we shall see in the following chapter.

1. This star is collapsing. It is maintaining the same temperature, but becoming less luminous. Therefore it must be getting smaller.
2. This star is also collapsing, in a different way. It is getting hotter, but it is not becoming any more luminous. This is only possible if the star is getting smaller.
3. This star is expanding. Though its temperature is falling, it is nevertheless becoming more luminous. It must therefore be getting larger at a very rapid rate. (This actually happens when a Main Sequence star evolves into a giant.)
4. This star is following a line of constant radius; its size is not changing. The star is cooling off, and as a result its luminosity is decreasing. (This actually happens to a white dwarf at the end of its life, when it has exhausted its fuel and shrunk to its minimum possible size.)

As an additional exercise, try reversing these four paths, and describe what must be happening to the star in each case.

The next question that reasonably follows is: If its gravitational force is so strong, what keeps the sun from collapsing? We know that the materials composing the earth, the solid rock and the molten core, are relatively incompressible, and this keeps the earth from collapsing under its own weight. But such an answer will not do for the sun. The sun is 300,-000 times as massive as the earth and the gravitational forces created are accordingly very much greater than the forces within the earth. Under such enormous pressure, all known solid, cold substances would be crushed into an extremely dense state. A mass as great as the sun can only resist collapse if it is a hot gas, for hot gases can exert enormous pressure. To understand the properties of the sun and other stars, therefore, it is necessary to know something about the properties of gases.

The Gas Laws

The pressure of a gas is a familiar property, encountered in many situations. It is the pressure of the trapped air in a balloon, for example, that keeps it from collapsing, despite the inward force of the stretched rubber skin. When air is permitted to escape, the pressure drops and the balloon shrinks.

Pressure is related to the density of the gas exerting the pressure. Consider the pressure in a tire. If it is 20 pounds per square inch, instead of the recommended 28 lbs/in², we take it to the gas station and put in some more air. Putting in air raises the density of the air inside the tire and thereby raises its pressure. This relationship was first enunciated by Robert Boyle. Boyle's law states that the pressure of a gas is proportional to its density if the temperature is unaltered.

Another familiar effect is that the pressure of a gas increases if the temperature is increased. Suppose the tires of a car are inflated to the regulation pressure of 28 lbs/in². Hard driving heats up the tires, and at the end of the day they are found to have a pressure of 32 lbs/in². The driver should not, however, let out air to readjust the pressure to normal, for when the tires cooled again the pressure would drop too low. The pressure of a gas is directly proportional to its temperature, provided its density is maintained constant. This is known as Charles's law.

There is another readily verifiable property of a gas. When a quantity of gas is compressed, without any energy being allowed to escape, it becomes hotter. A simple experiment demonstrates this effect. If the exit of a bicycle pump is closed and the pressure increased by pressing down on the pump handle, the barrel of the pump is found to get warmer. The compression of the gas inside the barrel has increased the temperature.

Notice that in this discussion we have not limited ourselves to any particular gas. All gases seem to behave in a very similar manner as long as we are dealing with them at low densities and at fairly high temperatures. Why should this be so? The answer can be found in the kinetic theory of gases.

The gas laws specify a set of simple relationships among the temperature, pressure, and density of most gases under normal conditions.

The Kinetic Theory of Gases

Around 1860 J. C. Maxwell (whose explanation of the nature of light was mentioned in Chapter 3) published a series of papers in which he explained the behavior of gases. In these papers he supposed gases to consist of extremely small, perfectly elastic particles, constantly in motion. The particles bounce off the walls of their container and off each other, but they do not react in any other way. Under these conditions the **average** energy per particle remains constant, although the energy of a particular particle may change every time it bounces off another particle. When the particles bounce against the walls of the enclosure they exert a pressure. The pressure is greater if there are more particles in the enclosure, which accounts for Boyle's law. Maxwell identified the average energy of motion per particle with the temperature of the gas. To increase the temperature is to increase the average energy of the particles. When this is done the particles bounce off the walls with more vigor. The higher the temperature, the greater the pressure, which explains Charles' law.

Maxwell also showed that if two different kinds of particles were present in the enclosure, the average energy of both particles would be the same. In a mixture of hydrogen and oxygen, for example, the hydrogen molecules and the oxygen molecules have the same average energy. But the energy of motion, known as kinetic energy, increases with the mass of the particles as well as their velocities. A heavy particle, in other words, has more kinetic energy than a light particle moving at the same speed. This is why it is more pleasant to be hit with a ping-pong ball travelling at 20 m/sec than a baseball travelling at that speed.

The exact formula for the kinetic energy of a particle with mass m and velocity v is $\frac{1}{2}mv^2$. The mass of an oxygen molecule is about 16 times greater than that of a hydrogen molecule. Evidently, then, a hydrogen molecule must travel with four times the velocity of an oxygen molecule to have the same amount of energy. This, we shall see, is the explanation for the lack of hydrogen in the earth's atmosphere. The high velocity hydrogen molecules have escaped from the earth's gravitational field, but the slower oxygen molecules have remained behind.

Maxwell's theory also explains the heating produced by compression. Consider an enclosure fitted with a piston that is moving inward, compressing the gas. The gas molecules in the enclosure bounce off the piston. Because the piston is moving, they gain energy each time they do so, like a tennis ball hit by a moving racket. The more energy they gain, the higher the temperature.

Hydrostatic Equilibrium

We can learn a lot about the interior of the sun merely from the fact that it has not collapsed under the force of its own gravity. Such a collapse is prevented by the pressure of the gases that make up the sun. Pressure, as we have seen, is related in a definite manner to density and temperature. Thus study of the pressure inside the sun reveals much about the densities and temperatures at various points of the solar interior.

Knowing the sun's mass and size, we can estimate the gravitational forces impelling it to collapse. A rough estimate indicates that the pressure we are likely to encounter inside the sun is as much as a billion times the air pressure at the earth's surface. This suggests that the temperature would also have to be at least several million degrees.

At these temperatures atoms are completely stripped of their electron shells and are reduced to bare nuclei. This has important consequences for the behavior of the solar material. The central density of the sun must be at least as great as its average overall density: 1.4 g/cm³. At such densities atoms are ordinarily packed so closely together that they interact in various complex ways, and so cannot be thought of as perfectly elastic spheres bouncing freely around in a large volume of space. The kinetic theory, in other words is not applicable to matter under these conditions. But when an atom is completely stripped of its electrons its volume diminishes by a factor of 10,000. The bare nuclei do behave very much like the atoms of a perfect gas. Therefore the gas laws are applicable to the matter inside the sun and other stars, even at densities hundreds of times that of water.

If the sun is to remain stable, the balance between gas pressure and gravity must be maintained at every point within it. This condition is called **hydrostatic equilibrium.** To understand how it works, it is convenient to divide the sun up into a series of concentric shells, as in Figure 6–9. (These divisions do not actually exist in the sun. We imagine them for convenience in calculation, just as we draw circles of latitude and longitude on the earth's surface for navigation.) Each shell has to support the shells outside it; without this support the outer shells would collapse inward. This means that the pressure in each shell must be sufficient to withstand the weight of all the shells above it. The pressure, therefore, increases as we proceed deeper into the star, reaching a maximum at the center.

Most stars consist of highly ionized gas, or plasma, which obeys the gas laws even at high stellar densities.

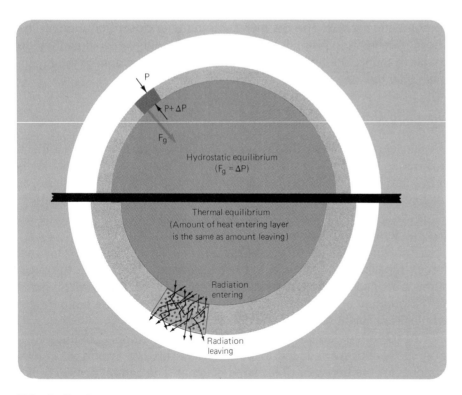

P

P+ΔP

F$_g$

Hydrostatic equilibrium
(F$_g$ = ΔP)

Thermal equilibrium
(Amount of heat entering layer
is the same as amount leaving)

Radiation
entering

Radiation
leaving

Figure 6-9
The pressure increases towards the center of a star, to balance the greater mass that has to be supported (hydrostatic equilibrium). And if the temperature of each layer is to remain constant, the amount of energy entering it has to be equal to the amount leaving it (thermal equilibrium).

What causes this pressure? We know from the gas laws that pressure depends both on temperature and density. The increasing pressure towards the center can be due to rapidly increasing density, or rapidly increasing temperature, or some combination of the two.

Consider one possible model of the mechanism of hydrostatic equilibrium. Suppose the density of the sun is constant throughout. The increase in pressure, therefore, would be due entirely to increasing temperature as we approach the sun's core. If this model is worked out in detail, the central temperature can be calculated to be about 2 million degrees, and the central pressure over a billion atmospheres. This is astonishing enough; but it turns out that if any other distribution of density is chosen, the model that results has an even higher temperature and pressure at the core. The values cited above, then, must be regarded as the **minimum** possible temperature and pressure for the heart of the sun.

How does a star actually adjust itself so that hydrostatic equilibrium is maintained? To understand this we might imagine what would happen if the temperature in the sun decreased by 10 percent. This would lower the pressure to a point where it could no longer balance the forces of the sun's own gravitation. Within a period of hours the sun would contract. But contraction would compress the gases that make up the sun, and both the density and temperature would rise. These increases would raise the pressure until it was again sufficient to balance the force of gravitational collapse. The sun would then once more be in hydrostatic equilibrium.

For a star to be in hydrostatic equilibrium, gas pressure must balance the inward force of gravity at every point.

Thermal Equilibrium

We can construct many different solar models in hydrostatic equilibrium. But only one of them can be the true one; how can we select it? To do this we must see what other conditions our model must satisfy. If we look at other physical laws we find that they impose additional restrictions on our choice of model. Eventually we may be able to disqualify all models but one.

One such law is that governing the flow of heat. It states that heat always flows from a hot region to a cooler one, tending to equalize their temperatures. We already know that the interior of the sun is much hotter than its surface. Therefore heat must be flowing between different layers of the sun, and this flow of heat must be outward from the hot core to the cooler surface, where heat energy is lost into space in the form of electromagnetic radiation. Moreover, in order for the sun to be stable this flow cannot be uneven; that is, the heat flowing into each layer must be exactly equal to the heat flowing out.

For a star to be in thermal equilibrium, the flow of heat into each layer must equal the flow of heat out.

As an analogy, we can think of people leaving a large department store at closing time by means of escalators. If the escalators leading down to the third floor are larger or faster than those carrying people from the third floor to the second, a bottleneck will develop. More people will be arriving on the third floor every minute than can leave. It will start filling up with people, and will soon be extremely crowded. If the same thing happened with the flow of heat in the sun, the temperature of the various layers would change, giving rise in turn to changes in pressure. The sun would have to expand or contract; it could not remain in stable equilibrium. Thus the right model of the sun must not only satisfy hydrostatic equilibrium, but must have a temperature distribution such that the flow of heat from layer to layer is uniform. This new condition is called **thermal equilibrium.**

HEAT TRANSFER Why is the sun's thermal energy lost little by little in the form of radiation, rather than all at once in a titanic explosion? The answer can be found in the way heat is transported inside the sun. There are three methods by which heat can travel: conduction, convection, and radiation. **Conduction** is the process whereby heat spreads directly by contact. When the end of a poker is thrust into a fire, for example, the handle soon becomes hot too. This transfer of heat is due to conduction. Metals are very good conductors of heat; most gases, by contrast, are poor conductors. Since gases are such poor conductors, conduction is not very important in the study of stars.

Convection is a method of heat transfer that depends on the motion of a gas or liquid. Take, for example, a small electric heater in the middle of a room. Air that comes into contact with the radiator gets hot, expands, becomes less dense, and rises towards the ceiling. Cool air from near the ceiling is displaced downwards, comes into contact with the radiator, and gets heated in turn. The circulation that is set up ensures that the whole room will be heated. Convective processes occur in certain regions of stars, especially where the temperature falls rapidly near the stellar surface. In some stars of extremely low luminosity convection may occur throughout.

When convection is not taking place, the only effective means of heat transfer is **radiation.** In this process heat energy is carried through space in the form of electromagnetic waves. Radiation is the process at work when we are warmed at a distance by a heat lamp, or when a piece of meat is cooked by an electric broiler. Although radiation is often the only mode of heat transport available inside a star, it is not usually very efficient at stellar temperatures. This is because stellar matter is highly ionized; almost all of the atoms are stripped of their electrons. As free electrons and bare nuclei swirl about at inconceivable speeds in the stellar interior, a nucleus will sometimes capture an electron to form an atom. When this happens radiation is emitted in the form of a single photon. Almost immediately the newly formed atom is broken up by another photon, however. Photons are thus being continually absorbed and emitted. An average photon can expect to travel no further than a few

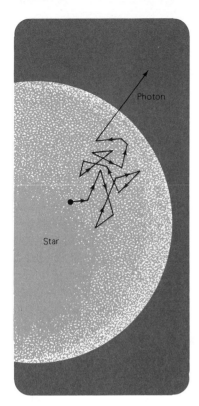

Figure 6-10
Photons created in the interior of a star are absorbed and re-emitted billions of times before they reach the surface. As a result, a journey that might take an unobstructed photon 2–3 seconds may take as much as a million years.

inches before being absorbed. It will be re-emitted almost immediately, but in a completely unpredictable direction.

The more frequently this happens, the more difficult it is for a photon to leave a star. To see why this is so, compare the situation to that of a heavy drinker who, overcome by remorse, tries to avoid further temptation by putting a healthy distance between himself and the bar where he has been drinking. He sets out in some direction away from the bar. Unfortunately, he is handicapped by the fact that he falls down every 10 feet or so. He immediately gets up and starts on his way again, but he is so drunk that he heads in a different direction each time. It is not hard to see that he will make very slow progress. He may cover a lot of ground, but most of it will be in highly inefficient zigzags and circles. It will take him a long time before he wanders far from his starting point. A photon at the center of the sun experiences similar difficulties. If unobstructed, it could reach the surface in less than three seconds. But it is absorbed and redirected at random every centimeter. Before it reaches the surface, some 70 billion centimeters away, it will travel for about a million years! This is why the thermal energy of the sun does not escape all at once.

OPACITY   **Opacity** is the ability of matter to absorb radiation. The opacity of a star's matter plays an important role in determining its internal structure. If the opacity of the material in the sun was lower, for instance, radiation would escape more easily and the sun would be more luminous. Because of the faster loss of energy, however, it would also become cooler. Pressure inside the sun would drop, and it would have to contract until equilibrium was restored. Suppose, on the other hand, that a star is in thermal and hydrostatic equilibrium and the opacity of a certain layer is suddenly increased. The greater opacity would hinder the flow of radiation outward, and radiation would back up. The inner layers would become hotter, and pressure there would increase. The star's hydrostatic equilibrium would be disturbed, and it would have to expand.

Calculation of the opacity of the material inside a star is very complex. It involves many factors, such as the kinds of atoms present, the temperature, and the pressure. In practice such calculations are done on high speed computers, and the results are made available in tables that are constantly being refined and improved as more sophisticated computation methods are developed.

MASS-LUMINOSITY REVISITED   Earlier in this chapter the mass-luminosity relationship was introduced simply as an observed fact. We were not in a position to understand why there should be a correlation between the mass of a star and the rate at which a star radiates energy away into space. Now, however, we can explain the relationship. The explanation was first worked out by the great English astrophysicist Arthur Eddington, a pioneer in this field who first developed many of the basic modern ideas of stellar structure in the 1920s and 1930s. The striking feature of Eddington's theory is that it shows the luminosity of most stars is determined solely by their mass and composition. The nature of the energy source on which the star draws to resupply its radiation losses does not matter. Stellar luminosity, in other words, is independent of the processes that produce stellar energy.

Eddington's calculations were quite complex, but the outline of his argument is very simple. Consider a star in equilibrium. In order to resist its own enormous gravitational forces and not collapse, its internal pressure must be very high. This means that its temperature too must be high. The more massive the star, the hotter it must be. At high temperatures a great deal of radiation is produced, as we have seen in Chapter 4. This radiation must trickle out of the star in the way described above: passing from atom to atom, constantly being absorbed and reradiated.

The amount of radiation trickling out of the star each second is its luminosity. The luminosity, therefore, does not depend on the nature of the stellar energy source, but only on the star's mass and composition. The mass determines how hot the star must be, which in turn dictates how much radiation is produced; mass and composition together determine the rate at which the radiation gets to the stellar surface and escapes into space. Eddington was able to clinch this argument by actually calculating just how the luminosity of stars should vary with their masses. He showed that the luminosity of a typical star should be proportional to roughly the $3\frac{1}{2}$ power of its mass ($m^{3.5}$). This is close to the observed relationship, and represented a great triumph for stellar theory.

A star, then radiates energy into space because it is hot, not because it necessarily has any way of replacing that energy. If the star has a suitable energy source to resupply its losses, well and good. But what if there is no adequate energy supply available? Can the star somehow economize? Can it cut back its radiation so as to conserve what energy it has? The answer is no; as long as the star is hot, it must continue radiating, at a rate determined solely by its temperature and the opacity of its material. A star in other words, radiates as much as it has to, not as much as it can afford to. In the following section we shall see just what energy sources a star can tap to balance its energy budget.

Figure 6-11
The work of Sir Arthur S. Eddington is the foundation of many branches of modern astrophysics.

STELLAR ENERGY

The sun emits 4×10^{33} ergs of energy every second. This is equivalent to 500,000,000,000,000,000,000,000 horsepower. What is the ultimate source of all this energy? At first glance that seems to be a very difficult question. We have seen that the nature of the source of solar energy does not affect the appearance of the sun or the rate at which it radiates. In fact, even if the sun were suddenly deprived of its source of energy entirely, this would have no observable effect for many thousands of years. How then is it possible to determine the process by which the sun constantly renews its energy supply?

AGE OF THE SUN Although the nature of the solar energy source does not determine the sun's rate of radiation, it does determine how quickly the sun runs out of fuel. The clue to the sun's energy source, therefore, lies not so much in the present appearance or luminosity of the sun, but in the fact that its appearance and luminosity have remained more or less constant over billions of years.

This fact is known from many different kinds of evidence. The sun has been observed accurately and extensively for only the last 75 years. During this time there has been no apparent change in its size or luminosity. We know from historical records, moreover, that the sun cannot have changed its luminosity by even 1 percent in the last 3000 years. A change as small as that would have greatly altered the ecological balance on earth through its effect on the climate. Furthermore, fossil algae have been found that are more than a billion years old. They could not have flourished if the temperature on earth was different by more than about 20 C from what it is now. It is clear, then, that the sun has maintained roughly its present luminosity for over a billion years.

Geophysical evidence, based largely on the rate of radioactive decay of certain rocks in the earth's crust, suggests that the earth is about 4.6 billion years old. The sun is thought to be at least as old as the earth, and possibly somewhat older. If the sun has maintained its present luminosity for a period of nearly 5 billion years, it has expended a vast amount of energy. Knowing the mass of the sun and its luminosity, it is easy to calculate that each gram of the sun's matter must have provided an average of 3×10^{17} ergs over that period.

ENERGY SOURCES The sun could not have maintained its luminosity for such an extraordinarily long time without some very generous source of energy. Most ordinary processes can be shown to be insufficient. Chemical energy, for example, can provide up to 10^{12} ergs per gram. But this is far too small for the sun's requirements. If the sun had had to rely on chemical reactions alone, it would have run out of fuel in about 300,000 years.

Gravitational energy is another possibility. Water at an elevation has potential energy due to its position in the earth's gravitational field. We can tap that energy by allowing the water to fall and turn the turbines of a hydroelectric power plant, thus converting its potential energy first into energy of motion, and then into electrical energy. Similarly, a mass like that of the sun, occupying a large volume, has gravitational energy which it can draw on by contracting towards its own center. It is thought that the sun was once a far larger, more diffuse mass of gas. Its **gravitational potential energy** was then greater than it is at present. To attain its present size it must already have cashed in some of that original gravitational energy. It could use up more of it by contracting further.

To see exactly how this process works, consider what happens when a star contracts. Its density must increase, so its internal temperature must also rise. The star has thus exchanged gravitational energy for thermal energy. When the star grows hotter, however, it will also radiate energy away faster. It can be shown that whenever a star draws on its gravitational energy, only half of it goes to heat up the star, while the other half is radiated away. You might say that, for a star, using gravitational energy is analogous to selling long-held corporate bonds. It increases one's spendable assets, but not by as much as the face value of the bonds, because there is a high tax on the transaction. Where gravitational energy is concerned, stars are always in the 50 percent bracket; the star gets to keep half in the form of higher temperature, but must give up half in the form of radiation.

We can look at the process from another point of view. Consider what would happen to a star if it radiated away a certain amount of energy and had no other source from which to replenish it. Its internal temperature would drop, and consequently so would its pressure. It would then have to contract until the temperature and pressure were again high enough to maintain equilibrium. But when the star gets smaller, its gravitational forces become stronger, since gravity increases inversely with the square of the distance. The old temperature and pressure will no longer be enough; equilibrium will not be restored until the star is hotter than it was before. Thus we have the paradoxical situation that, for a star living off its gravitational energy, the more energy it radiates away, the hotter it gets.

During the nineteenth century, physicists calculated how long the sun could have kept going solely by using its gravitational energy. Assuming that it was once a large, diffuse mass of gas that had contracted to its present size, they found that the process could have taken no more than about 20 million years. This number, large as it seems, met with objections from geologists, who even in those days already had good evidence that the earth was much older than that. It can be shown that each gram of the sun's mass had yielded up 5×10^{14} ergs of gravitational energy in the process of shrinking to its present size, but this is still not enough to account for the sun's age and luminosity.

There is also an astronomical reason for rejecting the idea that gravitational energy feeds the stars. It has to do with variable stars—stars whose luminosities fluctuate. Some stars of this type apparently pulsate in a very regular way, with periods that remain extraordinarily constant. It can be shown that the period of such stars is related to their density, very much as the pitch of a guitar string depends on its thickness. If these stars are living off their gravitational energy they must be slowly contracting. Since contraction involves a change in density, we would expect to see corresponding changes in their periods. The change would be small, but over time it would add up, just as a watch that is only slightly slow will eventually lose a very noticeable amount of time. A typical star of this type might have a period of about $3\frac{1}{2}$ days. A one-second change in the

Figure 6-12
One of the greatest physicists of all time was Albert Einstein. His theory of relativity transformed our notions of time and space. Among the ideas that followed from the theory was the equivalence of mass and energy.

period, which is only about .00033 percent, would add up to a full 20 minutes after 1200 periods, or 11½ years. Thus if such changes were occurring it would not be long before we could detect them. The fact that they have not been observed suggests that most stars are not contracting, and must have some other source of energy than gravitation.

Mass and Energy

What other source of energy might be available to stars? The answer began to emerge in 1905, with the publication of Einstein's Special Theory of Relativity. This theory presents a profound reinterpretation of the ideas of space, time, and energy that physicists had developed during the previous 250 years. One of the most revolutionary consequences of the theory is the idea that mass and energy are not completely independent entities, as scientists had assumed since the days of Newton. Einstein showed that a specific amount of energy could be considered equivalent to a specific amount of mass. The exact relationship is given by an equation which has become famous as the formula that made the atomic bomb possible:

The fact that the sun's luminosity has not changed much for billions of years suggests that it must be producing nuclear energy.

$$E = mc^2$$

In this formula E is the energy, measured in ergs; m is the mass in grams; and c is the velocity of light, 3×10^{10} cm/sec.

This relationship suggests that matter can be thought of as a concentrated form of energy, and that it might even be possible to transform a quantity of matter into the equivalent amount of energy. Moreover, since c is a very large number, and appears in the formula raised to the second power (that is, multiplied by itself), a very small amount of matter could make available a huge quantity of energy. Suppose, for example, we could change one gram of matter (about ⅓ the amount of matter in a penny) into energy. The amount of energy released would be

$$1 \text{ g} \times (3 \times 10^{10} \text{cm/sec})^2 = 9 \times 10^{20} \text{ ergs}$$

Expressed in more familiar units, this is 25 million kilowatt-hours: enough energy to keep 10 million hundred-watt bulbs burning all day.

Suppose the sun was living off its own mass, consuming some of its substance to provide itself with energy. If it could do this, it could shine for a very long time. Since it has a mass of 2×10^{33} g, it would have an energy reserve of

$$(2 \times 10^{33} \text{ g}) \times (3 \times 10^{10} \text{ cm/sec})^2 = 18 \times 10^{53} \text{ ergs}.$$

Tapping this energy reservoir at the rate of 12×10^{40} ergs/yr, as the sun now does, it could go on shining for about 15 trillion years.

The theory of relativity merely specifies how much energy is locked into every gram of matter; it does not say how to get it out. In fact, other laws of physics greatly restrict the possibility of transforming matter into energy, and the conditions under which such a transformation can take place are severely limited. This is certainly fortunate; otherwise you might find the pencil in your hand suddenly exploding with the force of a hundred H-bombs.

The difficulty of converting mass into energy is so great that before Einstein's theory scientists had thought it impossible to destroy matter by any means whatsoever. This idea was expressed in the law of conservation of mass, which stated that the amount of mass entering into any physical or chemical transformation must always equal the amount of mass that emerged. This is still true to a very good approximation in chemical reactions. Mass *is* lost in such chemical changes as the detonation of an explosive substance but so little that it is almost impossible to measure. If you weighed a stick of dynamite, and then weighed the products left over after it exploded (mostly hot gases), you would find that only about .0001 percent was missing. The conversion of just that much mass is enough to supply all the energy of the explosion.

Chemical events are highly inefficient. They only liberate a minute fraction of the mass-energy locked up in the reactants. In our century, we have begun to understand the far more powerful processes known as **nuclear reactions.** In chemical changes, only the outer electron shells of the atoms are involved. Nuclear reactions involve changes in the atomic nucleus. Nuclei of heavy atoms may break apart into lighter nuclei. This process, known as **fission,** supplies the energy of the atomic bomb, and of our nuclear reactors. Light nuclei may also be made to combine, forming a heavier nucleus. This is **fusion,** the process that takes place in the explosion of a hydrogen bomb—and in the energy production of the stars. The progressive building up of heavier elements from hydrogen is a very rich source of energy. The process can proceed all the way until iron is formed, at which point all the available energy has been extracted. About 99.2 g of iron can be made by fusion from 100 g of hydrogen. This means that .8 percent of the mass of the hydrogen is converted into energy during the process. Fusion is therefore about a million times more productive as an energy source then chemical reactions.

Most of this energy is released during the first step up the ladder of elements: the formation of a helium nucleus from four nuclei of hydrogen. This is in fact the process thought to be occurring in the sun and all Main Sequence stars. In the fusion process, about .71 percent of the initial mass of hydrogen entering the reaction is lost and converted into energy. How much energy does each gram of hydrogen produce in this way? The mass lost, .0071 g, is equivalent to $.0071 \times (3 \times 10^{10})^2 = 6.4 \times 10^{18}$ ergs. This is about twenty times greater than the 3×10^{17} ergs/g needed to have kept the sun going for the last 5 billion years. If the sun was once pure hydrogen, it has only used up about 5 percent of its fuel reserve in all that time.

A very small amount of mass is equivalent to a very large amount of energy.

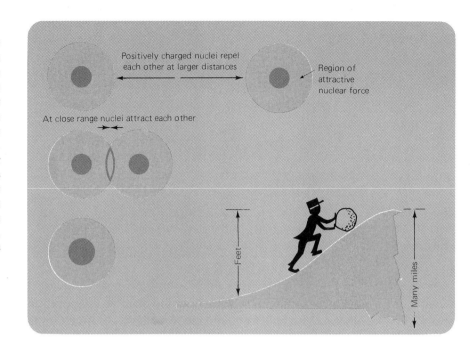

Figure 6-13
A man pushing a boulder to the edge of a cliff must invest a certain amount of energy. But when the boulder topples over, its fall makes a far greater amount of energy available — that is, it can be used to do work, such as turning a turbine or driving a pile at the bottom. Similarly, in the process of nuclear fusion, highly energetic collisions are needed to force positively charged nuclei together. But when the strong nuclear attractive force takes over and binds them together, far more energy is released than was necessary to start the process.

Rate of Energy Production

We have seen that the sun cannot budget its energy expenditure to match its energy production. If the sun lacked a source of energy sufficient to replace what it radiates away, it would have to cash in some of its gravitational energy by contracting. Yet we know that the sun has not had to do this in several billion years, for it has not changed appreciably over that time. Its energy production must therefore exactly equal what it loses through radiation. Such a perfect equilibrium could hardly be achieved, and maintained for so many billions of years, by accident. If the sun's energy is derived from nuclear reactions, how do they come up with just the right amount of energy to supply its needs?

Since the sun's energy loss cannot be regulated, it must be the energy production that is regulated instead. There must be some mechanism that adjusts the rate of energy release so that it always balances the rate at which energy is radiated out into space. The principle involved in such a self-regulating system is called **feedback.**

Feedback is a common phenomenon in our ordinary experience. Living things in particular make use of very complex feedback mechanisms. When your body overheats, for example, this condition is detected by certain sensors in your tissues. Electrical messages travel by way of the nervous system and chemical messages by way of the bloodstream to various organs. They respond in appropriate ways to restore the body to its normal condition. Your metabolism slows down, producing less heat;

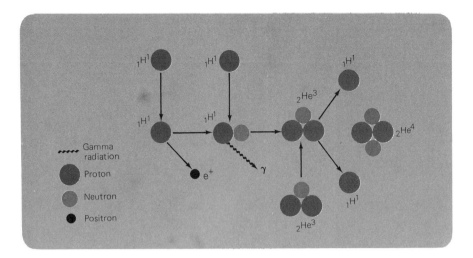

Figure 6-14
The proton–proton chain provides more than 90% of the sun's energy. Hydrogen is first transmuted into deuterium (heavy hydrogen) then into He³ and finally into He⁴. In the process, 0.7% of the mass is converted into energy. For any one atom, the entire process takes about 14 billion years; there are so many atoms in a star, however, that at any given instant, an enormous number are completing the cycle.

sweat glands release perspiration, which cools the skin by evaporation; and so on. Machines can be built to operate in a similar way. An example is the ordinary household thermostat. When the house gets too cold, a temperature-sensitive switch is activated that turns on the furnace. When a comfortable temperature is restored, the switch turns the furnace off again.

Something rather similar takes place within a star. To see how, we must look more closely at the fusion reactions that provide stellar energy. The nuclei of atoms are held together by a force that is very powerful at extremely close range. This strong nuclear force is opposed by an electrical force whose effective range is much greater. All nuclei are positively charged, and since like charges repel each other, electrical repulsion ordinarily keeps them apart. If two nuclei approach within about 10^{-13} cm of each other, however, the strong nuclear force overpowers the electrical repulsion. The nuclei are pulled tightly together to form a single heavier nucleus. This is the process of fusion.

But how can the nuclei get close enough to fuse if they are repelled by the electrical force? This is only possible if they crash into each other with high velocities. We have seen earlier that temperature can be considered a measure of the average energy of motion of atomic or sub-atomic particles. At high temperatures, these particles typically move at tremendous speeds, and the internal temperatures of stars reach millions of degrees.

Calculation of the rates of various nuclear reactions possible in stars has shown that there are two chief routes by which hydrogen can be converted into helium. The first, called the **proton-proton chain,** involves the progressive building up first of heavy hydrogen (deuterium), then of light helium, and finally of normal helium, in a three step process. (Figure 6–14). The second route, the **C-N cycle,** employs the

Most stars derive their energy from the fusion of 4 hydrogen nuclei into 1 helium nucleus. In the process, .7 percent of the mass is converted to energy.

nuclei of carbon and nitrogen as intermediaries. The carbon nucleus undergoes various changes (Figure 6–15) but finally returns to its original form — it is not used up. In the course of the various steps, meanwhile, four hydrogen nuclei have emerged as a single nucleus of helium.

The rate at which the proton-proton chain produces energy varies roughly with the fourth power of the temperature. Thus an increase of only 20 percent in the temperature more than doubles the rate at which energy is released ($1.2^4 = 2.1$). Doubling the temperature results in 16 times the energy production. The C-N cycle is even more critically dependent on temperature; the energy released increases with the 20th power of the temperature! Thus increasing the temperature just 10 percent increases the rate of energy production by more than 600 percent, and doubling the temperature increases it by a factor of a million.

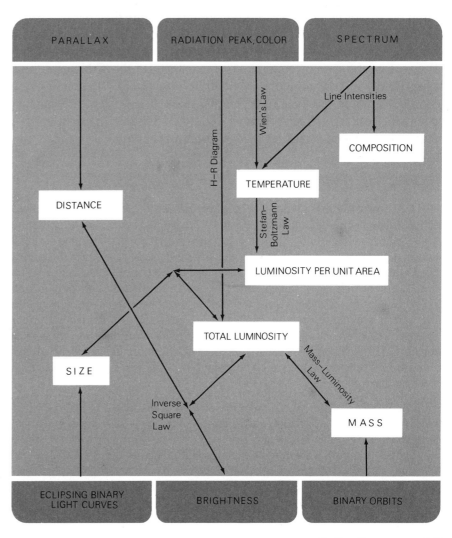

Figure 6-15
This chart shows some of the relationships among stellar properties that are described in the preceding chapters. The starting points are those things that we can observe directly: parallax, brightness, spectral line strengths, color, binary light curve, and binary orbits. Notice the two sets of triangular relationships (distance – brightness – luminosity and luminosity – size – luminosity/area). Knowing any two of either triad, we can easily find the third.

Because the C-N cycle is so strongly dependent on the temperature, it is more important than the proton-proton chain in hot stars. Main Sequence stars with internal temperatures even slightly higher than the sun's derive most of their energy from the C-N cycle. In the sun, however, about 92 percent of the energy is provided by the proton-proton chain and only about 8 percent by the C-N cycle, while cooler stars get virtually all of their energy from the proton-proton chain.

This temperature dependence explains how nuclear reactions can always supply exactly as much energy as a star radiates away. Think of a star that is producing energy through either of these two processes. If the process does not supply enough energy to balance the star's radiation losses, the star will have to contract. But when it does so its internal temperature, we have seen, must increase. Even a small rise in the temperature speeds up the nuclear energy production enough to meet the star's needs, and equilibrium is restored. This feedback mechanism explains another curious fact: all Main Sequence stars, despite their widely varying surface temperatures, reach roughly the same central temperature. The star Rigel, for example, is some 10,000 times as luminous as the sun. Because of the extreme temperature sensitivity of the C-N cycle, however, Rigel's internal temperature need be only about 70 percent higher than the sun's to produce 10,000 times as much energy. On attaining a central temperature of a few tens of millions of degrees, in other words, a star finds itself with an almost limitless source of energy.

The great temperature sensitivity of nuclear reactions enables stars to remain in stable equilibrium.

SUMMARY

Our modern understanding of the stars rests on the discovery of relatively simple relationships among observed stellar properties: mass, size, temperature, luminosity, and composition. In order to study these relationships, it is desirable to find representative samples of stars that will be typical of the stellar population of our galaxy. This presents certain difficulties. If we choose our sample from the brightest stars (which are the easiest to observe) we find that stars of high luminosity are greatly over-represented. If instead we choose relatively nearby stars the results are better, though faint stars are probably still somewhat under-represented. Luminous stars, though they are highly conspicuous, are actually relatively rare, whereas faint stars are extremely common. In our neighborhood, there are 10 times as many stars with only $1/10,000$ the sun's luminosity as there are stars like the sun.

When the masses of stars are compared with their sizes, it is found that, for most stars, the two factors are closely linked. Because of this, most stars have similar densities, on the order of the density of water ($1 \text{ g}/\text{cm}^3$). But there are two classes of stars that are notable exceptions. Red giants, though not unusually massive, are extraordinarily large, and therefore have very low densities; their outer layers would pass for a vacuum on earth. White dwarfs are extremely small, and thus very dense; a matchbox full of matter from such a star would weigh many tons.

When the luminosities of stars are compared with their masses, another important relationship emerges. The luminosity of most stars is roughly proportional to the 3½ power of their mass: $L \propto (m) \times (m) \times (m) \times (\sqrt{m})$. This enables us to estimate the relative lifetimes of stars of various mass. The fuel supply of a star can be assumed to be roughly proportional to its mass, and its rate of fuel consumption (its luminosity) is proportional to the 3½ power of its mass. The star's lifetime—its fuel supply divided by its rate of consumption—is thus inversely proportional to the 2½ power of its mass: $\propto 1/(m) \times (m) \times (\sqrt{m})$. Thus a star with 9 times the sun's mass will live only about ¹⁄₂₄₃ as long.

When stellar luminosity is plotted against spectral type (temperature), the resulting graph is called a Hertzsprung-Russell, or H-R, diagram. About 90 percent of all stars are found to lie along the Main Sequence band, running from the upper left of the diagram as it is usually drawn (hot, blue, luminous stars) to the lower right of the diagram (cool, red, faint stars). The white dwarfs, constituting about 10 percent of the stellar population, lie below and to the left of the Main Sequence band. Most of these stars are very hot, but their small size makes them dim. Red giants, comprising about 1 percent of the stellar population, lie above and to the right of the Main Sequence band; they are cool, but their great size makes them highly luminous. The existence of the Main Sequence makes it possible to estimate the luminosities of many stars solely from their spectral type. Knowing their luminosities, it is possible to determine their distances (by the inverse square law) and estimate their masses (using the mass-luminosity relationship).

The internal structure of the sun and other stars can be understood relatively easily for two reasons. First, stars consist largely of ionized atoms. They thus obey the ideal gas law, which states that the pressure of a gas is directly proportional to both its temperature and its density. Second, for much of its life a star is in equilibrium. The condition of hydrostatic equilibrium requires that the gas pressure at every point within the star exactly balance the force of the star's own gravity at that point. The condition of thermal equilibrium requires that the amount of heat flowing into and out of each layer within the star must be equal.

Within most stars, the chief method of heat transfer is radiation: the emission and absorption of electromagnetic energy. Because photons are constantly being absorbed and reradiated by atoms inside the star, their journey outward to the star's surface is very slow. Radiation from the sun's interior, for example, takes about a million years to reach the outer layers. The ability of stellar matter to absorb radiation is called its opacity. Since opacity determines the rate of energy transfer by radiation within the star, it plays a major role in determining the star's structure.

The rate at which a star radiates energy into space is determined solely by its temperature, which in turn depends on the star's mass and composition. In the case of the sun, an average star, this energy amounts to some 4×10^{33} ergs/sec. We know that the sun's luminosity probably has not changed significantly in several billion years. Energy sources such as chemical combustion or gravitational contraction could not supply so much energy for so long. The source of stellar energy is nuclear fusion: the combination of light atomic nuclei to form the nucleus of a heavier element. Stars are mostly hydrogen, and in most stars the chief energy-releasing process is the fusion of 4 hydrogen atoms into one helium atom. In this process .7 percent of the mass entering the reaction is converted into energy, in accordance with Einstein's equation $E = mc^2$. For each 141 g of hydrogen that is transformed into 140 g of helium, enough energy is liberated to light a million 100-watt bulbs for more than an hour and a half.

Fusion is only possible when nuclei overcome their electrical repulsion and approach each other very closely. To do this they must be travelling very rapidly. The needed velocities are attained when the temperature reaches about 10 million degrees, as it does in the central regions of stars. The fusion reactions that supply stellar energy are extremely temperature sensitive — a relatively small increase in energy production. Because of this, the fusion process is essentially self-regulating. A decline in energy production will cause the pressure in the star's interior to drop. The star will contract, becoming hotter, and thereby increasing energy production until equilibrium is restored. Similarly, an increase in energy production will cause the star to expand and cool, thus lowering energy output.

1. Why are very few of the brightest stars that we see in the skies amongst the nearest stars? Show that a star of 1000 times the luminosity of the sun is 1,000,000 times more observable than a star of solar luminosity. Then use the luminosity function shown in Figure 6–2 to show that such stars are likely to dominate our list of brightest stars.

2. If nearby stars cannot generally be identified by their brightness, how would you go about finding the nearest stars?

3. What is the mass-luminosity relationship? Does it hold for all types of stars? Is it possible for a star of 1 solar mass to have a luminosity 50 times that of the sun?

4. Why is it difficult to study spectrographic binaries in which the mass ratio of the two components is greater than about 3:1? (Consider the mass-luminosity relationship.)

5. How does the study of stellar densities divide stars into 3 classes? Is it possible for a star of one solar mass to have a radius much smaller than that of the sun? What will be the luminosity of such a star?

6. You are a cosmic engineer asked to provide illumination for a certain region of space. You are given 10 solar masses of hydrogen to make into stars. How should you divide up the hydrogen if you want to illuminate the vicinity a) brightly? b) for a long time?

7. Assume that the lifetime of the sun is 10 billion years. What will be the lifetime of a star of 4 solar masses?

8. What do we need to know about a star to plot it on the H-R diagram?

9. A white dwarf is a star that has used up its energy reserves and is slowly

cooling down. It does not change its radius as it cools. Plot its path on the H-R diagram.

10. How can the H-R diagram be used to find the distance of a star?
11. Why are most of the stars we observe with the naked eye blue?
12. Using the kinetic theory of gases, give two reasons why the atmosphere of Jupiter contains hydrogen while that of the earth does not.
13. Explain the two equilibrium conditions satisfied by a stable star. What role does opacity play in determining stellar structure?
14. Why do we no longer believe that the sun's energy is provided by gravitational contraction, as was proposed in the nineteenth century? What is the sun's energy source?
15. Why does nuclear fusion depend strongly on temperature?
16. For how many years could a 100-watt bulb burn, using the energy produced by the annihilation of 1 gram of matter? (A 100-watt bulb uses 10^9 ergs/sec; 1 year $= 10^7$ sec.)
17. Assuming that $1/500$ of the sun's mass is converted into energy during its entire lifetime, estimate the lifespan of the sun.

CHAPTER SEVEN

LIVES AND DEATHS OF THE STARS

The most intriguing problem in the study of stars is their evolution: how they are born, how they age, and how they die. Much of what we know or what we can guess about the most fundamental questions of astronomy—the age of the universe as a whole, the way in which our solar system was formed, the possibility of intelligent life elsewhere in the cosmos—depends on our understanding of the life histories of stars. Indeed, one of the things we have learned from studying stellar evolution is that everything in the universe is connected. The very atoms of our bodies were made billions of years ago in the hearts of giant stars—stars that blew themselves apart in violent explosions long before our own sun was born.

By their very nature, stars must change with time, for they are constantly radiating energy away into space. The nuclear reactions that replenish this energy can keep a star burning for a very long time, but they cannot make it immortal. Eventually the supply of nuclear fuel must run out. Even before that point is reached, however, the processes by which stellar energy is produced begin to alter the star's internal structure. After a while these changes become large enough to produce observable effects—changes in the star's size, its temperature, its luminosity. The star, in other words, evolves. Eventually, it must die. We shall see, however, that stars end their lives in various ways, some of them quite spectacular.

171

The astronomer studying stellar evolution, Sir John Herschel once wrote, is like a person trying to deduce the entire life history of a tree after spending just one hour in a forest. An hour is a very small fraction of the lifespan of a tree. But a year, or even a hundred years, is an even smaller fraction of the lifespan of the average star. On the vast time scale of stellar evolution, the entire history of modern astronomy represents only a brief instant. Our observations, therefore — even when made over a period of decades or centuries — are no more than snapshots. They cannot tell us the story of a star's life, any more than a single movie frame can tell us the complete plot of a long, complex film.

We do have one great advantage in our investigation of the stars: there are so many of them. Presumably we see them in all stages of their life cycles — young stars, mature stars, dying stars. But how can we know **which** stars are young, or old? This is a problem that no amount of mere observation will resolve. Our position is very like that of a scientist from another planet who tries to understand the human race from a handful of photographs. The pictures show infants in cribs, children, teenagers, adults, old people, and coffins. They also reveal that some people are black, others brown, still others yellow or pink. But without some knowledge of human physiology and development he cannot interpret this mass of evidence. He might well come to the conclusion, for example, that people come in various sizes, but that all are born white and darken as they grow older, becoming first yellow, then brown, and finally black. Or he might conclude, with equal plausibility, that people develop under the earth in boxes, are born with the assistance of gravediggers and embalmers, and grow up to be babies.

Similarly, we observe stars of all sorts: large and small, hot and cool, luminous and dim. It is not immediately apparent from these observations which of these properties are intrinsic (like the color of your skin) and which change with time (like your physical strength). Are red giants, for example, a special kind of star right from birth, or are all stars red giants at some time in their lives? If we could follow a sample of stars from birth to death, we could easily answer such questions. But this project would require millions of years, and perhaps billions. Instead, we must rely heavily on theory — on our understanding of stellar structure and the way that stars produce energy.

Since stellar lifetimes are so long, we cannot observe the process of stellar evolution, but must reconstruct it with the help of theory.

Models of Stars

If we fully understood all the relevant physical principles, it would be possible to construct a theoretical model of a star that would match the real star as a good map matches the terrain. In actual practice, theoretical models match stars well, but not perfectly. Indeed, it is often the slight discrepancy that gives us a clue as to what is wrong with the physics we are using. Such clues help us to improve our understanding of stellar interiors, and thus to refine our models further.

CONSTRUCTING A MODEL In the previous chapter we discussed briefly some of the physical principles that are applicable to the study of stellar interiors. The gas laws, for example, giving the relationship of pressure, temperature, and density, must be obeyed everywhere inside the star. The way in which energy passes outward through the stellar gases is determined by their composition, temperature, and pressure in complex but calculable ways. So is the rate of nuclear fusion at each point within the star's central regions.

In addition to these general physical laws, certain specific conditions must be satisfied if the star is to be stable. The star must be in hydrostatic and thermal equilibrium, and the rate at which its nuclear reactions provide energy at its center must equal the rate at which energy is radiated away into space at its surface. It is possible to state these conditions in a very brief and precise way—a kind of shorthand—by putting them in the form of mathematical equations. The equations summarize the relationships that must exist among the physical variables (pressure, temperature, density, energy production, and so forth) at each point within the star. It turns out that all of the essential relationships can be expressed in four equations, known as the **equations of stellar structure.** Solving these equations will tell us the physical conditions everywhere within the star—in other words, it will give us the model we seek.

What do we mean by solving these equations? The actual equations of stellar structure are too complex in their mathematical language to be given here, but the principle is the same one that we use in much simpler situations. If a man buys 10 pounds of coffee for \$20, it is clear that the price of coffee must be \$2/lb. Suppose, however, you are told that a man paid \$13 for 2 lb of coffee and 3 lb of tea: $2C + 3T = \$13$. You cannot tell from this information what coffee costs per lb; it depends on the price of tea, and vice versa. If you know either one, you can easily find the other. But to find the price of coffee, you do not have to be given the price of tea directly—merely some other relationship between the two. You may know, for example, that tea always sells for 50 percent more than coffee in this particular store: $T = 1\frac{1}{2}\,C$. Then it is not hard to show that coffee must cost \$2/lb, and tea \$3/lb. No other set of values will fit both the relationships that were given. (Similarly, we saw in Chapter 5 that the masses of binary star components could be found because we had two equations—one giving the sum of their masses, and another the ratio of their masses.

The equations of stellar structure are difficult to solve, however, because they express an intricate network of relationships. The variables are all interdependent, and each affects the others in complex ways. A change in the pressure, for example, affects the condition of hydrostatic equilibrium. It also influences the temperature and density through the operation of the gas laws. It will probably change the opacity, and quite possibly the rate of energy production at that point.

The study of stellar evolution depends on the construction of models of stellar interiors.

If the equations can be solved, they will give us a set of values for each of the physical variables at every point within the star. But with equations as complex as those of stellar structure, it is hard to be sure that a solution such as we are seeking actually exists. Some sets of equations may have several solutions; in other cases, it may not be possible to find any solution at all.

THE RUSSELL-VOGT THEOREM   The mathematical problem represented by these equations has been closely studied. The result of this analysis is the **Russell-Vogt theorem.** It asserts that, for a star in hydrostatic and thermal equilibrium living off nuclear energy, the equations of stellar structure have one and only one solution. Moreover, the solution depends on just two factors: the distribution of the chemical elements within the star, and its mass. Given the mass and composition of a star, in other words, we can build a model that will tell us virtually everything about the star's structure.

We can look at this result in a more physical way. The mass and composition of a star are its most basic properties; once set at birth, they are not easily altered, as are size or temperature. The Russell-Vogt theorem asserts that, when a certain amount of matter of a certain sort becomes a star, it can only become *one kind* of star. Its luminosity, the way in which its mass is distributed, the pressure and temperature at every point are all determined. Only two developments can affect the star's structure: a change in its total mass, as sometimes happens in close binary systems (see Feature), or a change in the way elements are distributed within the star.

To see what this means for stellar evolution, consider what happens when hydrogen is transmuted into helium by nuclear fusion in the core of a Main Sequence star. As the process goes on, helium starts to accumulate at the star's center. When about 15 percent of the hydrogen has been changed to helium the star rapidly expands, grows more luminous, and becomes a red giant. All along the mass has remained virtually constant, since less than 1 percent is lost during fusion. The change in configuration is caused by the changing distribution of elements within the star.

While the Russell-Vogt theorem has never been proven in a completely general way, and exceptions are known to exist, it nevertheless remains a very useful general principle. In 1931 the Danish astronomer Bengt Strömgren used the Russell-Vogt theorem as a point of departure to establish a valuable new result. He showed that stars of identical composition but different masses would all lie along a diagonal on the H-R diagram. That diagonal coincides very closely with the Main Sequence, where most stars are in fact found. This suggests that most stars are quite similar in their makeup. (If they were identical they would all lie on a narrow line rather than a band of some width, as we actually observe.) Strömgren's is in effect the "explanation" of the Main Sequence.

The Russell-Vogt theorem asserts that a star will evolve only as the result of a change in its mass or the distribution of elements within it.

The Russell-Vogt theorem helps us to understand the existence of the Main Sequence.

COMPUTATION OF MODELS

COMPUTATION OF MODELS   Once the equations of stellar structure are formulated, the Russell-Vogt theorem tells us that they can be solved. Doing so, however, may still require thousands of separate calculations. Until twenty years ago, the process took a year or more of difficult, tedious work. Nowadays, high-speed electronic computers can do the same task in minutes, with greater accuracy than any human could achieve, and the computation of a stellar model has become almost a routine chore. There is nothing mysterious or miraculous about what these computers do, however. They do exactly what a human being could, in principle, do—they merely do it faster. The value of their results still depends upon the quality of the data they are given. Supplied with incorrect data, a computer will simply give you the wrong answer more quickly.

In order to solve a mathematical problem a computer employs a program which tells it how to go about its calculations. The program has to be written by a human being, of course, but once it has been written it can be used to solve a variety of similar problems. Also, only a couple of instructions may be needed in order to tell the computer to carry out a series of, maybe, 100,000 calculations. Computers, however, are expensive and the computation time available to an investigator is limited; therefore, all programs attempt to produce exact solutions with the minimum of calculations. There are many tricks and shortcuts involved and a good computer program is almost a work of individual artistry.

There are several ways in which the equations of stellar structure can be solved. The scheme described below is a commonly used one that is particularly suited to the study of stellar evolution. As a first step, the star is divided up into 50 or more concentric shells. Each shell of radius r is characterized by its temperature, T; its pressure, P; the amount of mass lying within it, M; and its luminosity, L (the amount of energy produced every second by the matter inside it). The four equations are set up so as to describe the relationships among these four variables, and how they change from shell to shell.

To construct a model it is necessary to find values for $P, T, M,$ and L in each shell that satisfy the equations. It turns out that only these four quantities are needed for a complete stellar model; other factors, such as density and opacity, can be easily calculated from them. With 50 shells, however, there are still 200 values to be determined—not a simple task.

One of the shortcuts often used in solving the equations of stellar structure is a technique discovered by Newton. The data are first fed into the computer along with an approximate solution, based on estimation or crude calculation. The computer modifies this solution to fit the actual data more closely, and comes up with a much more accurate solution. These values can in turn be fed into the computer again, as the rough basis for a still better solution. The process can be repeated till a very precise solution is obtained. This solution, however, is only as good as the physics and the data that went into the equations. Newton's method is like a meat grinder that grinds as fine as you wish, but the hamburger it produces is no better than the meat that you start with.

This process of successive refiniment of a solution is called **iteration.** Iterative methods are very useful when calculating a sequence of models that differ from each other only slightly. For instance, an accurate model of a Main Sequence star with a mass equal to that of the sun can be taken as the rough solution for a 1.1 solar mass star. This model could then be refined in only a few iterations, because a 1.1 solar mass star does not differ too much from a star of 1 solar mass. The refined model could then be used as the rough solution for a star of 1.2 solar masses, and so on.

Stellar Evolution Calculations

To study stellar evolution it is necessary to calculate a sequence of models which represent the same star at successive points in time. We begin with a single model of a star. From the star's luminosity we know the rate at which it is losing energy. We also know where the energy is being produced, since this can be understood from the theoretical model. After a certain interval of time the chemical composition of the star has changed slightly—some of the hydrogen near its center has been converted into helium by nuclear fusion. In order to find out how the structure of the star will change as a result, a model of the evolved star is calculated using the Newtonian scheme. The old model is used as the rough solution, and the changes produced by fusion are included in the data. After a few iterations a good model of the slightly evolved star emerges. The procedure is repeated in order to get a succession of models of the star as it ages.

An easy way of displaying the results obtained in this way is to plot the successive positions of the evolving star on the H-R diagram, showing how its temperature and luminosity change with time. Such theoretical H-R diagrams can then be compared with those derived from observation. If the two coincide, we have strong evidence that stars do actually evolve in the way that our models predict, and our understanding of stellar interiors is sound.

In theory it is best to have as many models as possible to mark the star's evolutionary path. Economically, however, it is impractical to multiply models without limit, and much thought is spent on devising means by which a given range of time can be spanned by the fewest number of models. A star such as the sun, for example, changes very little during its 10 billion year stay on the main sequence. The entire period may be accurately represented by 5 models, the age of each model differing from the next by about 2 billion years. The next phase of evolution is more rapid; the star becomes a red giant in a few million years. This phase of evolution may be spanned by about 50 models, differing in age by less than a million years.

The evolution of a star is reconstructed by calculating a series of models that show how the star changes over a period of time.

The time interval from one model to the next is called the **time step.** The need to adjust time steps in order to have the most efficient computation may be compared to the most efficient way of tracking down a fugitive. As long as the fugitive, in his getaway car, is on the turnpike it may be necessary for the police to have reports only every 10 minutes, but when he enters the city it is necessary to know which way he turns at every block. When M. Schwarzschild and R. Harm carried out some of the pioneering stellar evolution calculations in the 1950s and early 1960s they used time steps ranging from 1 billion years down to 2 minutes.

ORIGIN AND INFANCY OF STARS

Until recently star formation was one of the least understood phases of stellar evolution. Just fifteen years ago a leading scientist in the field, Geoffrey Burbidge, could remark that if it were not for the fact that stars do exist, it would be easy to prove that they could not. Even today there are some uncertain elements in our theories of how stars are born. But better physical models, calculated with the aid of modern computers, and a wealth of new observational data, obtained chiefly in the infrared and radio regions of the spectrum, have made the origin of stars far less mysterious than it seemed only a couple of decades ago.

This is fortunate, for star formation is an important subject. We would like to understand how stars come into being as a part of our overall attempt to trace their evolution. Beyond this, however, we find that our ideas about star formation shed light on other astronomical problems: the range of stellar masses, for example, and the tendency of stars to occur in binary and multiple systems. The rate at which stars form may also play a critical role in determining many of the structural features of galaxies. And perhaps most interesting of all, our understanding of how stars form may tell us a great deal about how planetary systems are created, and how common they may be throughout the universe.

Young Stars and Interstellar Clouds

When we examine hot, highly luminous stars of spectral types O and B, we find that they tend to occur in clusters. These clusters are very often associated with vast clouds of interstellar gas and tiny solid particles known as dust. (Figure 7–1) Stars of low luminosity, by contrast, are found widely scattered throughout the galaxy. The interpretation of these observations hinges on the fact that, as we saw in Chapter 6, the most highly luminous stars have the shortest lifetimes. Some hot blue stars, such as Rigel (in Orion) and Deneb (in Cygnus), are so luminous that they can hardly have maintained their present rate of energy production for more than a few million years. These stars, therefore, must be very young; it is fairly certain that they were not yet shining when man first walked the earth.

Young stars are often found near the clouds of gas and dust from which they formed.

PORTFOLIO

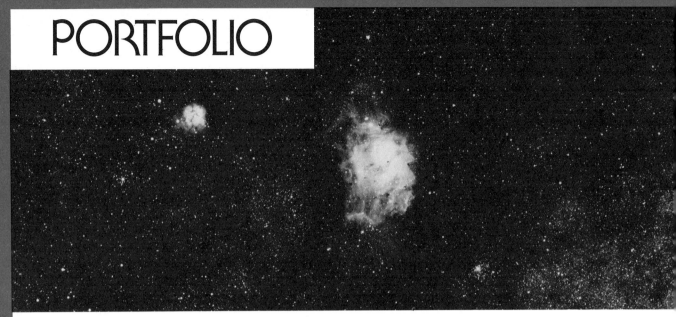

Above, star fields of the Milky Way with M 20 and M 8. These are both regions of star formation. In the photographs below, we have M 8 and a detail within M 8. The bright regions of gas have been excited by ultraviolet radiation from the hot, luminous, short-lived stars embedded within them. Complex instabilities have produced dark regions of dark, dust laden gas. Numerous globules are visible as dark spots.

Herbig-Haro objects are peculiar nebulas, all located in dusty clouds amidst T Tauri stars. Three examples are seen at right. Herbig-Haro objects are believed to be, in some way, precursors of stars, but the theory of star formation is so uncertain that it is not possible to explain them in any definite manner.

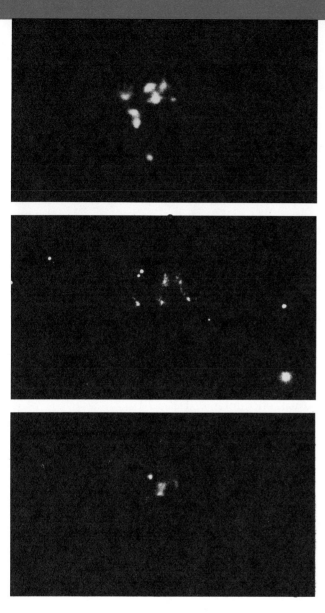

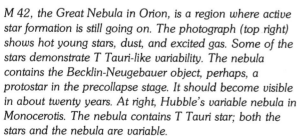

M 42, the Great Nebula in Orion, is a region where active star formation is still going on. The photograph (top right) shows hot young stars, dust, and excited gas. Some of the stars demonstrate T Tauri-like variability. The nebula contains the Becklin-Neugebauer object, perhaps, a protostar in the precollapse stage. It should become visible in about twenty years. At right, Hubble's variable nebula in Monocerotis. The nebula contains T Tauri star; both the stars and the nebula are variable.

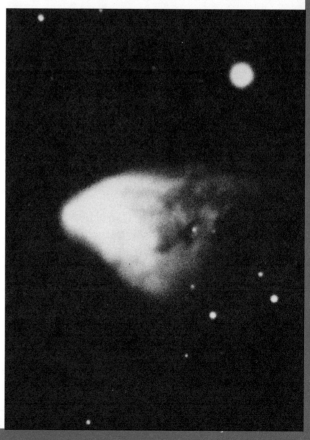

From these facts, and from other evidence that we will encounter later in the chapter, astronomers are convinced that all stars originally form from interstellar clouds. Highly luminous stars, in their brief lifetimes, do not have time to travel far from their birthplaces, and so we find them still closely associated with the clouds from which they first condensed. Stars of low luminosity, though they too were presumably born in the same way, live long enough to travel far from their places of origin. They leave the clouds in which they formed behind them, and gradually drift away from the clusters in which they spent their infancy.

BEGINNING OF STAR FORMATION Reduced to its essence, the process of star formation is extremely simple. A diffuse interstellar cloud collapses in upon itself, becoming hotter and denser, until it finally becomes a star. Exactly how and why this process takes place, however, are complex questions which have been much studied, and we do not yet have all the answers.

A collapsing cloud of gas and dust, on its way to becoming a star, is called a **protostar.** During its condensation, the radius of the protostar decreases from about a tenth of a parsec (3×10^{12} km) to roughly that of the sun (7×10^5 km), and the physical conditions within it change dramatically. The density of the matter in the original cloud increases from about 10^{-18} g/cm³ to about 1 g/cm³—a billion-billion-fold increase. At the same time, the gravitational energy released in this collapse

Figure 7-1
A region of dust and gas, in which many hot, young stars are embedded. Another such region is shown in the chapter opener photo. Massive, hot blue stars have very short lifespans, and do not have time to wander far from their places of birth. They are thus generally found amidst the dust and gas from which they are formed.

raises the temperature of the protostar from about 20 K to 10,000,000 K — the temperature at which nuclear fusion begins, and the protostar becomes a star.

The problem of why a protostar begins to collapse was first analyzed by James Jeans in 1919. He considered what would happen within a vast cloud of gas of uniform density. Each particle of this gas exerts a gravitational force on every other particle. If the density of the gas is uniform throughout, however, we would expect that at every point within the cloud, the forces acting in many different directions would cancel each other out. Since there would be no net force in any direction, it might seem at first thought that the gas should remain uniform and diffuse forever.

But a gas consists of particles in ceaseless motion, and as a result of these motions the density at every point is constantly fluctuating in a random way. At one instant it may be below average at a given point; the next moment it may be more than average. When the density increases at a particular point, other particles in the vicinity are more strongly attracted by the increased gravitational force. They start to fall in toward the point of concentration, increasing the density and the gravitational force still further. The process accelerates as more and more matter is drawn in. Eventually all the particles in the vicinity have condensed into a relatively small volume. Jeans demonstrated, in other words, that a cloud of gas does not remain uniform and diffuse indefinitely, but tends to condense into a number of discrete blobs.

According to the latest versions of Jeans' theory, star formation begins when clouds of dust and gas collide with each other in the arms of the galaxy. In the regions where this takes place, the density of the dust and gas becomes high enough for large blobs of about 120 solar masses, called **globules,** to form. As the globules collapse they become still denser, and tend to break up into smaller blobs with masses equal to that of a star. These are the protostars.

By ordinary earthly standards, the density at which star formation begins — about 10^{-18} g/cm^3, or about 10,000,000 hydrogen atoms per cubic inch — would be an excellent vacuum. But it is still much higher than the average density of matter between the stars. One unfortunate result of the relatively high density required for star formation is that the process is difficult to observe at visible wavelengths. Interstellar dust particles are very effective at blocking out visible light. Long before the clouds have reached the protostar stage they have become quite opaque.

Globules do tend to radiate at radio wavelengths, however, and protostars, as they collapse, soon become hot enough to emit infrared radiation as well. Working with radio and infrared telescopes, astronomers have recently observed several objects embedded in the Orion nebula and the Taurus dark cloud that are probably protostars in the process of becoming stars.

Figure 7-2
Small random fluctuations in density in a diffuse cloud of gas can cause the cloud to condense into a number of discrete globules. Whenever the density rises locally, gravitational forces attract more matter to that region. It is thought that the process of star formation begins in this way, as regions of gas collapse to form protostars.

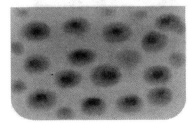

ROTATION   We know that the galaxy rotates slowly, with the sun completing a circuit of the galactic center once every 200 million years. A cloud belonging to the galaxy should share in this slow rotation. When such a cloud collapses it preserves its angular momentum (Chapter 5). The smaller it becomes, therefore, the faster it must spin. It would seem, then, that by the time a protostar finally collapsed into a star, the star's surface would have to be travelling at speeds approaching that of light!

Long before this happened, of course, the centrifugal force would become so great that the collapse would be halted. Unless the excess angular momentum could somehow be dissipated or transferred to another object, the protostar would never be able to become a star. It has been suggested, however, that the collapsing protostar might break up into two or more objects in orbit about each other. The angular momentum of the system would then be taken up largely by the orbital motion of the components. This theory could explain why most stars seem to be members of binary or multiple star systems.

Protostars

The progress of a protostar towards becoming a star can be divided into two phases. In the first, the gravitational contraction takes place unopposed to any significant degree by the pressure within the protostar. This is the **dynamic stage,** and is very brief. As the protostar continues to shrink, however, the pressure inside it builds up. Eventually it becomes great enough to halt the rapid collapse. Hydrostatic equilibrium is reached, and further contraction proceeds slowly, made possible only by the loss of energy from the star's surface. This phase of collapse is known as the **thermal stage.**

DYNAMIC STAGE OF CONTRACTION   A typical cloud, with a mass about equal to that of the sun, has a radius of roughly .1 pc when it starts to collapse. The temperature of the cloud is low—only about 20 K (−424 Fahrenheit). Though much energy is released by the process of gravitational contraction, it is quickly radiated away at infrared wavelengths. The gas of the protostar, because of its low temperature and density, is very transparent to this radiation, which thus escapes easily into space. As a result, the temperature of the protostar does not rise during this phase of its collapse. The dynamic stage lasts, typically, several hundred thousand years. It begins slowly, but steadily accelerates, with much of the contraction taking place in the final few years.

During the first stage of contraction, the protostar radiates energy away into space, and its temperature does not increase.

THERMAL STAGE OF CONTRACTION   Eventually the density of the protostar becomes high enough so that the radiation released within it is trapped—absorbed and reabsorbed so that it can only trickle out slowly. With radiation no longer carrying off all its thermal energy, the protostar starts to heat up. This further increases the opacity of the gases, so that more and more radiation is trapped within the protostar. The soaring temperature causes a rise in pressure, which tends to oppose the process of collapse. It has been shown that the central region of the protostar reaches hydrostatic equilibrium first, while the outer regions are still collapsing inward. When these outer layers collide with the central region, violent shock waves are produced, which travel through the protostar generating still more heat.

During the second stage of contraction, the density of the protostar becomes great enough to trap radiation, and its temperature rises.

Eventually most of the protostar reaches hydrostatic equilibrium. It is still a fairly large object at this stage, with a diameter some 200 times that of the sun, or about the size of Saturn's orbit. It contracts slowly as energy is radiated away into space from its surface. As we saw in Chapter 6, the process of contraction liberates energy, half of which is radiated away and half of which is retained as heat. When the temperature of the star reaches 1800 K, hydrogen molecules (H_2) begin to dissociate into hydrogen atoms. This process absorbs most of the energy released by gravitational contraction, so that the temperature of the protostar no longer rises. Since the protostar is now unable to heat up any further, hydrostatic equilibrium can no longer be maintained, and a second collapse sets in. This phase is very rapid, and lasts only a few decades. When it is completed the protostar has a radius about 50 times that of the sun and a luminosity about 200 times greater. The material inside the protostar is completely ionized, and the surface temperature is high enough to make it a visible object. It is now almost a star.

WATCH THIS SPACE!   Astronomers have recently discovered an object in the Orion nebula that they believe is a protostar with a radius about 1500 times that of the sun. It is a strong emitter of infrared radiation. Calculations have shown that, if the theory outlined above is correct, this object should become a visible star in only 20 years or so. The infrared radiation amounts to a call to "watch this space"; a new star is preparing to make its debut!

Coming of Age

When a protostar reaches hydrostatic equilibrium, it becomes possible to study it using the techniques developed for stable stars. One of the astronomers to undertake such studies was the Japanese C. Hayashi. Hayashi found that for part of the period of prestellar collapse, protostars evolve almost vertically downward on the H-R diagram (Figure 7–3). Less massive stars reach the lower reaches of the Main Sequence directly in this way, taking 10 million years or more for the process. Stars more massive than the sun turn off to the left before joining the Main Sequence, and evolve much faster. A star of 100 solar masses, for instance, may require only about 10,000 years to arrive at the Main Sequence.

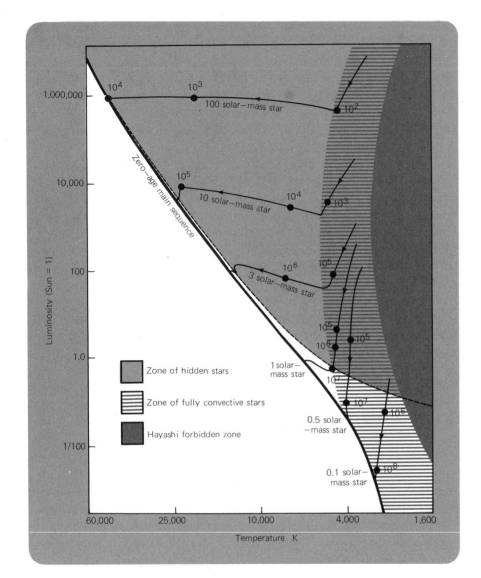

Figure 7-3
The evolution of a series of proto-stars of different mass, contracting towards the Main Sequence. These paths are derived from theoretical calculations. Notice how the stars avoid the Hayashi forbidden zone, and how their paths change when they leave the region where convection is dominant. the more massive stars evolve most rapidly, but for all stars, the final stages of the process are the slowest. Stars lying above the dashed line on the diagram will probably be hidden from our view by infalling gas and dust. The most massive stars will never be visible at all.

Hayashi's work led to another interesting result. His calculations showed that there is a near-vertical line on the H-R diagram, to the right of which no stable stars can exist. This region is known as the **Hayashi Forbidden Zone,** and it has been confirmed by observation, for there seem to be no stars with surface temperatures below about 3000 K. We do sometimes observe stellar objects that seem redder than the Hayashi limit allows, but these are never stars in equilibrium. Usually they are collapsing rapidly, or else blowing off their outer layers so that they are surrounded by shells of relatively cool gas. The Hayashi law is like a very strict zoning ordinance; it states that the right-hand edge of the H-R diagram must be considered off limits for the construction of stars.

LIMITS OF THE MAIN SEQUENCE We can see now why the Main Sequence appears to contain no stars with masses less than about $1/12$ that of the sun. Stars of such low mass never become hot enough to ignite the nuclear reactions that enable Main Sequence stars to burn for so many millions of years. Before this temperature is reached, such stars attain densities so high that the matter in their interiors becomes degenerate (see Box). Radiation from the surface continues to cool the star, but now the cooling does not lead to further contraction, because the pressure of degenerate matter does not vary with the temperature. Degeneracy pressure continues to support the star against collapse as it radiates away its heat into space, becoming steadily less luminous until it finishes its abortive life as a cold, dark, dead chunk of matter, more like a superdense planet than a star.

We do not usually find extremely massive stars on the Main Sequence either, for a rather different reason. When a protostar with a mass greater than 150 times that of the sun collapses, its central regions condense much faster than its outer layers. Equilibrium is soon reached in the center, and the star arrives on the Main Sequence still surrounded by a cloud of matter that has not yet completed its collapse. The luminosity of such a star is very high, and the pressure of radiation from the stellar core against these outer layers stops their infall and eventually disperses them. The mass of the central star, it has been calculated, cannot exceed about 100 solar masses; the rest of the protostar's mass never gets a chance to become part of the star, and is lost into space.

Whereas stars with masses greater than 100 solar masses probably do not exist, stars with masses between 60 and 100 solar masses probably *do* exist, but are not directly observable. Such stars, when they first reach the Main Sequence, will be surrounded by cocoons of dust and gas that obscure them from our view. We have seen that the Main Sequence lifetime of a massive star is very short. For stars of greater than about 60 solar masses, it is actually less than the time required to disperse the dense clouds of dust around them. Such stars would thus never be visible to us. The surrounding material, however, absorbs the star's intense radiation and reemits it in the infrared. Such objects have been detected by infrared telescopes in recent years.

Stars with less than 1/12 the sun's mass never become hot enough to ignite fusion reactions; stars with more than 60 times the sun's mass are hidden by clouds of dust for their entire lifetimes.

MATURITY AND AGE

It may seem odd that the phase of a star's evolution which is by far the longest—its stay on the Main Sequence—is the phase about which there is least to be said. But this is in fact the case. Whereas the births and deaths of stars are full of dramatic incidents and rapid changes, the middle lives of stars are curiously placid and uneventful. It is almost as if the star, having once tapped the rich energy source of nuclear fusion, settles down to live happily ever after—or at least for a very long time.

The tranquility of life on the Main Sequence is due in large part to the fact that, as we saw at the end of the previous chapter, nuclear energy production is essentially a self-regulating process. Thus for most of its stay on the Main Sequence the star can remain in almost perfect equilibrium. This equilibrium begins to be disturbed only when the accumulation of helium — the end product of hydrogen fusion — becomes great enough to cause a change in the star's structure. Even in the most massive stars, which consume their fuel rapidly, this may not happen for millions of years. During its birth pangs and death throes, by contrast, the star is usually not in equilibrium, and often changes its structure and appearance dramatically in short periods of time.

Life on the Main Sequence

Once a star has begun using nuclear energy, its temperature and luminosity no longer change very much; the star remains in one place on the H-R diagram for many millennia. The points at which stars of different mass begin using nuclear energy all lie along a narrow band stretching diagonally across the H-R diagram. This is the **Zero Age Main Sequence** (ZAMS). Massive stars are the hottest and most luminous; they lie on the upper left-hand portion of the band. Stars of lower mass, cooler and less

Figure 7-4
Evolution away from the Main Sequence. Slow at first, the process accelerates rapidly at a certain point (the little zig-zag on each track). The more massive stars leave the Main Sequence in a few million years; stars like the sun take billions. Notice once again how the stars turn away from the Hayashi zone, causing their tracks to "funnel." The figure at right shows that stars differing only a little in mass evolve at very different speeds. A star of 1.2 solar masses may just have begun to evolve, while one of 1.3 solar masses may already have reached the red giant region, and a 1.4 solar-mass star is well on its way to becoming a white dwarf. Consequently, in any particular cluster, the stars of the horizontal branch and red giant region all have nearly the same mass.

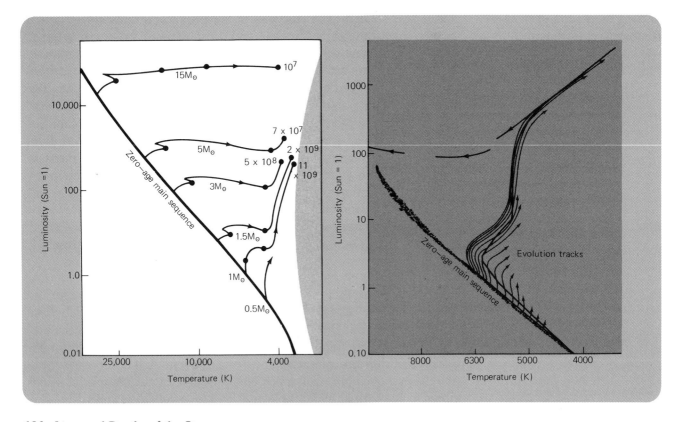

luminous, occupy the lower right region of the band. The term "Zero Age" is used because, as we shall see, stars undergo some evolution away from their initial positions on the H-R diagram even before they leave the Main Sequence entirely.

The precise position of the ZAMS also depends to some extent on the composition of the stars composing it. On the whole, stars can be roughly divided into two classes, called Population I and Population II. A fuller study of their differences belongs to the following chapter. At this point it is necessary only to point out that the two classes differ in their content of helium and heavier elements. About 2 or 3 percent of the mass of a typical Population I star will consist of elements heavier than helium, while in a Population II star the abundance of these elements will be only a tenth or even a hundredth as great. This difference in composition affects the opacities of Population II stars, and thus their luminosities compared to Population I stars of the same mass. As a result, the ZAMS for Population II stars lies slightly below that for Population I stars.

The Main Sequence lifetime of stars ranges from about 10^6 years for a star of 50 solar masses (about the upper limit for stars that we can observe) to perhaps as much as 10^{12} years for the faint, red stars on the toe of the Main Sequence with masses only a tenth that of the sun or less. (There are some such stars, formed early in the history of the galaxy, which have still not begun their evolution away from the Main Sequence.) During its stay on the Main Sequence the star undergoes no drastic change in its appearance. After a while, however, it does begin a slow evolution away from its Zero Age position on the H-R diagram.

MAIN SEQUENCE EVOLUTION It is easy to see why the star must evolve slightly even while it is still on the Main Sequence, with a plentiful supply of fuel to burn. Most of the star's energy production takes place in a relatively small region at its center, for only there is the temperature high enough to make nuclear fusion possible. Studies of model stars have shown that there is very little mixing of material among different layers within the star. Thus the helium "ash" that is formed by the fusion of hydrogen accumulates in the star's core.

We saw in the last chapter that when four hydrogen nuclei are fused to form one helium nucleus, less than 1 percent of the mass entering the reaction is lost as energy. But four nuclei and four electrons have been replaced by one nucleus and two electrons. The pressure of a gas at a given temperature depends on the number of particles in a unit volume. When eight particles are replaced by three, therefore, the pressure in the energy-producing core of the star drops. The core must contract, becoming hotter, until the pressure is again high enough to restore hydrostatic equilibrium.

The increased radiation from the star's core has an effect on the outer layers. They expand and become less dense, like a cake rising in the

Stars spend most of their lifetimes on the Main Sequence, undergoing relatively little change in luminosity and temperature.

oven. In the process they grow cooler, and thus each square meter of the star's surface becomes less luminous. The increase in the total surface area, however, is more than enough to offset this, and so the total energy output of the star actually increases. The star, in other words, appears both redder and brighter; it drifts slightly upward and to the right on the H-R diagram.

AGES OF CLUSTERS   This evolution is naturally most rapid for massive stars. In a cluster, therefore, in which all the stars are the same age, the massive stars "peel away" from the ZAMS first, followed by less massive stars. It is possible to estimate the age of a cluster from the point where its observed Main Sequence has begun to branch off from the ZAMS. If this point is high up on the Main Sequence, only the most massive stars have begun to evolve, so the cluster must be quite young. If the turn-off point is well down the Main Sequence, stars of moderate or even low mass have already started to evolve, so the cluster must be fairly old.

LEAVING THE MAIN SEQUENCE   As a Main Sequence star grows older, helium accumulates in its core, and hydrogen continues to burn in a shell around it. Until the temperature becomes very high, the helium itself cannot undergo fusion into still heavier elements. With two units of

It is possible to estimate the age of a cluster from its Main Sequence turn-off point.

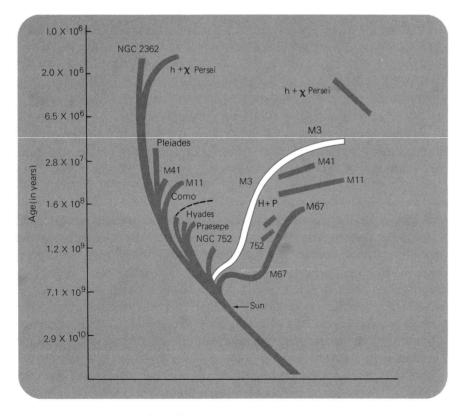

Figure 7-5
An H-R diagram for a typical globular cluster. The Main Sequence is intact up to the turn-off point of a star slightly more massive than the sun. After leaving the red giant region, the stars are strung out along the horizontal branch. All the stars in globular clusters are believed to be about equally old, but some, being more massive, evolve faster than others.

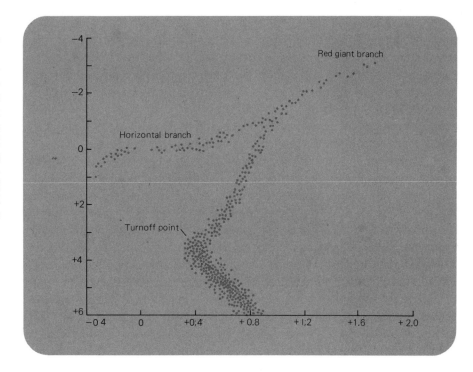

Figure 7-6
An H-R diagram for ten open clusters and one globular cluster (M 3). The age of each cluster can be estimated from the location of its turnoff point on the Main Sequence; the ages are calibrated on the scale at right. Notice the Hertzspring gap between the Main Sequence and red giant region for the young clusters. M 67 is a much older cluster; its pattern is similar to that of the globular cluster M3, since all globular clusters are known to be very old.

electrical charge, helium nuclei repel each other four times as strongly as do hydrogen nuclei, and temperatures above 100 million K are needed to overcome this repulsion. Because there is no energy source within the helium core, heat flows inward to it from the hydrogen burning shell until the temperatures of the two regions are identical. The helium core in this situation is isothermal — that is, the same temperature throughout.

When the mass of the isothermal core exceeds a critical value — about 12 percent of the star's total mass — the star ceases to be stable and starts to readjust its configuration very rapidly. The process resembles the Main Sequence evolution we have just described, but it is more drastic and much faster. The core shrinks; its density, temperature, and energy production all increase. At the same time, the outer envelope of the star expands markedly. Its surface temperature drops, but the area so increases that the star's total luminosity increases. The star evolves upward and to the right of the Main Sequence, and becomes a red giant.

When the helium core reaches 12 percent of the star's mass, the star begins to evolve into a red giant.

The Red Giant Stage

The time required to reach the red giant region depends once again on the star's mass, with the most massive stars evolving most rapidly. It also depends on the star's most of energy production. As we saw in Chapter 6, Population I stars of low mass burn hydrogen through the proton-proton chain, while more massive stars (above 1.2 solar masses) depend primarily on the C-N cycle. In Population II stars there is so little carbon and nitrogen, the necessary intermediaries for the C-N cycle, that most of these stars also use the proton-proton chain. Since the C-N cycle is far more temperature-sensitive than the proton-proton chain, the energy-producing region in stars using it tends to be concentrated in a small region at the center, where the temperature is highest. This concentration sets up a convective zone around the core in these stars, not usually present in stars relying on the proton-proton chain.

It has been found that the presence of such a convective core speeds up the evolution of a star, so that it reaches the red giant much more rapidly than a star lacking a convective core. The difference is clearly evident if we look at the H-R diagrams of various clusters. Clusters of stars in which most of the members are using the proton-proton chain usually have many stars strung out between the Main Sequence band and the red giant region. This is true of Population II clusters, and also of older Population I clusters, in which stars of low mass are just starting to reach the red giant stage. But in young Population I clusters, in which only the massive stars that use the C-N cycle have so far started to become red giants, we seldom see any stars between the two regions. The process of evolution is so fast that stars are rarely caught in the middle. They change from Main Sequence stars to red giants virtually overnight, by astronomical standards. This empty region on the H-R diagrams of young clusters is called the **Hertzsprung gap** (Figure 7–6).

Within each cluster, however, the most massive stars will leave the Main Sequence first, followed by stars of progressively lower mass. Thus the Main Sequence burns down like a fuse, throwing off stars into the red giant region. In all but the very youngest clusters, therefore, the top of the Main Sequence is missing. The stars that formerly occupied that region are now red giants.

Since the most massive stars become red giants first, all but the youngest clusters are missing the upper parts of their Main Sequences.

As a star evolves into the red giant region, its core becomes hotter and hotter. When the temperature reaches about 100 million K, the helium nuclei are travelling fast enough to overcome their electrical repulsion and can approach each other closely enough for fusion to take place. When this happens, the transmutation of helium into carbon begins. The chief way in which this takes place is called the **triple-alpha process,** because three helium nuclei, also called alpha particles, combine to form a carbon nucleus. Another reaction that also takes place at these temperatures is the fusion of carbon and helium to form oxygen.

The Helium Flash

In stars of moderate mass (between about .5 and 5 solar masses) the onset of helium burning is a sudden and dramatic event. In evolving to the red giant stage, these stars develop cores of enormous density—so great that the electrons in the core become **degenerate** (see Box). Now, one of the properties of degenerate matter is that it is a very good conductor of heat. When even a very small amount of helium fusion begins to take place, the heat that it produces spreads immediately throughout the entire core, igniting the rest of the helium. The process resembles a flash fire. Within seconds, all the helium of the star is burning furiously, liberating tremendous amounts of heat.

If the core were a normal gas, the sudden increase in temperature would increase the pressure; the core would have to expand and thus cool down. Because the gas is degenerate, however, the increased temperature does not affect the pressure; it only increases the rate at which the helium burns, which in turn raises the temperature still further. This runaway process is called the **helium flash.** At its height, the luminosity of the core may exceed a billion times that of the entire sun.

When the temperature of the helium core reaches about 350 million degrees, the gas becomes non-degenerate again. It starts to expand and cool. The entire sequence of events has taken only a very short time, perhaps a few minutes. The vast outpouring of energy from the star's core during the helium flash never reaches our eyes directly. It is absorbed by the outer layers of the star, which change their configuration in response. Within a few days the star has returned to a point lying slightly to the left of its old home on the Main Sequence—hotter, in other words, but not appreciably more luminous than it was before.

When the helium flash is halted by the expansion and cooling of the core, helium continues to burn, but only in the center, where the temperature is highest. Hydrogen continues to burn in a shell around the core.

Figure 7-7
The structure of a red giant. The small dot in the center of the left panel is the core, shown in detail at right. As hydrogen burns in a shell, matter from the hydrogen-rich envelope flows in through the shell and is deposited in the helium core. In the central regions the density is so high that matter is slightly degenerate. Conduction is a significant means of energy transport in such stars, along with radiation, and (in the outer layers) convection.

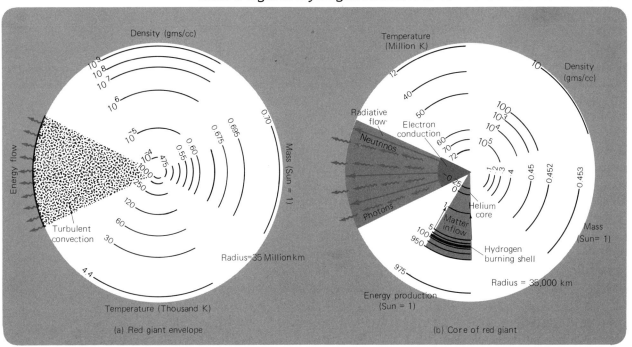

(a) Red giant envelope

(b) Core of red giant

Degeneracy

The gas laws that were discussed in Chapter 6 describe the behavior of real gases fairly well, but only under a limited range of conditions. The properties of gases at high densities have to be analyzed by using the principles of quantum mechanics. We have already encountered these principles in Chapter 4, where it was mentioned that an atomic electron can possess only certain discrete amounts of energy. Moreover, there is a law, called the Pauli exclusion principle, asserting that no two electrons in an atom can be in exactly the same energy state. This law was not very important in the examples that we discussed there, for we were concerned chiefly with hydrogen, which normally has only a single electron. But the exclusion principle is essential for understanding the energy states of more complex atoms.

There is a similar principle that applies to electrons which are not part of any atom, but merely confined to a finite volume of space—the interior of a star, for example. Each electron in such a volume has a total energy, of which its energy of motion generally makes up the greatest part. These energy states too are quantized; only certain discrete states are possible, and no two electrons can occupy the same energy state.

When the electron density is low, there are always enough unoccupied energy states so that the electrons can gain or lose energy quite freely. The restrictions imposed by the exclusion principle are of little importance. A very different situation arises, however, when the electron density becomes very high. Under these conditions, the number of permitted energy states toward the bottom of the energy ladder may be much smaller than the number of electrons that would ordinarily occupy them.

This has important consequences if the electron gas is allowed to cool down by giving up energy. As the electrons lose some of their energy of motion, they start to fill up the lower energy states. But the number of such states is soon ex-

The structure of such a helium-burning star is shown in Figure 7–7. Once the star has settled down after the cataclysm of the helium flash, it resumes a course of evolution somewhat similar to that which first led it away from the Main Sequence. It moves upward and to the right on the H-R diagram, in the direction of the red giant region once more.

Now, however, the star evolves much more rapidly than before. Its high luminosity means that it is radiating away a lot of energy each second. But the sources of energy available to it are much poorer. When helium is fused into carbon, for example, it produces only about $\frac{1}{10}$ the energy that would result from the fusion of an equal mass of hydrogen into helium. Subsequent nuclear transformations are still less fruitful sources of energy. It is clear, therefore, that once a star has started to consume helium the end is near.

hausted. If an electron occupies the lowest possible energy state, for example, no other electron can cool down to that state; it can fall no lower than the second allowed state. The next electron can give up even less of its energy, for the lowest vacant energy state is the third from the bottom, and so on. Thus the trillionth electron will be unable to cool down below the trillionth energy state. This means that it will still be traveling at a very high velocity, and thus exerting a high pressure.

Electrons trapped in such artificially high energy states are called **degenerate.** No heat can be extracted from these electrons, for there is no way they can give up any of their energy and slow down. Their temperature is thus effectively zero, yet the pressure they exert can be very great. This pressure, which does not depend on temperature, but only on the density of the gas, is known as **degeneracy pressure.** Electron degeneracy pressure starts to become significant in stars when the electron density approaches 10^{29} electrons/cm^3. This corresponds to a mass density of about 10^6 g/cm^3, about a million times that of water.

It is also possible for neutrons to become degenerate, though the pressure required is about 100 million times greater than for electron degeneracy. One interesting aspect of neutron degeneracy is that it allows neutrons to achieve a stability that they do not normally possess. In free space, a neutron soon decays spontaneously into an electron and a proton, releasing a certain amount of energy in the process. This does not happen, though, if the neutrons are packed so densely that they become degenerate. In order for the neutron to decay, the electron produced must have some unoccupied energy state available to receive it. But when the density is so high, all the lower energy states are quickly filled, and only extremely energetic ones are vacant. The decay process cannot provide the electron with this much energy, and so the neutron simply does not break up.

Approaching the End

When helium burns into carbon and oxygen, a new core consisting of these elements starts to form in the center of the old helium core. As more and more helium gets used up, the carbon-oxygen core grows, and the star moves back into the red giant region. When the temperature in the center of the star gets high enough, it is possible for the carbon to ignite and be fused into still heavier elements such as neon and silicon. (There is evidence that carbon ignition may cause the whole star to explode like a bomb; we shall take this matter up when we discuss supernovas.) With each new fusion reaction, the internal structure of the star becomes more complex; successive cores form, each surrounded by a shell in which a different nuclear reaction is taking place.

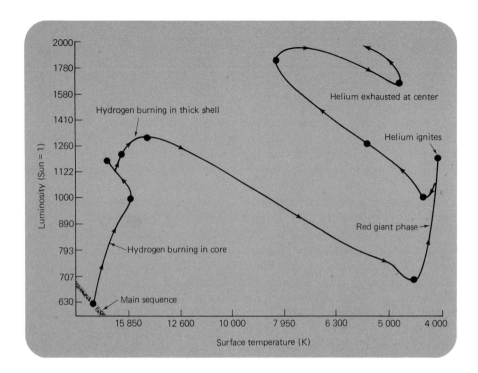

Figure 7-8
The evolutionary track of a star five times as massive as the sun. The Main Sequence lifetime of such a relatively massive star is about 65 million years — less than 1/100 that of the sun. The total time required for the star's evolution after leaving the Main Sequence is on the order of 50 billion more years. By the time the star exhausts its supply of helium, its life is almost over. Though it may go on to other nuclear reactions, it will likely end as a supernova before too long.

The rates of these various reactions have a complex effect on the appearance of the star. As each switches on in turn, its size and luminosity change abruptly, and the star executes many sudden stops, turns, and swerves on the H-R diagram (Figure 7 – 8). The star's evolution in this phase can only be studied by patiently calculating the effects of each energy source, and the many complex factors involved, such as the mixing of material through convection, make the results rather uncertain. As a general rule of thumb, however, it can be said that each time a new energy source is ignited, the star moves to the left on the H-R diagram, back toward the Main Sequence. Each time another isothermal core develops, the star moves to the right on the H-R diagram, into the red giant region again.

If the star is very massive, the core will become hot enough for many different nuclear reactions to be ignited, one after another. Each successive reaction produces heavier and heavier elements, and each provides the star, briefly, with a new source of energy. But the amount of energy available from these reactions diminishes steadily until iron, with a nucleus containing protons and neutrons, is reached. This is the end of the line as far as nuclear energy is concerned. The creation of elements heavier than iron by nuclear fusion does not release any more energy; in fact, it consumes energy. When iron is produced in the core of a massive star, the star has exhausted its fuel — there is nothing left that it can burn.

Not all stars get even this far. Less massive stars never attain central temperatures high enough for the fusion of heavier elements. Rather

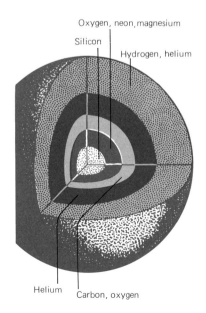

than running out of fuel, these stars lack the means of igniting the fuel they have. In fact, stars with less than half the sun's mass cannot even begin burning helium; when the hydrogen in their central regions has been depleted, they are through. But whether the star stops at helium or runs through all the nuclear reactions up to iron, the final result is the same: the end of its career as a user of nuclear energy. When this happens, the star has only gravitational energy to fall back on. It starts slowly to collapse, much as it did when it was still a protostar on the way to becoming a star, not yet hot enough to initiate nuclear fusion. Like some people, stars in their old age regress; they revert to their infancy.

DEATH AND TRANSFIGURATION

As an aging star collapses, its density increases. The gravitational energy released by its collapse keeps it extremely hot. As we have already seen, a star cannot cut back its radiation to fit its impoverished state; as long as it is hot, it must continue to radiate, dissipating the remains of its dwindling supply of thermal energy. We would expect a dying star, therefore, to be small, dense, hot, and (because of its small surface area) dim.

White Dwarfs

We do not have to look far for observational evidence that this is indeed the way many stars end their lives. The description just given fits perfectly a class of objects familiar from the previous chapter: white dwarfs. Clearly a white dwarf is a star that is radiating its remaining heat energy away into the void, cooling slowly toward oblivion.

When Eddington first studied white dwarfs in the 1920s, one aspect of their fate intrigued and alarmed him. He realized that these stars had obviously used up their nuclear resources and were collapsing, attaining very small sizes and high densities. They could not expand back to normal stellar density unless they were supplied with energy, which was highly unlikely; in fact, they were still losing energy through radiation. But how far could they collapse? Would they continue to dwindle until they became mere points?

The atoms of a white dwarf, of course, are completely ionized and can pack very closely together. If they were packed so tightly that the nuclei touched, the density would be about $10^{15} g/cm^3$; on earth, a cubic inch of such matter would weight about 15 billion tons. But long before this point is reached, other factors come into play. At a density of $10^6 g/cm^3$, the electrons in the star's core become degenerate.

At about the same time that Eddington was investigating white dwarfs, his colleague R. H. Fowler was studying the theoretical properties of degenerate matter. Eddington quickly realized that electron degeneracy pressure could halt the collapse of white dwarfs and allow them to cool

Figure 7-9
The concentric cores of a highly evolved star. Different fusion reactions take place at the interfaces of each shell.

down. The cooling would not affect the degeneracy pressure, so the star would not alter its size in the process. Its radius would remain constant, but as it slowly lost its thermal energy it would grow redder and dimmer. In time the star would be reduced to a mere stellar ember, glowing faintly in the infrared. Eventually it would reach oblivion as a black dwarf—a cold, dead, extraordinarily dense chunk of largely degenerate matter, leaving no trace of its existence save the gravitational force it might happen to exert on a nearby companion.

Following Eddington, astronomers began to construct theoretical models of dwarf stars. It was found that white dwarfs should obey a very definite mass-radius relationship, but one quite unlike that for Main Sequence stars. The more massive the star, the smaller it must become. A white dwarf of about .8 solar masses will have a radius only $1/100$ that of the sun; a dwarf of 1.2 solar masses will have a radius only $1/200$ that of the sun, or 3500 km—not much larger than the planet Mars.

The theoretical mass-radius relationship for white dwarfs is shown in Figure 7–10. It is evident from this diagram that as the mass of the star approaches about 1.4 times that of the sun, its radius approaches zero. In physical terms, this means that the pressure in the interior of the star is so great that even electron degeneracy cannot resist it. The star cannot exist as a white dwarf, and collapses. The greatest mass that a white dwarf can have, roughly 1.4 solar masses, is known as the **Chandrasekhar limit,** after the eminent Indian (now American) astrophysicist who predicted it.

Mass Loss

The existence of the Chandrasekhar limit has important implications for the late phases of stellar evolution. One thing it tells us is that most stars must somehow lose part of their mass during the final phase of their lives. We know this because, as we have already seen, it is the most massive stars that evolve most rapidly. Most of the stars in our galaxy that have already completed their evolution began with masses well above the Chandrasekhar limit. (There are very few clusters so old that stars of less than 1.4 solar masses have started to leave the Main Sequence.) Yet the number of white dwarfs in the galaxy is very great—enough, it would seem, to account for virtually all the stars that have already completed their course of evolution. This means that stars which began with masses greater than the Chandrasekhar limit have ended up as white dwarfs with masses below the Chandrasekhar limit!

Evidently, then, most stars expel part of their substance at some point late in their life cycles. This supposition is confirmed by observation. Many red giant stars seem to be losing mass continuously. They are surrounded by thin shells of gas and dust that they have apparently cast off. Studies of these ejected shells reveal that they are often expanding

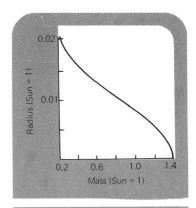

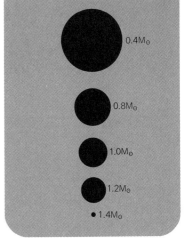

Figure 7-10
The mass-radius relationship for white dwarfs.

away from the star's surface at high velocities. Large parts of the outer envelopes of red giants may be blown away in this manner. The small, hot, dense cores that remain become white dwarfs. The expanding shells of gas are often visible as **planetary nebulas,** so called because their spherical form made them look like planets to early telescopic observers (Portfolio). A small but important minority of stars may also lose much of their mass, or even blow themselves to bits, in violent explosions.

It is important to realize that the mass expelled by a star late in its life is not necessarily identical in composition with the matter from which the star originally formed, millions or even billions of years earlier. During the course of the star's evolution beyond the Main Sequence, part of the star's original hydrogen and helium have been transmuted by nuclear fusion into heavier elements. The matter lost from stars enriches the interstellar clouds of gas and dust with elements beyond hydrogen and helium, and new stars that are born from the clouds are richer in these elements than stars of an earlier generation. As we shall see in subsequent chapters, this is the explanation for the two stellar populations, referred to briefly above. It also explains how the iron and silicon that make up our planet, and the carbon, oxygen, nitrogen, sulfur, phosphorus, and dozens of other elements that compose our bodies, were created. If it were not for massive stars that lived and died more than five billion years ago, neither the book that you are holding, nor the hand holding the book, would exist.

Some stars lose mass by ejecting shells of gas that we observe as planetary nebulas.

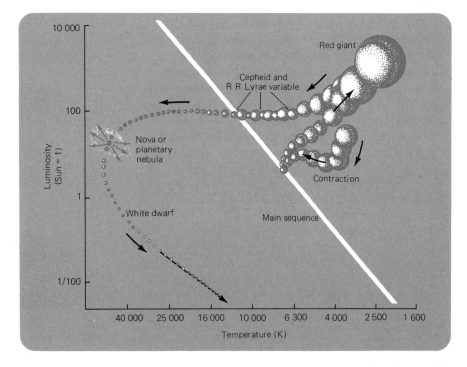

Figure 7-11
The complete evolutionary path of a star slightly more massive than the sun. The sizes of the star shown on the figure are not meant to be literally correct, but only suggestive. If these were drawn to scale, the actual size of the star in its red giant phase would be several feet. (Cepheid variable stars are discussed in Chapter 9.)

Mass Exchange in Binary Systems

The Russell-Vogt theorem tells us that a star should change only if either its mass or its makeup is altered. The mass of a star generally remains pretty nearly constant throughout most of its life, at least until the end of the red giant phase. When astronomers study stellar evolution, therefore, they usually need consider only the effects of nuclear fusion on the internal composition of the star. But there are some instances when the mass of a star **does** change significantly. This often happens in close binary pairs, where matter lost from one star may fall onto the surface of its companion. In such systems, the evolution of both stars is greatly complicated by the transfer of mass between them.

This exchange of mass can produce peculiar and even violent effects, such as strong X-ray emissions and repeated explosive flare-ups — phenomena which had been observed, but not previously explained. But the study of mass exchange was not undertaken in an effort to understand these effects; that they can now be explained is a bonus. The field actually grew out of the attempt to explain a paradoxical situation that is often encountered in the study of eclipsing binary systems.

For many decades, binary systems had been known in which the more massive component is a Main Sequence star, while the less massive component is a stunted giant known as a **subgiant**. Algol, mentioned in Chapter 5, is such a system. We know from the theory of stellar evolution that Main Sequence stars eventually expand and evolve into giants. We also know that the more massive the star, the shorter its stay on the Main Sequence, and the more quickly it becomes a giant. Systems such as Algol are thus very odd, since in them it is the **less** massive component that seems to be in a more advanced stage of evolution.

To understand this situation it is necessary to analyze the gravitational influence of both stars in the space around them. Each star is surrounded by a region in which its gravitational force is dominant. The boundary of this region is called the **Roche limit.** Any matter within the Roche limit of a particular star is bound to that star and cannot escape. Matter outside

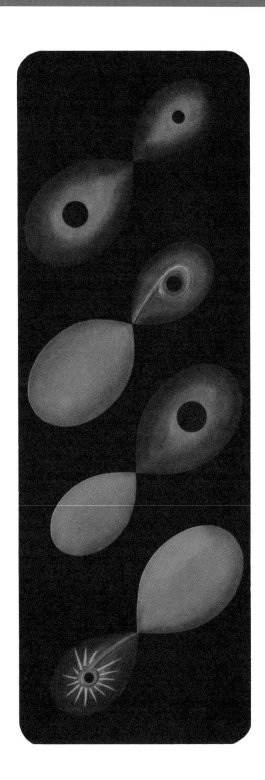

a) The more massive star evolves more rapidly, and expands into a red giant.

b) As the star overflows its Roche limit, some of its matter falls onto the surface of the companion, speeding its evolution.

c) The giant star has settled down as a subgiant filling its Roche limit; the companion star, now the more massive of the two, starts to evolve rapidly into a red giant.

d) The subgiant has completed its evolution and is now a white dwarf; its companion, now a red giant, overflows its Roche limit, and matter falls onto the surface of the white dwarf, triggering nova outbursts.

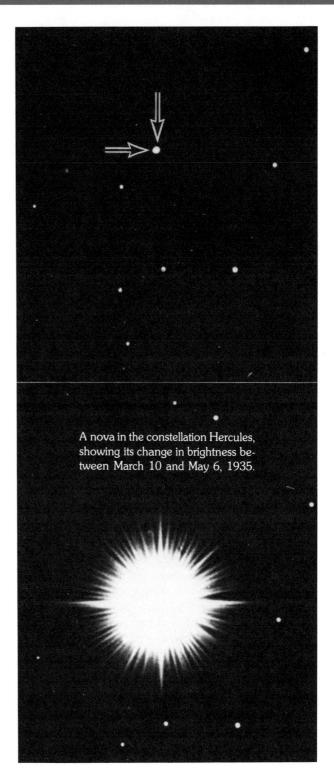

A nova in the constellation Hercules, showing its change in brightness between March 10 and May 6, 1935.

the Roche limit, however, is free of the star's gravitational "sphere of influence," and may be lost. Notice that the Roche limits of binary components are in contact. This means that matter from one star may pass through this point of junction into the domain of the other star, and can then actually fall onto its surface.

In the case of Algol and similar systems, the less massive star seems to fill its Roche region completely. This has led astronomers to suggest that the less massive star was once the more massive component. Being more massive, it consumed its supply of hydrogen more rapidly than its companion, and evolved into a red giant. As it expanded, it overflowed its Roche region, and matter from its outer layers was captured by the other star. Eventually enough mass was transferred in this way for a reversal of mass to take place: the less massive star became the more massive of the pair, and vice versa.

This theory at first met with great skepticism, since it implied that, to explain some known binary systems, as much as 80 percent of one star's mass might be transferred to its companion. Moreover, there were very few binary systems in which such a transfer could be seen taking place. The process had to occur so quickly that the binary systems were unlikely to be caught in the act. In 1960, however, new theoretical studies lent strong support to this model. It was shown that when a star expands rapidly on the way to becoming a giant, it indeed exceeds its Roche limit and loses its outer layers. When this happens the star expands even further, but because of the loss of mass, its gravitational attraction diminishes, and so its Roche limit contracts. The process accelerates as more and more mass finds itself outside the Roche limit. The star goes through a phase of rapid mass loss, in which almost all of its outer envelope is transferred to the companion.

When only the helium core and a much reduced hydrogen envelope remain, the star finally becomes stable again. It is then a subgiant, filling its Roche region almost exactly. After the period of rapid mass loss, which may last only a few thousand years, there is a slower phase in which a trickle of gas may continue to flow from the subgiant onto the surface of the other star. This is the phase we are likely to observe; and in fact we do sometimes see close binary pairs at this stage, surrounded by an envelope of very rarified gas.

Among the peculiar phenomena which this model enables us to understand are **novas**—stars that experience recurring explosive outbursts. Novas suddenly brighten by a factor of about 10,000, shine with this greatly increased luminosity for a few days, and then slowly decline in brightness over a period of years. These outbursts are not as violent as supernova explosions, and evidently do not disrupt the star very greatly.

In the last decade it has been found that most, if not all, novas seem to occur in binary systems where one component is a white dwarf. When matter from the companion star falls onto the surface of the dwarf, the tremendous surface gravity compresses and heats it. Nuclear reactions are suddenly ignited in the infalling matter, helped by the great density of the material and the availability of much carbon and nitrogen, in the highly evolved white dwarf, to catalyze the C-N cycle. The result is rather like what happens when lighter fluid is squirted onto the hot coals of a dying charcoal fire. There is an explosive flare-up in which the outer layers of the white dwarf are blown off with considerable violence. As the companion star loses more of its mass to the white dwarf, the cycle may repeat itself many times.

It seems likely that both of the processes described above may take place within the same binary system at different stages in its evolution. To see why this should be so, consider what happens when a member of a binary system transfers a large part of its outer envelope to its companion. Having lost much of its mass, including most of its unburned hydrogen fuel, the star quickly reaches the white dwarf stage. Meanwhile its companion, having gained mass, begins to evolve more rapidly. It soon expands into a giant, and overflows its Roche limit. The transfer of mass now starts to take place in the opposite direction. The outer layers of the giant fall onto the surface of the dwarf, triggering nova flare-ups.

An event similar to that which produces novas may be responsible for the mysterious X-ray radiation from some stars. A number of stars are strong emitters of **non-thermal** X-rays—that is, the X-rays are not part of the star's normal black-body spectrum. At least two such sources seem to be members of close binary systems. It has been suggested that the X-ray emissions result from nuclear reactions triggered in gas falling onto a relatively small surface region of a white dwarf or neutron star.

Still a third class of objects that may be involved in the process of mass exchange are **Wolf-Rayet stars.** These stars are very hot and luminous, with surface temperatures of about 100,000 K. Their spectra exhibit broad emission lines, suggesting the presence of extended, hot, turbulent atmospheres. Strong lines of carbon and nitrogen are especially conspicuous. These stars are also considerably less massive than might be expected from their high luminosities, which average about 10,000 times that of the sun.

Wolf-Rayet stars are not yet fully understood, but it has been noticed that many of them are members of spectroscopic binary systems. It is thought that they may be stars that are undergoing a phase of rapid mass loss of the sort previously described. This process appears responsible for the extended, turbulent regions of gas that surround these stars, and which in turn probably account for their unusual luminosity. The high concentrations of carbon and nitrogen presumably represent material from the core that has been brought to the surface during the rapid loss of the star's outer layers.

Evolved giant stars appear to lose their outer layers
before becoming white dwarfs. The result is an
expanding shell of gas, a planetary nebula, surrounding
the star which has collapsed into a compact, extremely
hot object. Such a central star is clearly seen in two of
the three photographs; the ultraviolet light it emits serves
to excite the nebula.

Planetary nebulae have various shapes. These three are
called the "Bubble," the "Ring," and the "Dumbell".
The complicated shapes are a consequence of
interactions with magnetic fields, or with the interstellar
medium.

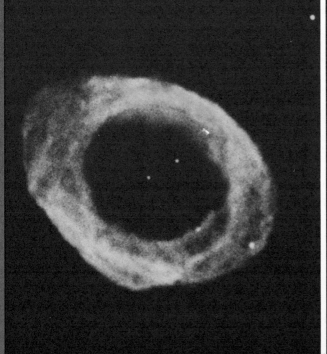

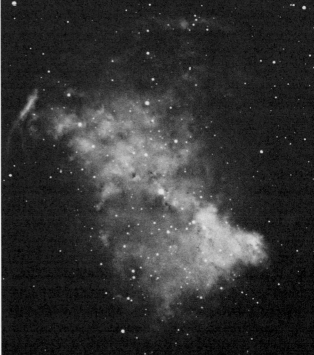

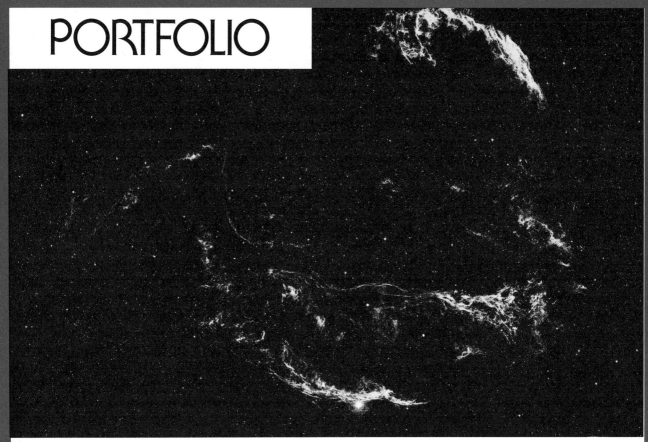

Above, a supernova remnant in Cygnus. The brighter part is the "Veil" nebula. The luminous gases have been heated by shock waves generated in collisions with the interstellar medium. Notice the greater density of star images inside the circle of expanding gas. This is evidence that the explosion has swept away the obscuring interstellar dust in that region.

Also, contrast this object with a similar object in Cassioeia, below left. The supernova remnant in Cassiopeia is greatly obscured by intervening dust clouds and only a portion of it can be seen in the photographic negative. The radio map, superimposed on the negative, reveals the more complete structure. Below right are photographs, in blue and red light.

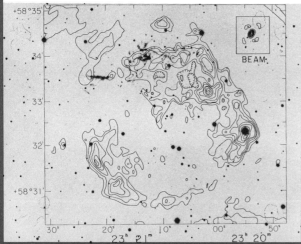

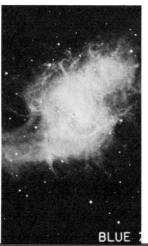

BLUE

RED

Supernovas

On at least seven occasions in the last two thousand years, new stars have appeared, and for a few days outshone all others in the heavens. In 1054, for example, the Chinese observed a "guest star" in the constellation Taurus; in 1572 Tycho found a new star in Cassiopeia; and in 1604 Kepler observed another in Serpens. All three of these stars, at their peak, were bright enough to be seen during the day.

These temporary stars are called *supernovas,* and they represent the most spectacular of stellar events, the explosion of an entire star. When such an explosion takes place the luminosity of the star increases a billionfold. If the supernova occurs within our galaxy, a faint or invisible star suddenly becomes one of the brightest objects in the heavens. The high luminosity persists for a few days, and then declines slowly over a period of several years. Usually the star remains bright enough to be visible to the naked eye for many months. After it has faded from view, a rapidly expanding cloud of gas can be observed telescopically at the site of the original explosion.

The Crab nebula, for example, is located in the same region of the sky as the supernova seen by the Chinese in 1054 (Figure 7–12). It is possible to analyze the motions of this cloud of gas and trace them backward in time, as if we were running a film in reverse. When this is done, we find that the cloud must have begun its expansion from a single point about the middle of the eleventh century. It seems certain, therefore, that the Crab nebula is the remnant of the supernova of 1054.

Modern observations of supernovas within our galaxy are not available because of the rareness of the event. It is estimated that one supernova occurs within the galaxy every 25 to 75 years. Since interstellar matter obscures much of the galaxy from our view, however, the observed frequency is much lower—it averages about one every 300 years or so. The most recent, known as Kepler's supernova, was seen in 1604. Each year we observe many of these stellar explosions in other nearby galaxies. Some of them are so bright that they outshine the entire galaxy in which they occur. But extragalactic supernovas are too distant to be studied intimately.

Our models of supernova explosions therefore rest as much on theoretical deduction as on observation. There are several ways, however, in which we can check our theories against empirical evidence. One is to compare the observed frequency of supernovas with that predicted by theory. Another is to compare the known chemical composition of the galaxy as a whole with the abundances of various elements that our theory predicts should be formed during supernova explosions. We can also compare the velocity of expansion of known supernova remnants, such as the Crab nebula, with the velocity that our models would lead us to expect. Whether a star becomes a supernova depends upon its mass. It is

Figure 7-12
A supernova in another galaxy. It is not uncommon for such stellar explosions to outshine the entire galaxy in which they occur.

thought that stars of low mass (3.5 solar masses or less) are not liable to these explosions. After leaving the Main Sequence, they manage to lose enough to get down to the Chandrasekhar limit (1.4 solar masses) without cataclysmic violence, and end their lives peacefully as white dwarfs. Some of them apparently do this by shedding their outer layers, as previously described.

A very different fate may await stars of intermediate mass — 3.5 to about 8 solar masses (the precise upper limit is not yet known with certainty). Many of these stars apparently do **not** lose enough mass during their lifetimes to become white dwarfs. In the later stages of stellar evolution, they develop cores of carbon and oxygen. Helium burns in a shell around this core, while hydrogen fusion continues to take place in another shell outside the helium-rich region. As the carbon-oxygen core builds up, its density reaches $10^9 g/cm^3$, and the temperature rises steadily. At about 2 billion nuclear reactions involving carbon and oxygen start to occur within the core. The situation is similar to the helium flash, described earlier, but with one crucial difference. The carbon and oxygen core is so dense that it remains degenerate even when the temperature soars to billions of degrees. Thus there is none of the expansion and cooling that eventually damps the helium flash. Once carbon and oxygen start to burn, there is nothing that can halt the runaway process. The core explodes, blowing the entire star to bits. The shock wave from the exploding core travels outward, heating and compressing the stellar gases, and initiating many nuclear reactions. These produce heavy elements, which are spewed out into interstellar space.

In stars of 10 or more solar masses, a somewhat different process is believed to operate, though with no less violent results. These stars do not develop dense, degenerate cores. Carbon and oxygen ignite and burn smoothly to produce heavier elements such as neon, sulfur, and silicon. The internal temperature keeps rising until it reaches the vicinity of 5 billion degrees. At this point, however, the nuclear fusion reactions undergo a major change: they start to produce large numbers of neutrinos.

Neutrinos are particles that interact very weakly with matter; they can travel through hundreds of light-years of solid steel with very little likelihood of interference. Unlike photons, which must struggle for millions of years before escaping from the interior of a star, neutrinos reach the surface within seconds of their creation. Pouring from the star in vast numbers, they carry away a considerable amount of energy. Pressure inside the core drops; it contracts, and becomes hotter. The increasing temperature causes more neutrinos to be produced, and still more energy to be lost. Within seconds, the star's core collapses.

At the inconceivably high pressure within the collapsed core, the atomic nuclei themselves are crushed and obliterated. Subatomic particles such as protons and electrons can no longer maintain distinct identities; they are squeezed together to form neutrons. The entire core of the

Stars that do not lose enough mass to become white dwarfs are liable to explode with great violence as supernovas.

star is compressed to a small sphere, perhaps 20 km in diameter, of densely packed, degenerate neutrons. The outer layers of the star, deprived of their pressure support, start to fall inwards—only to meet the core as it bounces back in its attempt to achieve some stable equilibrium position. The result of this collision is a violent shock wave that travels swiftly through the star. The carbon-oxygen layer is heated to the ignition point, and explodes with such violence that the remainder of the star's envelope is blown off into space at a high velocity.

As we saw earlier, elements heavier than iron are not ordinarily created within a star, for the necessary fusion reactions absorb rather than liberate energy. A supernova explosion such as we have been describing, however, provides enough energy for the synthesis of elements beyond iron. The heaviest metals found in the galaxy are created in this way.

Supernova explosions enrich the interstellar material with heavy elements, which are incorporated into the next generation of stars.

Neutron Stars

We have seen how electron degeneracy pressure can support a star against collapse. The same role can also be played by neutron degeneracy pressure. It is possible to calculate equilibrium models of stars that derive most of their pressure support from degenerate neutrons. Such stars are called **neutron stars.** Neutrons do not become degenerate, however, until the density reaches about 10^{14} g/cm^3. This is some 10 to 100 million times higher than the density inside a white dwarf. A teaspoon of matter from such a star would weigh a billion tons. When the first theoretical models of neutron stars were made, in the 1930s, there was no way of knowing whether any stars of this sort actually existed, or how they might come into being.

Observation has now settled the first question fairly decisively. As we shall see in the following Interlude, we have discovered a class of objects, known as pulsars, which are almost certainly neutron stars. The second question has also been clarified by the advance of our knowledge of stel-

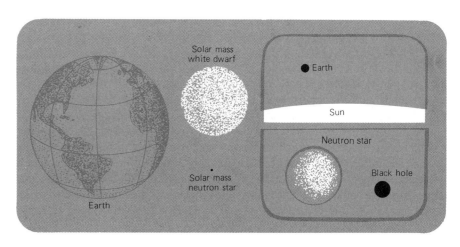

Figure 7-13
The relative sizes of the earth and sun in comparison with a white dwarf, a neutron star, and a black hole, each of 1 solar mass.

lar evolution. Neutron stars appear to have an upper limit for their mass, similar to the Chandrasekhar limit for white dwarfs. The exact value of the neutron star limit is not known at present, for our understanding of subatomic interactions at the densities found within a neutron star is still very incomplete. If the neutron star limit is greater than the white dwarf limit of 1.4 solar masses, no unusual process is needed to account for the existence of neutron stars. A star that is too massive to become a white dwarf will simply contract further until it becomes a neutron star.

If the neutron star mass limit is less than 1.4 solar masses, however, the creation of a neutron star is not so easy. A collapsing star with a mass low enough to become a neutron star would also have a mass low enough to beome a white dwarf. When it reached the white dwarf configuration, it would attain stability, and stop contracting. If this is the case, the only way to produce a neutron star is through external compression, such as the core of a massive star experiences during a supernova explosion. Which of these scenarios is responsible for the neutron stars of our galaxy is still uncertain.

Black Holes

It seems impossible to imagine that there could be anything odder or more "final" than a neutron star—a ball of neutrons some 20 km in diameter, with a mass equal to that of our sun, and a density a thousand

Balancing the Energy Budget

One of the most powerful concepts in physical science is the law of conservation of energy, discussed briefly in Chapter 6. The great advantage of this law, as of all conservation laws, is that we do not have to analyze all the details of a physical process to draw up an energy budget for it. We know that, regardless of how complicated the intervening steps, the final energy of the system we are studying must be equal to its initial energy, less whatever energy is lost from the system along the way.

We have now traced the life cycle of a star from its beginning as a diffuse cloud of gas, to its possible end, after billions of years, as a neutron star. Remarkably, despite all the complexities of its evolution, it is possible to construct a balanced energy budget for such a star over the entire course of its life.

In the beginning, the star consists of hydrogen atoms—that is, protons and electrons. As it evolves, much of the hydrogen is fused to form successively heavier elements: helium, carbon, oxygen, and so on up to iron (if the star is massive enough). But we need not concern ourselves with these intermediate steps. We need only know the final stage, in which all the elements are broken down again, and the original protons and electrons are fused to form neutrons.

A black hole is formed by the complete gravitational collapse of a star that is too massive to become a neutron star.

trillion times that of water. Yet as the distinguished British biologist J. B. S. Haldane has said, "The universe is not only stranger than we imagine; it is stranger than we *can* imagine." It now appears that the destiny reserved for some stars is indeed stranger than most of us can easily imagine: to disappear from our universe entirely by falling into a hole in space that it has itself created. Such a star becomes a **black hole.**

A star that is too massive to become a white dwarf or a neutron star will continue to collapse. How far can it collapse? The answer—strange as it seems to common sense—is that we now know of no force that could stop such a collapse. The collapse, in other words, can go on indefinitely. The entire mass of a star—3×10^{30} kg or more—could shrink to the size of a baseball, to the size of a bacterium, to the size of an atom. . . .

The idea of a physical object being compressed to an infinitesimal size is quite alien to our everyday notions. We tend to think of neutrons as solid, spherical objects, like billiard balls, having a definite size. Of course you could compress a billiard ball until it was pulverized to dust— but even the dust would still occupy some volume. Though you might be able to pack it more tightly than when it was part of the ball, it is hard to imagine compressing it indefinitely.

But on the subatomic level, these analogies are misleading. Particles such as neutrons do not have solid surfaces of the kind we associate with

Neutrons represent a *higher* energy state than separate protons and electrons. (In fact, as we saw earlier, a free neutron will soon decay into a proton and an electron, releasing energy.) Moreover, during the course of its life the star has given off a vast amount of energy in the form of electromagnetic radiation. The energy budget appears unbalanced—the star, having lost energy into space, seems to have ended up with more energy than it started with. Where did it get enough energy to keep it radiating for billions of years, and still leave it with a net surplus?

The answer is simple. The hydrogen from which the star formed was originally dispersed over a very large volume of space. It thus possessed a tremendous amount of gravitational potential energy, like water behind a dam, or a rock at the edge of a high cliff. The star draws on this store of energy in its infancy, contracting until it is hot enough to begin nuclear fusion, and again at the end of its career, when it collapses to become a neutron star. In fact, it has been calculated that the difference in gravitational energy between a diffuse cloud of gas, 6 trillion km in diameter, and a ball of neutrons, 20 km in diameter, accounts for precisely the missing amount of energy: the energy radiated away, plus the energy surplus of neutrons compared to protons and electrons.

macroscopic objects. It is more useful to think of subatomic particles as regions of space in which certain forces operate strongly. The "radius" of a particle is simply the distance at which one of these forces is strong enough to repel another approaching particle. Theoretical calculations have indicated, however, that no known repulsive force is strong enough to withstand the attractive force of gravitation, once enough mass is packed together in a small enough volume. This means that there seems to be no limit to the ultimate compressibility of matter, and no minimum radius at which the collapse of a star must halt!

What is left when a star virtually vanishes in this way? To answer this question, it is important to realize that, although the star's size decreases without limit, its mass remains undiminished. All that mass continues to exert gravitational forces on any matter near it. In fact, the surface gravity of the star becomes more and more intense as its radius diminishes. This is a consequence of the inverse square law of gravitation. A white dwarf and a red giant, for example, may have the same mass; nevertheless, the surface gravity of the dwarf is enormously greater, simply because the star is so much smaller, and its surface is so much closer to its center of mass. If the white dwarf's radius is $1/1000$ that of the giant, its surface gravity will be 1,000,000 times greater.

As a collapsing star contracts, therefore, the force of gravity at its surface increases. This is reflected in the **escape velocity**—the velocity which a body leaving the star's surface must have in order to break free of its gravity and continue on into space forever. The escape velocity on the earth's surface is 11.2 km/sec. If you fired an artillery shell away from earth at that velocity, it would never return . The escape velocities of several sorts of stars are shown in the accompanying table. Notice that the escape velocity at the surface of a neutron star is about $1/3$ the speed of light.

LIGHT AND GRAVITATION   What would happen if a star, continuing to collapse beyond the minimum possible size for a neutron star, shrank so far that the escape velocity at its surface exceeded the speed of light? Einstein's general theory of relativity predicts that even light is subject to gravitation. Though this effect is normally too small to detect, the prediction has been verified; in 1919, during a solar eclipse, scientists measured the bending of light from distant stars as it passed close to the sun. If the escape velocity at the surface of a collapsing star reached 300,000 km/sec, therefore, light itself would be unable to leave the object.

We can look at this conclusion in another way. Relativity also predicts that light loses energy moving against a strong gravitational field. As we might expect from the quantum theory (Chapter 4), this loss of energy takes the form of a decrease in frequency and an increase in wavelength. The wavelength of light climbing up through a strong gravitational field will therefore be shifted toward the red end of the spectrum, just as if the source were receding from us and its light were being red-shifted by the Doppler effect. In the case of a white dwarf, this gravitational red shift can actually be measured; it is equivalent to the Doppler shift we would find if the star were moving away from us at about 20 km/sec.

If the gravitational field of a collapsed star were intense enough, the light leaving it would be red-shifted so far that its wavelength would be infinitely long and its frequency infinitely low. In other words, it would lose all its energy in the fight against the star's gravitation. No light would emerge from such a star; it would be totally dark, like the "hole in the heavens" that Herschel imagined he saw. Such an object is a **black hole.**

The radius which an object of a given mass must reach in order to become a black hole is called its **Schwarzschild radius,** after the German astrophysicist who first predicted its existence. The Schwarzschild radius, in kilometers, is roughly equal to 3 times the mass of the black hole, expressed in solar masses. Thus a black hole created by the collapse of a star twice as massive as the sun will have a radius of only about 6 km — a very tiny object, by astronomical standards. It is interesting to note that once this radius is reached, the size of the star that created the black hole in the first place is no longer important. The star itself will presumably continue to shrink, although (since no light from within the Schwarzschild radius can reach us) we will be unable to observe the process further. Its gravitational field, however, will be unaffected.

If light cannot escape from a black hole, neither can any physical object, whether it be a planet, a space ship, or an electron. Anything that happens to fall within the Schwarzschild radius is lost from our universe. Not only will it never emerge, but we will never be able to see it or learn anything about what becomes of it. For this reason, the Schwarzschild radius is also known as the **event horizon**; anything that takes place inside it is forever hidden from us. Indeed, when an object falls into a black hole, for all practical purposes it ceases to be. The only evidence of its existence will be the properties that it contributes to the black hole itself. There are just three of these: its electrical charge, its angular momentum (for it is believed possible for rotating black holes to exist, with peculiar characteristics all their own), and its mass (which will serve to increase the radius of the black hole slightly).

Because of their intense gravitational fields near the Schwarzschild radius, black holes have excited a great deal of speculation recently. Some writers have imagined the black hole as a kind of cosmic vacuum cleaner, swallowing up all the matter in its vicinity. The truth is slightly less dramatic. In reality, the ability of a black hole to attract matter at a distance is exactly the same as that of a star of equal mass. If the sun were replaced by a black hole of identical mass, the planets would continue in their orbits just as before, with no more tendency to fall into it than they now have to fall into the sun. Only at close range does the black hole show its remarkable properties.

The escape velocity at the edge of a black hole is so great that not even light can escape.

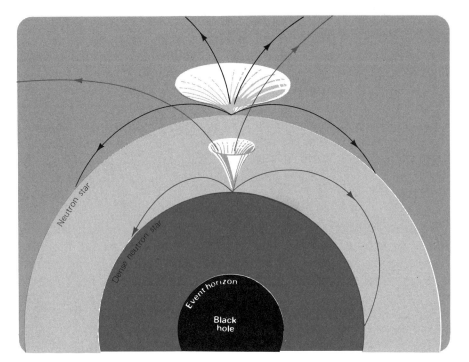

Figure 7-14
The cones represent the escape paths of photons leaving a neutron star. Though we are accustomed to thinking that light always travels in straight lines, the gravitational force of the neutron star is so great that it bends the photon paths. Only photons leaving along paths that lie inside the cones can escape; photons leaving at other angles fall back. The denser the neutron star, the stronger its surface gravity, and the narrower the cone of escape. At the event horizon of a black hole, gravity is so strong that no photons can escape, regardless of their path.

TIME AND GRAVITATION   Among the many odd properties of black holes, perhaps the oddest is their ability to distort the flow of time. This is a consequence of Einstein's general theory of relativity, which predicts that, to an outside observer, time will appear to run more slowly in a strong gravitational field. A space traveler falling into a black hole would detect nothing peculiar about his ***own*** experience of time. He would find himself approaching and passing the event horizon in a reasonable period (if either he or his clock could somehow survive the wrenching tidal forces created by the black hole's enormous gravity). As distant observers watching this fall, however, we would see something very strange.

We would expect the unfortunate voyager to fall ever faster, accelerated by the gravitational attraction of the black hole. Yet the opposite would seem to happen. The closer the voyager came to the event horizon, the more slowly he would appear to fall. Nor would his rate of fall be the only phenomenon affected in this way, for every temporal process would exhibit the same slowing down. The voyager's clock would seem to run more slowly, his heartbeat would decrease proportionately, the very atoms of his body would appear to vibrate less rapidly. (In fact, this is just another way of looking at the gravitational red shift described earlier; the slowing of time causes the frequency of electromagnetic radiation to decrease.) In fact, as the voyager neared the event horizon, time would become so stretched out that it would seem to take him forever to reach it. He would appear to spend all eternity at the edge of the event

Nothing that passes the event horizon can ever leave a black hole; but because of the slowing of time near the event horizon, we can never see anything fall in.

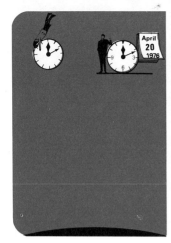

Figure 7-15
To a person falling into a black hole, the process will take a finite time — perhaps a few minutes. But the powerful gravitational field near the event horizon causes time to flow more slowly, as witnessed by an outside observer. No matter how long we watch, we will never see the conclusion of the unfortunate traveller's fall.

horizon, constantly inching infinitesimally closer — but no matter how many eons we watched, we would never see him fall in! This is one of the most tantalizing of the many paradoxes that black holes present us with — not only can we never see anything leave a black hole, we can never see anything enter!

Anything that falls within the event horizon of a black hole is lost to our universe — but black holes themselves, if they exist, should be very much part of our universe. Though we cannot see them, we ought to be able to detect their existence by their influence on their surroundings. The possibility of doing this is discussed in the following Interlude.

SUMMARY

Because stars are hot, they must radiate energy away into space. Because they are constantly losing energy in this way, they must evolve. When stars are not replacing that energy through nuclear fusion, they must fall back on their store of gravitational potential energy, contracting and growing hotter. This is what happens before stars become hot enough to begin nuclear fusion, and again after they have exhausted their supply of nuclear fuel. In between, when stars are making good their energy loss by nuclear fusion, they do not need to draw on gravitational energy. Nevertheless they still must change, for the process of fusion is continuously altering their internal structure.

The time scale of stellar evolution is so long that we can rarely witness any of these changes taking place. We do, however, see many stars of all ages. This fact, along with our understanding of stellar energy processes, and our ability to construct models of stars, enables us to predict the course of a star's evolution. Models of stars are created by solving (usually with a computer) the equations of stellar structure. The Russell-Vogt theorem assures us that, for a star of given mass and composition, there is only one possible solution to these equations. This means that the star's structure is determined completely by its mass and composition, and will change only as one of these factors does. To study stellar evolution, astronomers calculate a model of a star, and then see how that model will change as a result of the fusion process. A series of such models gives us a picture of the star's evolution throughout its lifetime.

Bright, hot stars, which we know must be young, are often found in clusters, and the clusters are generally associated with clouds of interstellar gas and dust. All stars are believed to form from the condensation of such clouds. When the density of the interstellar material becomes great enough at any point, gravitational attraction holds the particles together and attracts more material from the surrounding areas. A cloud collapsing in this way, called a protostar, eventually becomes hot enough for nuclear fusion to begin. At this point, it becomes a Main Sequence star, its position on the H-R diagram depending on its mass.

Because stars consist mostly of hydrogen, and the fusion of hydrogen to helium is such a rich source of energy, the Main Sequence lifetime of a star is by far the longest part of its life. As helium accumulates in the stellar core, however, the star begins to evolve slightly away from its original Main Sequence position. When about 12 percent of the star's mass has been converted into helium, the star's evolution accelerates abruptly. Its surface becomes cooler, but also larger and thus more luminous. The star becomes a red giant.

Eventually, the temperature at its core may rise high enough for the fusion of helium into carbon and oxygen to begin, and subsequently the fusion of carbon and oxygen into still heavier elements. The more massive the star, the higher its central temperature, and the more transformations can be initiated. But while the star is radiating away more energy than ever, successive fusion reactions provide less and less energy. Once iron is reached, no more energy can be extracted from nuclear fusion.

When all nuclear reactions have been exhausted, most stars become white dwarfs—small, hot stars that slowly cool as they radiate away energy. When they reach a certain density, the electrons of a white dwarf become degenerate. The star is then supported against collapse by degeneracy pressure, and does not contract any further. The more massive the white dwarf, the smaller it must be. No white dwarf can exist with a mass greater than 1.4 solar masses. This suggests that most stars lose an appreciable part of their mass late in life—for example, by shedding their outer layers to form planetary nebulas.

Stars of moderate and high mass are liable to explode with great violence, blowing themselves to bits. When such a supernova explosion takes place, the luminosity of the star increases a billion-fold or more, and it may for a time outshine an entire galaxy. An expanding nebula, known as a supernova remnant, usually marks the site of past supernova explosions. The stellar core, compressed by the explosion into a rapidly spinning ball of degenerate neutrons only a few km in diameter, may also be left behind. We observe these neutron stars as pulsars—sources of regular pulses of radio waves.

If enough mass is concentrated in a small enough volume, the process of gravitational collapse cannot be halted by any known force. If a star reaches this condition, it will continue to collapse indefinitely. At a certain radius, its gravita-

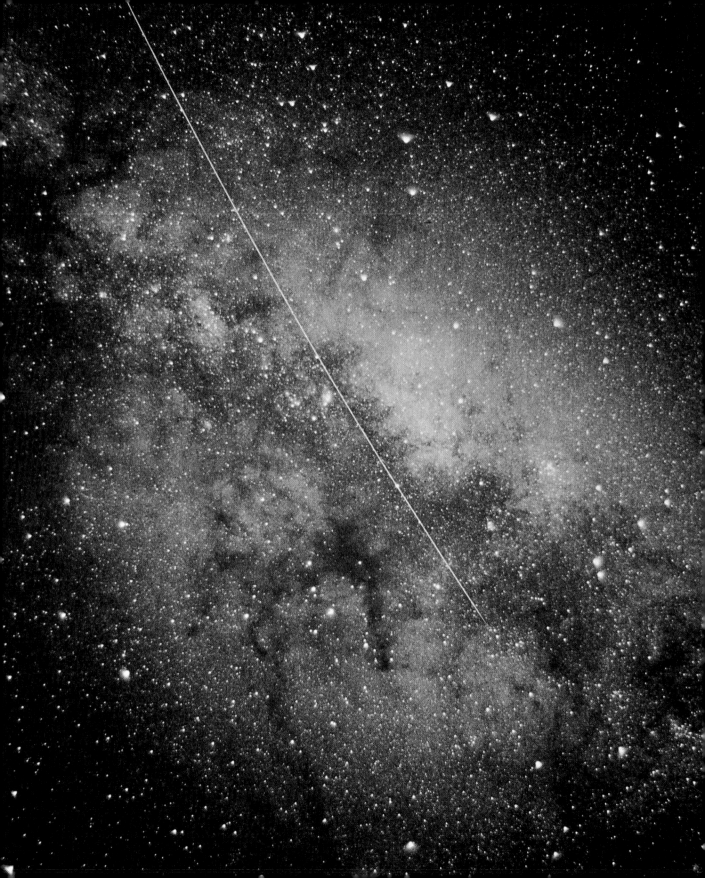

M 13, in Hercules, is the brightest globular cluster in the northern sky. The stars are all old, members of Population II. They travel in randomly oriented, elongated orbits about the center of the cluster, while the cluster itself travels in a similar orbit about the galactic center. Notice how the central regions are far too densely packed to be resolved into individual stars. Some estimates put the number of stars in such a cluster at a million. (U.S. Naval Observatory photograph.)

The Pleiades, in Taurus, form a bright, open cluster. Besides the six stars visible to the naked eye, there are about 150 others in the group. This photograph shows the filamentary structure of the reflection nebula that surrounds the stars. The light from these clouds of interstellar dust is merely that of stars, reflected and scattered by the tiny dust particles. It is therefore bluer than the original starlight. It is not clear whether the nebula is material left over from the formation of the cluster, or merely a dusty region through which the stars are passing. (Hale Observatories.)

Preceeding page: the Milky Way in Sagittarius, looking toward the center of our galaxy. The galactic center itself cannot be seen because of obscuration by clouds of interstellar dust, as evidenced by the dark areas in the photograph. The diagonal streak is caused by the highly reflective communications satellite Echo I, which crossed the field during the 30-minute exposure. (U.S. Naval Observatory photograph.)

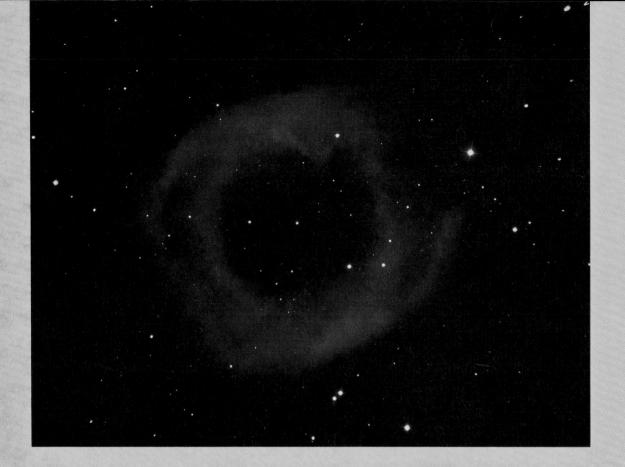

NGC 7293 is a nearby planetary nebula in the constellation Aquarius. This object, which has a diameter about half that of the full moon, may be no more than 150 parsecs away. The red color is due to emission lines of hydrogen and once-ionized nitrogen. The shell of gas is expanding at 25 km/sec. In long exposures, a fainter second envelope can be seen surrounding the one visible here. The central star has a surface temperature of about 60,000 K. Having lost its outer layers, it is probably on the way to becoming a white dwarf. (Hale Observatories.)

M 27 is another planetary nebula, known as the Dumbbell nebula. The green light from this object is due to forbidden lines of twice-ionized oxygen; once-ionized oxygen adds blue-violet tints. Here, too, longer exposures reveal the presence of a fainter, outer shell of gas. The temperature of the star at the center is about 50,000 K. (Lick Observatory.)

M 42, the Great Nebula in Orion, is a region of recent star formation. This vast cloud of gas is excited by hot, luminous young stars. The ultraviolet radiation from these stars is re-radiated by the nebula in the visible range. This region contains CO and OH radicals; it has been mapped by radio observations. Infrared telescopes have also detected what appear to be protostars—stars still in the process of formation. (U.S. Naval Observatory photograph.)

Eta Carinae, an unusual mass of gas and dust, with embedded stars, in the southern hemisphere. This object has undergone unusual changes over the years. A rather faint "star" to observers in 1677, it became as bright as Sirius in 1837, and has now faded below naked eye visibility. Some astronomers believe it is a region of star formation; others have suggested that it may be a star about to become a supernova. (Courtesy of Donald H. Menzel, from his book, *Astronomy;* Harvard College Observatory, Boyden Station photograph.)

M 31, the Andromeda galaxy, a spiral galaxy in the Local Group. Somewhat larger than our own galaxy, it comprises several hundred billion stars. The light by which this photograph was taken began its journey to us more than 2 million years ago, about the time the first human beings walked the earth. The difference in color of the inner and outer regions reflects the distribution of the two stellar populations: old, reddish stars of Population II in the nuclear bulge, and young, hot, luminous blue stars of Population I in the spiral arms. The two satellite galaxies are dwarf ellipticals. (Hale Observatories.)

M 82 is an exploding galaxy. Some violent eruption appears to have hurled streamers of dust and gas many thousands of light-years from the galactic center. The redness of the central regions is due to hydrogen emission lines and starlight reddened by its passage through dust. (Lick Observatory.)

M 51, the Whirlpool galaxy, a beautiful spiral in Canes Venatici. Dust and bright blue stars can be seen in the spiral arms. The companion system, NGC 5195, is an irregular galaxy also rich in dust and gas. Notice how one of the spiral arms appears to connect the galaxies. The evolutionary relationship between the two is not understood. (U.S. Naval Observatory photograph.)

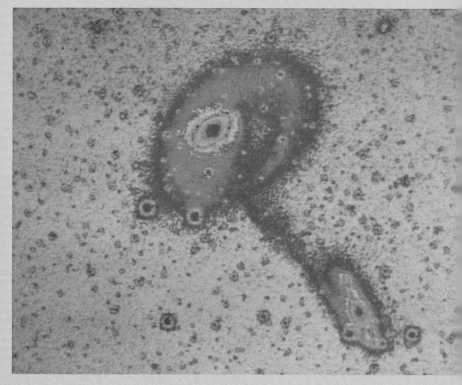

A pair of interacting galaxies, photographed with the 4-meter telescope at Cerro Tololo, Chile. The larger of the two systems is about twice the size of our own galaxy. A collision seems to have taken place; star formation in the huge ring has apparently been triggered by the passage of the smaller galaxy through the disc of the larger. This false-color image is designed to emphasize areas of neutral hydrogen. (Kitt Peak National Observatory.)

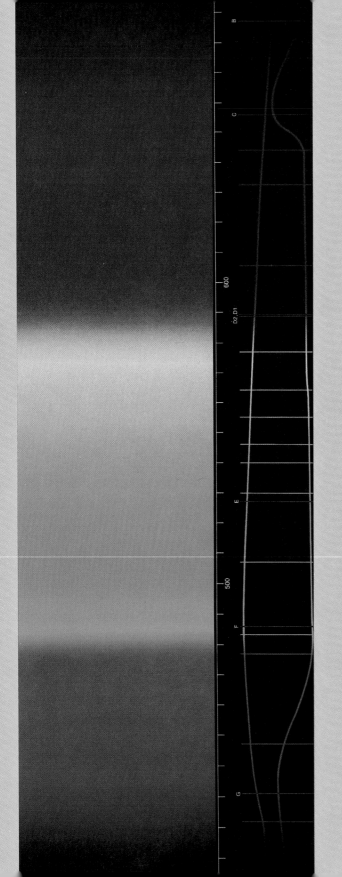

The visible spectrum, a tiny segment of the entire known range of electromagnetic radiation. The wavelengths of visible light range from 400 nm (violet) to 700 nm (red). Underneath the spectrum is a curve representing the distribution of wavelengths in sunlight reaching the earth's surface. The dark lines marked by letters are some of the most prominent Fraunhofer lines — absorption lines created by the gases of the chromosphere. Line C, for example, is the Balmer alpha line of hydrogen at 656.3 nm, while the double D lines are due to sodium, and the green line at 530 nm represents 14-times ionized iron in the solar corona. The lower curve indicates the absorption of various wavelengths by the chlorophyll of green plants, which use the energy of sunlight to make sugar from carbon dioxide and water. Solar radiation is thus the source of all the food on earth, as well as nearly all the energy. The retina of the human eye contains cells of three different sorts, each with its maximum sensitivity at a different wavelength: 477, 540, and 577 nm. (Courtesy of Eastman Kodak Company.)

tional field becomes so strong that nothing can escape it, not even light. Such an object becomes a black hole. We can never see or communicate with anything that passes the threshold of a black hole.

EXERCISES

1. Why do we seldom see stars evolving? Why do we believe that they evolve at all? Do we have any empirical evidence for stellar evolution?
2. What do we know about stars that enables us to construct models of their interiors?
3. What is the Russell-Vogt theorem? Why is it important? How does it shed light on the existence of the Main Sequence?
4. How do we use models of stars to study stellar evolution? What is the time-step? Why are time-steps of widely different lengths used in studying the evolution of a star?
5. What evidence about star formation do we get from studying stars of spectral types O and B?
6. How do protostars form from interstellar material? Describe briefly the stages through which a protostar passes before joining the Main Sequence.
7. Why is it difficult to observe star formation with ordinary telescopes? How can this limitation be overcome?
8. Why do the masses of Main Sequence stars fall within definite limits?
9. Why do we find such a substantial majority of all stars on the Main Sequence? Why is a star's stay on the Main Sequence relatively uneventful?
10. How does the extent of its Main Sequence on the H-R diagram tell us the age of a cluster?
11. An earth-like planet 100 AU from a star of luminosity 10,000 times that of the sun would have about the same temperature as earth. Why is it unlikely that intelligent life could exist on such a planet?
12. What is degeneracy pressure? Mention three instances of its importance in the study of stellar evolution.
13. What are white dwarfs? Why don't they contract indefinitely as they cool?
14. What is the Chandrasekhar limit? What does it tell us about the evolution of stars beyond the Main Sequence?
15. What is a supernova? Is the sun likely to become one? Explain.
16. Why do some stars have higher abundances of heavy elements than others? Where do these heavy elements come from?
17. How does a neutron star differ from a white dwarf? How is it created? What keeps it from further collapse?
18. Why do some stars have higher abundances of heavy elements than others? Where do these heavy elements come from?
19. Why cannot elements heavier than iron be built up through gradual fusion within stars? How are such elements produced?
10. How does a neutron star differ from a white dwarf? How is it created? What keeps it from further collapse?
21. What is a black hole? Why should they be avoided?

pulsars & black holes

Photographic record of a pulsar in the Crab nebula.

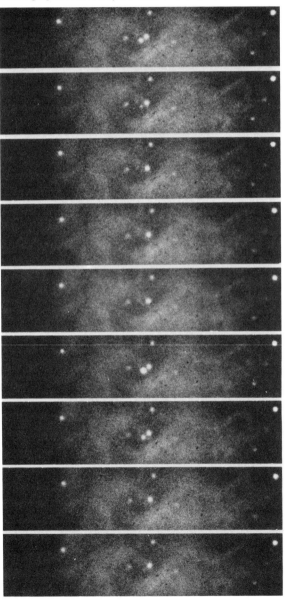

Of all the objects to be seen in the heavens, supernovas—stars which suddenly flare into incredible brightness, only to fade and disappear within a matter of months have always seemed to astronomers to be one of the most remarkable. To pre-Galilean astronomers, who believed that the heavens were incorruptible and unchanging, the phenomenon was not only spectacular but frightening. European observers did not even report the great supernova of 1054 although it surpassed Venus in brightness. It was left to the Chinese and American Indians to preserve records of the event.

In more recent times, it was suggested that a supernova must be a star in its final death throes—a stellar explosion of almost unimaginable scope and violence. But what caused the explosion, and what became of the star afterward, remained entirely unknown. The first hypothesis offering an insight into the nature of supernova came in 1934. Fritz Zwicky and Walter Baade published a paper in which they speculated that "supernovas represent the transitions from ordinary stars into **neutron stars,** which in their final stages consist of extremely closely packed neutrons." At this point, no one had knowingly observed a neutron star, either through an optical telescope or by any other means. Calculations by Zwicky and Baade, based on contemporary knowledge of the nature and behavior of matter, suggested that such objects **could** exist, but this was far from proof that they **did** exist. The concept of a neutron star remained purely a conjecture, but it received additional support in 1939, when J. Robert Oppenheimer and his associates showed mathematically that the reduction of a star to a sphere of closely packed neutrons was possible within the framework of the quantum theory, then still relatively new. But the final test of the neutron star hypothesis—the identification of an acutal neutron star—did not come until almost 30 years later.

The break came, not as the result of an attempt to validate the neutron star concept, but as an entirely unexpected offshoot of another project. Many important discoveries in the history of science have occurred in this manner. In 1967, Jocelyn Bell and Anthony Hewish, of Cambridge University, England, were conducting research on the scintillation of radio sources. Scintillation is a rapid fluctuation in the intensity of a radio signal from a point source, such as a QSO(Chapter 9). It is caused by interstellar gas, in much the same way that the earth's atmosphere causes visible stars to "twinkle."

Soon after they began their study, they were astonished to find a source that emitted radio waves in extremely short, regular pulses. The bursts of radiation were of precisely .016 seconds duration, and occurred at intervals of 1.33730115 seconds. Later, other sources were discovered, pulsing at different intervals. The investigators were at a loss to explain the origin of these bursts. One theory, proposed half-seriously at the time, was that they emanated from cosmic lighthouses—artifical signalling devices set up by a space-faring extraterrestrial civilization as an aid to navigation.

Pulsars, as the radio sources later came to be called, did not turn out to be artificial signals, but they did prove to be something almost as interesting. In 1968, astronomers at the National Radio Astronomy Observatory in Green Bank, West Virginia, located a pulsar in the center of the Crab nebula. The Crab nebula had already been shown to be the remnant of the supernova of 1054. Thus, in this one case at least, a connection between pulsars and supernovas was indicated. Could the pulsating object in the center of the nebula be the fabulous neutron star which Zwicky, Baade, and Oppenheimer had predicted three decades before?

The evidence seemed to support this conclusion. Scientists often test a hypothesis by playing devil's advocate—that is, by trying to find alternative explanations that might also account for the phenomenon. In the case of the Crab Nebula pulsar, such attempts proved futile. Astronomers recognize only three types of objects which might produce regular bursts of radio noise: a pulsating variable star, which expands and contracts; a rotating star with a bright spot; and a double star system with one star regularly eclipsing the other. None of

these objects, however, would be able to pulse as rapidly as the new found radio sources. Only an object of the size and density predicted for a neutron star—about 20 kilometers across, the size of a large city—would be able to oscillate or spin with such rapidity.

One of the difficulties in explaining pulsars was that the radio emissions could not at first be identified with any visible object. Astronomers reasoned that those pulsars most likely to radiate at visible wavelengths were those that contained the greatest amount of energy. According to the generally accepted models for pulsars, a part of their energy would be stored in their rotation. As they radiated away energy, therefore, they should slow down. This line of reasoning suggested that the pulsar that pulsed fastest should be the youngest and the most luminous.

The fastest pulsar known was the Crab nebula pulsar, which blinks on and off 30 times a second. Astronomers at Lick Observatory attempted to locate the optical counterpart of the Crab nebula pulsar. They used a special camera, something like a movie camera, which could be adjusted to admit light for short intervals every $1/30$ second. It could therefore be synchronized with the pulsar, taking pictures of it when it was "on" and when it was "off," as in the accompanying photographs . With this device, they found a faint star in the center of the nebula which flickered optically in the same rhythm as the bursts at radio noise. The star had been observed before, but the flickering had gone unnoticed, since it was too rapid for the human eye to detect. The discovery of the pulsar in the

INTERLUDE

Crab nebula also solved another long-standing astronomical problem. The nebula itself shows unmistakeable signs of receiving energy from some source, for it emits synchrotron radiation at both visible and X-ray wavelengths. This radiation indicates the presence of high velocity electrons in the nebula.

How a neutron star emits its pulses of radiation is not yet fully understood. There seems little question, though, that the phenomenon is almost certainly related to the star's rapid rotation. That neutron stars spin at high speeds is not surprising. If a neutron star really represents the collapsed core of a star that has undergone a supernova explosion, as we now believe, it will retain most of the star's original angular momentum. Though the star may not have been rotating very rapidly to begin with, the conservation of angular momentum requires that it spin faster as its radius decreases. During its collapse, the radius of the star will diminish by a factor of many thousands, and its rate of rotation will increase by a similar factor.

The rapid rotation accounts for the regularity of the pulses and their short period, but not their origin. The currently favored theory is that there is some sort of "hotspot" on the surface of the star which sends out a beam of electromagnetic energy, rather like the rotating searchlight on top of a police car. Each time the beam sweeps past the earth, we observe it as a pulse of light and radio noise. The fact that the earth is in the path of the beam is purely a matter of chance. There are doubtless many pulsars whose beams miss the earth and which are therefore invisible to us.

Although the pulsations of neutron stars are remarkably regular, careful monitoring of their emissions with radio telescopes have detected some significant changes in rotation rate. As we have seen, the rate at which all pulsars rotate is continually diminishing, due to their loss of energy by radiation. Each year, for example, the rotation rate of the Crab nebula pulsar slows by some 44 billionths of a second. In 1200 years, it will rotate only 15 times per second; by this time, the optical pulsations will be readily detectable. The gradual

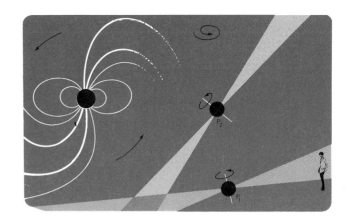

slowdown does not always occur smoothly, however. Occasionally, certain pulsars, including the Crab, have shown tiny but abrupt increases in their rotation rates, after which the slowing down has resumed again. The sudden speed-ups have been called "glitches." But in order to comprehend what a glitch is and why it occurs, we must first understand something about the structure of neutron stars.

As we saw in Chapter 7, a neutron star is the product of a drastic stellar collapse, halted only when the star is stabilized by neutron degeneracy pressure. At this point, the interior of the star has a density of about 10^{14} g/cm^3, roughly that of an atomic nucleus. In fact, it is possible to think of a neutron star as essentially a huge nucleus. With dwarfs, with their densities of 10^6 g/cm^3, are mere cotton candy by comparison.

According to the most generally accepted model, the outer few meters of a neutron star consist of a gas so dense that it possesses metallic properties, much like the innermost layers of a large gaseous planet such as Jupiter. Beneath this gaseous layer is a solid shell a few kilometers thick. It has a crystaline structure and is unbelievably hard—about 10^{17} times stiffer than steel. Beneath the solid shell is a superfluid core. A superfluid is matter which flows without friction or viscosity. The only example of a superfluid produced on earth is helium cooled to a temperature just above absolute zero.

An object containing 1.4 or more solar masses compressed into a sphere with a diameter of only 20 kilometers would naturally possess a gravitational field of overwhelming strength. A 70-kg man standing on the surface of a neutron star would weigh approximately 7.5 billion tons. Of course, he would be squashed quite flat. In fact, not only something as mushy as a human being, but even the star's super-dense and super-hard crystaline shell would be affected by the enormous gravity. Gravitational forces would make the star an almost perfect sphere, with very little surface irregularity. Mountains might exist on a neutron star, but they would be no more than a few centimeters in height. The star's shell, hard as it is, may occasionally develop cracks caused by strains in the interior. Although these starquakes would produce surface movements of no more than a millimeter, their effects would be far more cataclysmic, in relation to the star, than the most violent tremors on earth. In order to conserve angular momentum, the star would have to increase its rate of rotation in response to the minute decrease in its radius. Such events are believed to be the mechanism which causes the glitches observed by pulsar-watchers on earth.

The search for neutron stars seems to have met with success. We have every reason to believe that neutron stars are formed in supernova explosions

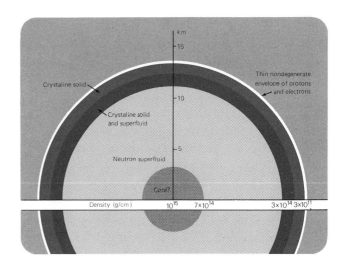

(Opposite page) Only by chance do we pick up the beam of a pulsar. (Above) Structure of a neutron star. (Below) A small quake in a neutron star causes an observable glitch.

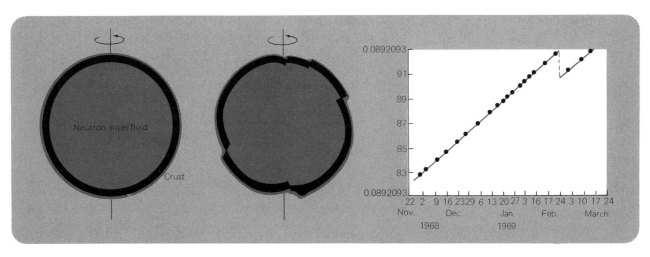

(though it is possible that they may be created in other ways as well), and that we observe them as pulsars. But what of the neutron star's even more exotic relative, the black hole?

As we saw in Chapter 7, supernova remnants beyond a certain size (the critical point is generally agreed to be not more than three solar masses) ought to develop gravitational forces which cause them to shrink indefinitely. When such an object shrinks to a size where its escape velocity exceeds the speed of light (the Schwarzschild radius), it should become a black hole. In theory, then, a black hole is just as possible as a neutron star.

The two objects have had a remarkably similar history in astronomical thought. Black holes were first analyzed as a theoretical possibility by Karl Schwarzschild during World War I. (It is said that the calculations were performed with the paper resting on a gun-butt on the western front.) Schwarzchild's work was essentially an exercise with the general theory of relativity, then quite new; it was not undertaken with the primary aim of helping to discover a novel class of astronomical objects. The next step, too, was purely theoretical. Just as he had done for neutron stars, Oppenheimer and his colleagues analyzed the concept of a black hole and showed that, from the point of view of quantum mechanics, there was nothing impossible about the idea.

The task of finding a black hole, however, is far more difficult than that of finding a neutron star. Neutron stars can be detected by the electromagnetic energy they emit. Actually, they are among the more conspicuous objects in the heavens. The difficulty astronomers had in finding them was due to the fact that they did not know quite what to look for. Black holes, on the other hand, by their very nature emit no radiation of any kind. Since they do not give off visible light, there would be no point in looking for them with optical telescopes. Trying to see a black hole in space would be like looking for the proverbial black cat in a dark cellar, except that the cellar is billions of cubic light years in volume. In fact, there is no point in searching for them directly at any wavelength. Fortunately,

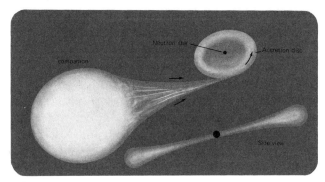

The binary star system of Epsilon Aurigae.

however, we may be able to deduce their existence from indirect evidence: their effects on entities in their vicinity, such as other stars or clouds of interstellar gas.

One binary star system which astronomers think may contain a black hole is Epsilon Aurigae. The visible component of Epsilon Aurigae is a very large star of at least 17 solar masses. It is so large that if it were to change places with our sun, the planet Mercury would find itself revolving inside the body of the star. The two components of Epsilon Aurigae move in an orbit whose period is 27 years. At one point in the orbit, the visible star is hidden by its invisible companion, creating an eclipse which lasts for months. The light curve of the eclipse, however, has certain peculiarities that are impossible to explain in terms of a large companion star. It seems to suggest, rather, an eclipse by a disc of matter—perhaps surrounding a central black hole. This theory is supported by calculations of the mass of the invisible object, which indicate that it must be at least 8 times that of the sun. If so, it evidently cannot be a white dwarf or a neutron star.

Because of these alternative explanations, we cannot say with certainty that Epsilon Aurigae contains a black hole. We can only say that it might contain one. A few other systems have been observed, however, where the case is clearer than in that of Epsilon Aurigae. The principal difficulty which remains with respect to these systems is in estimating the mass of the primary from its spectrum. As more examples accumulate, however, our doubts about the existence of black holes are slowly disappearing.

In recent years, a new tool has evolved to help astronomers in their search for black holes. This is the technique of X-ray astronomy. A black hole in close proximity to clouds of dust or gas (such as the gaseous outer layers of a companion star) would be continually attracting great quantities of this diffuse matter due to its enormous gravitational force. The black hole is very small compared with the size of its companion. As the gas funnelled into the black hole, it would be greatly compressed, causing it to reach very high temperatures. The hot gas would emit X-rays strong enough to be detected from earth.

The difficulty in locating X-ray sources among the stars is that X-rays do not penetrate the earth's atmosphere, and so can only be detected from a rocket or a satellite. Rockets have been used with some success to find X-ray sources, but their great limitation is that they remain above the atmosphere for only a few minutes before falling back to earth. In 1969, the first X-ray satellite observatory was launched from Kenya bearing the name *Uhuru,* the Swahili word for "freedom." From its position above the atmosphere, *Uhuru* sent back more information about X-ray sources than had been gathered in all the previous rocket flights put together. One of the sources which it located is a binary system in the constellation Cygnus known as Cygnus X-1. There appears to be a very good chance that Cygnus X-1 contains a black hole.

Analysis of the spectrum of the visible component of Cygnus X-1 shows that it is a giant blue star of about 30 solar masses. Judging from its motion, it seems to be in orbit with an invisible companion of at least 5 solar masses. The companion's mass is too great for it to be a white dwarf or a neutron star. Moreover, the peculiar X-ray emissions coming from the binary system match the pattern predicted for an extremely small, dense object into which gas is funnelling at very high temperatures. Other models have been proposed to explain

Cygnus X-1 but none of them seem to be quite as convincing as the hypothesis that the system contains a black hole which is slowly draining away the outer layers of its giant companion. Thus, while the evidence is not totally conclusive, we can say at this point that Cygnus X-1 is **probably** a black hole. At least, it is rather difficult to explain the observations that have been made in any other coherent way. But whether Cygnus X-1 is a black hole or not, the search for black holes will go on until theory and observation are eventually brought into harmony.

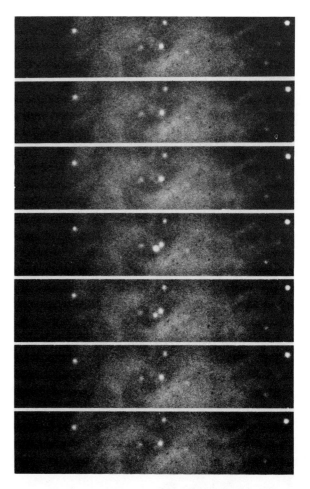

CHAPTER EIGHT

THE MILKY WAY

Today, it is possible to pass one's life in a city and never see the Milky Way. But on a clear, moonless night, away from the lights and the smog, there is no mistaking the broad, irregular, luminous belt that arches across the heavens. The Greeks called this hazy band *galaxies kyklos*—"milky circle." From this expression we derive, not only our Milky Way, but also the word *galaxy*. For the Milky Way is nothing but our own galaxy, seen from within. Indeed, when we do not refer to it simply as "our galaxy," or "the galaxy," we sometimes call it the Milky Way galaxy. But regardless of what we call it, it is our home.

The Milky Way galaxy is a system of some hundred billion stars, of which the sun is a very unremarkable member. Since we cannot see our galaxy from outside, our most vivid image of it is derived from photographs of other galaxies which seem to be similar in structure. The form of our galaxy is that of a flattened disc, some 30,000 pc in diameter, surrounded by a less conspicuous halo of more widely scattered and fainter stars. The halo has the shape of a spheroid, or flattened sphere. In the plane of the disc, it extends about as far as the disc itself, but at right angles to the plane, it is only about $1/2$ or $1/3$ as large. The galactic disc flares out at its center into a thick bulge that echoes the shape of the halo. The central plane of the disc, where the density of stars is greatest, is divided into a series of pinwheel-shaped spiral arms, which are such a prominent feature in photographs of many other galaxies. It is in one of these arms that our solar system is located.

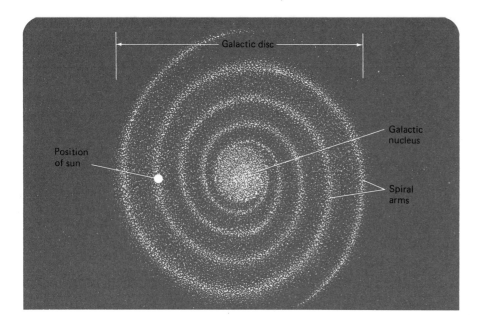

Figure 8-1
Two views of our galaxy, represented in semi-schematic form. The diameter of the galactic disc is about 30 kpc. The sun is located in a spiral arm, near the central plane, some $3/5$ of the way out from the center. The halo may be somewhat more flattened than depicted here.

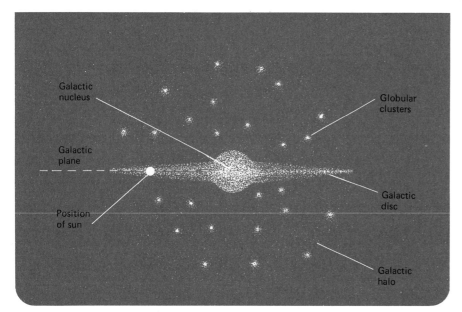

The first hint as to the size and structure of our stellar system came in 1610, when Galileo turned his primitive telescope toward the Milky Way. He found that, as a few people had already speculated, the hazy band of light is in fact composed of a multitude of stars, most of them too faint and too closely spaced to be seen individually with the unaided eye.

THE GALACTIC DISC

Shape of the Disc

The fact that the Milky Way girdles the entire sky suggested to some astronomers that this band of stars was in fact nothing but the stellar system of which we are part, seen from within. As an analogy, imagine that you are standing in a forest and trying to get your bearings by observing the trees around you. In your immediate vicinity, the trees seem far apart, with large empty spaces between them. The farther into the distance you look, however, the denser the distribution of trees appears. At the limits of your vision, all you can see is an almost solid wall of trees. But this does not necessarily mean that you happen to be standing in a clearing, and that the trees at greater distances are really closer together. They will appear this way even if they are evenly distributed throughout the forest; it is merely an effect of perspective.

Suppose, though, that you think that you are near the edge of the woods, and wish to orient yourself. If the trees seem most densely packed in one particular direction, that is probably the direction of the center of the forest. On the other hand, if the trees seem unusually sparse in another direction, that is probably the way to the forest's edge. William Herschel used this line of reasoning in his attempts to determine the shape of our stellar system. He developed the technique that he called "star gauging"—counting the stars in a number of small, representative areas of the sky. Herschel found that the stars were densest in the vicinity of the Milky Way—that is, along what we now call the plane of the galaxy. As he moved away from the plane, toward the galactic poles, the number of stars dropped off steadily. Herschel concluded that our stellar system extended farthest in the direction of the Milky Way, and was least extensive at right angles to the plane of the Milky Way. On the basis of these observations, he developed his model of our stellar system, which he pictured as a flattened, lens-shaped disc, with a thickness about 1/5 of its diameter.

Because his star counts were about the same in all directions along the Milky Way, Herschel concluded that we must be near the center of the system—otherwise we would see more stars in one direction (toward the center) and fewer in the opposite direction (toward the nearest *edge*). In this, however, he was mistaken, for he had no way of knowing that the galactic plane contains thick clouds of light-absorbing interstellar dust. The effect of this dust is to limit our visibility through the plane, in all directions—so much so that the high density of stars toward the center of the galaxy is obscured. Our situation is rather like that of the person in the forest, described earlier, trying to get his bearings in the midst of a heavy fog. The fog keeps him from seeing more than a few feet in every direction, making it impossible for him to determine in which direction the density of trees is greatest or least.

Herschel's star counts enabled him to deduce that we are situated in a disc-shaped stellar system.

Interstellar dust limits our visibility in the plane of the galaxy, making it hard to determine our distance from the galactic center.

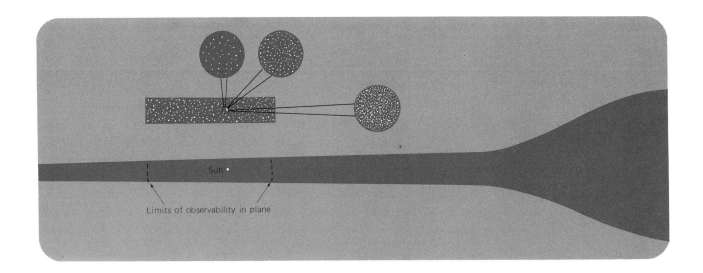

Sun •

Limits of observability in plane

POSITION OF THE SUN   This picture of our galaxy remained fundamentally unchanged for nearly 150 years. As late as the 1920s, many astronomers were still thinking in terms of a sun-centered galaxy, although its diameter had been increased to about 12 kiloparsecs. (A kiloparsec, abbreviated kpc, is 1000 parsecs; thus 12 kpc = 12,000 parsecs, or some 39,000 light-years.) The role of obscuring interstellar dust was not fully understood until 1930. Yet by 1917, a more accurate picture of the galaxy—one in which the sun was **not** located at the center—had begun to emerge through the work of Harlow Shapley.

Shapley studied the distribution of globular star clusters—dense, spherical clusters which typically contain between 100,000 and 1,000,000 members. It had been observed that these clusters seemed to avoid the disc of the galaxy, and were generally found in the space above and below it. Shapley used a convenient yardstick that enabled him to determine the distances of many globular clusters: **RR Lyrae stars.** These are short-period variable stars—that is, stars whose luminosity varies in regular cycles of less than 24 hours. All RR Lyrae stars are about equal in their luminosity, so that the observed brightness of such a star enables us to find its distance quite simply by the inverse square law. In this way, Shapley was able to map the distribution of globular clusters about the galactic disc. He found it to be roughly spherical—but the center of the sphere was located, not near the sun, but far away in the galactic plane.

Astronomers, like other scientists, usually expect to find symmetrical forms in nature. It seemed odd that the galaxy should be lopsided in this way, with the globular clusters distributed about a point near the edge of the galaxy. Moreover, if the globular clusters moved in orbits about the galactic center (as we now know to be the case), we would certainly not expect them all to be concentrated toward one side of the galaxy. A more logical conclusion was that astronomers had somehow erred in placing

Figure 8-2
Herschel deduced the shape of the galactic disc through "star gauging"—counting the stars visible in various representative regions of the sky. He assumed that the galaxy was most extensive in the direction where he saw the densest concentrations of stars. But Herschel did not realize that visibility in the galactic plane is limited by interstellar dust, which absorbs starlight. Since he saw equal numbers of stars in all directions in the plane, he concluded that we are near the center of the galaxy.

We are not situated near the center of the galaxy, as Herschel thought, but about 30,000 ly from it.

Figure 8-3
A mosaic of the Milky Way. The dark rift that runs through the center is caused by interstellar dust, concentrated most heavily in the central plane of the galaxy.

the sun at the center of the galactic disc—it was really out toward the edge of the system. In fact, this view is the correct one. We now believe that the diameter of the galactic disc is about 30,000 pc, and that the sun is some 9000 pc from the center. (In light-years, the convenient approximations to remember are 100,000 ly and 30,000 ly, respectively.)

Rotation of the Galaxy

In the 1920s, astronomers started to learn something about the dynamics of the Milky Way as well as its shape. By studying the radial velocities of a large number of stars, Jan Oort showed that the stars of the galactic disc, including the sun, move in nearly circular orbits about the center of the system. Oort's findings are summarized, schematically, in Figure 8–1. Notice that the velocities in the neighborhood of the sun obey Kepler's second law—stars farther from the galactic center travel more slowly (like the outer planets of the solar system), while those nearer the galactic center travel more rapidly (like the inner planets).

If we plot the rotation curve for the galaxy as a whole—that is, the velocity of the stars against their distance from the galactic center—we find that the rotation is Keplerian for the stars in the outer regions of the disc. At a point near the center, however, the character of the galactic rotation changes sharply. For the inner regions, the curve is almost a straight line, indicating that the velocity of a star is proportional to its distance from the center. The farther a star is from the hub of rotation, in other words, the *faster* it moves, so that all the stars have equal periods of rotation about the center. This is the way that a solid body, such as a merry-go-round or a phonograph record, rotates—but just the opposite of what we would expect if Kepler's second law were being obeyed.

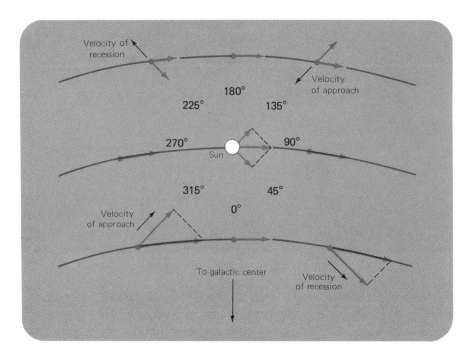

Figure 8-4
Oort's method of analyzing the rotation of the galaxy, based on his studies of the radial velocities of a large sample of stars. In our part of the galaxy, stars farther out from the galactic center than the sun orbit more slowly than the sun does. We are always overtaking stars ahead of us, which thus have radial velocities toward us, and outracing those behind, so that their radial velocities are away from us. But just the opposite is true for stars slightly closer to the galactic center than the sun. Those ahead of us are pulling away, and have radial velocities of recession; those behind are catching up, and have radial velocities of approach.

The situation in the galaxy, however, is somewhat more complicated than in the solar system, which Kepler's laws were formulated to describe. We saw in Chapter 5 that the period of an orbiting body depends on *two* factors: the radius of its orbit, and the mass of the body about which it is revolving. Most of the galaxy's mass is concentrated toward the center. Thus for a star near the periphery of the disc, the mass about which it is revolving is essentially that of the galaxy as a whole. It is not significantly affected by the radius of the star's orbit, for regardless of the size of the orbit, the vast majority of the galaxy's mass is still within it. The period of the star, then, will depend almost entirely on its orbital radius, in accordance with Kepler's third law: $P^2 \propto a^3$.

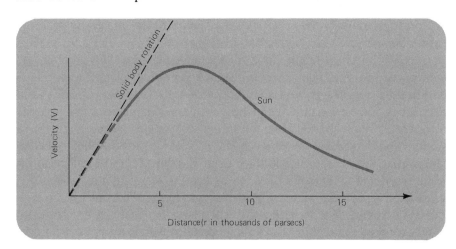

Figure 8-5
The rotation curve of the galaxy. Most of the galaxy rotates in accordance with Kepler's laws, with stars farther from the center traveling more slowly. But near the center, the galaxy rotates like a solid body, with the stars nearest the hub traveling most slowly.

Once we get very near the galactic center, however, an appreciable part of the galaxy's mass lies *outside* the orbits of the stars. In this situation, it is no longer possible to calculate the gravitational force acting on the star as if the entire mass of the galaxy were concentrated at the center. Some of the galaxy's mass is now exerting its attractive force in the opposite direction, away from the galactic center. The smaller the radius of the star's orbit about the galactic center, the smaller the fraction of the galaxy's mass that contributes to the star's orbital velocity.

The smaller orbital radius and the lower mass pretty much cancel each other out, so that all the stars in this region have similar periods, regardless of their distance from the hub of the galaxy. Thus the central part of the galaxy rotates as if it were a solid body. (Very close to the center, in the region known as the galactic nucleus, this scheme too breaks down. We do not know much about the rotation of the stars in the nucleus, and what we do know is not well understood. As we shall see, there are many mysteries about this region.)

The velocity of the sun about the galactic center is not easy to determine with precision. The best estimates give a figure of roughly 250 km/sec. At that rate, the sun has probably made about 20 full circuits of the galactic center since it was formed some 4.6 billion years ago.

Most stars circle the galactic center in orbits that obey Kepler's laws; near the center, however, the galaxy rotates in the manner of a solid body.

The Spiral Arms

Though Herschel thought the galactic disc to be relatively uniform, we now know that this is not the case. The disc is about 1 kpc thick in the vicinity of the sun; it becomes thicker towards the center, and balloons out into nuclear bulge resembling a somewhat flattened sphere with a radius of about 3 kpc. The density of stars increases as we approach the central **galactic plane,** which runs through the disc somewhat like the filling between the two pieces of bread of a hero sandwich. Though the plane is not as sharply distinct from the rest of the disc as this analogy might suggest — there is no clearly marked dividing line — it is usually thought of as a separate region, with a thickness of about 500 pc, because it contains one of the galaxy's most characteristic structural features: the spiral arms.

Trying to see the spiral arms of our galaxy is a bit like trying to see your own nose. You can see the noses of many other people quite clearly, and the glimpse you can get of your own suggests it is probably rather similiar. Still, without a mirror, its hard to be certain. Similarly, we can observe the spiral structure of thousands of external galaxies, but our vantage point for observing that of our own galaxy is just about the worst imaginable. The sun lies in, or very near, the plane of the galaxy, so that we are looking at the spiral arms edge-on. Thus we have to look through each arm to see the one behind it. Moreover, the plane in general, and the spiral arms in particular, have the greatest concentration of obscuring interstellar dust, which acts as a kind of galactic smog to hinder observa-

The central plane of the galaxy contains the spiral arms, with their rich concentration of hot, young, blue stars, gas, and dust.

The Mass of the Galaxy

How do we determine the mass of the galaxy? Estimating the masses of the various components — stars, dust, and gas — would be quite difficult, if we did not have some independent means of finding at least an approximate total that they should add up to. Fortunately, we can calculate the mass of the galaxy as a whole with a fair degree of accuracy simply by applying Kepler's third law to the rotation of stars in the disc. Let us take our sun, and assume that it moves in a circular orbit about the galactic center. We will also assume that most of the galaxy's mass lies inside the sun's orbit, and can be treated as if it were concentrated at the center of that orbit. (The fact that the rotation curve for the outer region of the disc is basically Keplerian suggests that this is a safe assumption.)

We saw in Chapter 5 that Kepler's third law,

$$(m_1 + m_2) \ P^2 \propto a^3$$

can be used to describe the mutual orbital motion of two masses, m_1 and m_2, whose average separation is a and whose orbital period is P. To apply this equation to the movement of the sun around the galactic center, we can choose our units in such a way that this proportionality becomes an equation — that is, the constant of proportionality is 1. We do this by considering the case of the earth in orbit about the sun. Its period is 1 year, and its average distance from the sun is 1 AU. Since the earth's mass is tiny compared to that of the sun, $m_1 + m_2$ is essentially the sun's mass. We can then write Kepler's law as

$$m_{(\text{solar masses})} = \frac{a^3_{(\text{AU})}}{P^2_{(\text{years})}}$$

Now we can substitute the values for the sun's orbital radius and period about the galactic center, and solve for the mass of the galaxy, in solar masses. We have seen that the period of the sun, traveling at about 250 km/sec in an orbit some 9 kpc in radius, is about 230 million years. (As an exercise, you might try verifying this, remembering that the circumference of a circle is $2\pi r$, and that there are about 3×10^7 seconds in a year, and 31 trillion kilometers in a parsec.) A parsec is about 206,000 AU, so the sun's orbital radius of 9 kpc is equivalent to 1.9×10^9 AU. Substituting these values in the equation, we have

$$m_{(\text{sun})} + m_{(\text{galaxy})} = \frac{(1.9 \times 10^9)^3}{(2.3 \times 10^8)^2} = \frac{6.4 \times 10^{27}}{5.3 \times 10^{16}} = 1.2 \times 10^{11}$$

Thus the mass of the sun plus that of the galaxy is somewhat more than a hundred billion solar masses. Of course, the sun's mass is a negligible part of the total, so this figure can stand for that of the galaxy as a whole. It is possible to perform these calculations in a more refined manner, but the results are not very different — probably no more than 200 billion solar masses. We shall see in the next chapter that this is a fairly typical mass for a galaxy of our type.

Figure 8-6
NGC 4565, a galaxy probably much like our own, which we see edge-on. In this photo we can observe the shape of the disc and the nuclear bulge. The dark strip along the edge is caused by dust in the galactic plane (compare Figure 8-3).

tion. Consequently, much of our information about the arms has come from radio observations (described in greater detail later in this chapter), for radio waves can pass through dust with little absorption.

Their heavy concentration of interstellar dust is one of the distinguishing features of the spiral arms. Another is the relatively high density of interstellar gas. These two components of the interstellar medium are usually found together. And, as we saw in Chapter 7, they are also generally associated with clusters of young stars—especially hot, bright, blue giants and supergiants of spectral types O and B. These stars have such short lifetimes that they seldom have time to travel very far from the clouds of gas and dust in which they formed. Thus the spiral arms are the star nurseries of the galaxy; they are the regions where star formation is still proceeding most intensely. It is primarily because of their rich population of type O and B stars that the spiral arms show up so prominently in photographs of other spiral galaxies.

THE DENSITY WAVE THEORY   For many years, the existence of the spiral arms in our own and other galaxies was something of a mystery. If the entire galaxy rotated as a solid body, as the central regions do, the persistence of the spiral arms would not be surprising. If you paint a spiral pattern on a phonograph record, you would expect to find it still there no matter how many revolutions the record made. But given the Keplerian rotation of most of the galaxy, with stars closer to the center having shorter periods than those near the periphery, it is not hard to show that an initial spiral pattern would not be stable. After a few revolutions, the spiral arms would "wind up" and disappear.

The error in this analysis lies in the assumption that the spiral arms are permanent structures which always contain the same stars. But this is not the case. The spiral arms are regions where something unusual is taking place: regions where the density of gas, dust, and stars is unusually high, and where the formation of new stars is proceeding at a particularly rapid rate. This does **not** imply, however, that their population is constant. A maternity hospital can be defined as a place where there is a high concentration of doctors and nurses, and where women are constantly giving birth—but not always the same women, or even the same doctors and nurses. In fact, it has been shown that the spiral arms do rotate about the galactic center, but their velocity is only about half that of the stars in the galaxy. Thus the stellar population of the spiral arms is always changing as individual stars overtake and pass through the spiral arms.

As an analogy, think of a highway which is being resurfaced by a road crew. The workmen move slowly down the highway, fixing one stretch of road after another. Wherever they are working, there is a slight bottleneck, cars must slow down briefly, and traffic backs up. Seen from a helicopter, therefore, the bottleneck would appear to be moving

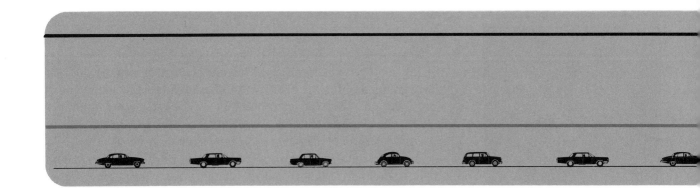

slowly along the road. From road level, however, we can see that the traffic jam never contains the same cars for very long; cars are constantly passing through it and leaving it behind, to be replaced by others catching up.

We saw in Chapter 3 that a wave is a moving disturbance. The spiral arms apparently are density waves—waves of compression. They persist because material passing through these regions is slowed slightly by the increased density, like the cars reaching the traffic bottleneck discussed previously. Thus the spiral arms are self-perpetuating.

As clouds of dust and gas overtake and enter the high-density zones, they are compressed. In these areas, the density of the interstellar medium becomes high enough for the rate of star formation to be substantially increased. The hot, young stars that are produced in turn excite large clouds of interstellar gas to incandescence. In fact, when we look at other galaxies, we find the greatest concentrations of young O and B stars along the edges of the spiral arms—just as this model would lead us to expect. The original cause of the density waves remains unknown. But whatever it is, it must be very common, for as we shall see in the following chapter, there are a great many galaxies with spiral arms.

Figure 8-7
A density wave—a moving region of greater than average density—caused by a highway bottleneck.

The spiral arms represent density waves propagating through the material of the galactic plane.

The Nucleus

We have mentioned the nuclear bulge at the center of the galactic disc. At the heart of this large central region lies the galactic nucleus. The nucleus is quite small—perhaps 5 pc in radius, or about the size of a globular cluster. (Recall that the nuclear bulge has a radius of two or three **thousand** parsecs.) The stars in the nucleus are quite crowded by the standards of our own immediate neighborhood. The average distance between nearest neighbors may be as little as a few hundred AU, compared to the 270,000 AU separating the sun and Alpha Centauri. This does not mean that stars are jostling each other like people in a crowded bus. A separation of 270 AU—$\frac{1}{1000}$ the distance of the sun from Alpha

Centauri—is still some 40 billion km. Nevertheless, an inhabitant of a planet circling a star in the center of the galactic nucleus could expect a rather spectacular display each night after sunset. His skies would hold thousands of stars brighter than the planet Venus, and probably a dozen or so brighter than the full moon. (Would the science of astronomy flourish under such conditions?)

We know very little about the galactic nucleus, but everything we do know points to it as the scene of extraordinary activity. Since dense concentrations of interstellar matter obscure it from our view at optical wavelengths, most of our information is derived from radio, infrared, and X-ray observations. The nucleus radiates strongly in all three regions of the spectrum, and some of the radiation is non-thermal (Chapter 9), indicating the presence of highly energetic electrons and magnetic fields. Moreover, clouds of gas have been detected that seem to be moving away from the galactic center at great velocities, as if they had been ejected by some violent event in the nucleus. We shall see in the following chapter that the processes taking place within the nucleus of our own galaxy, whatever their nature, appear to be repeated on a much larger scale in the nuclei of other observable galaxies.

THE HALO

So far we have dealt chiefly with the disc of our galaxy. It is in the disc that our own solar system is located. It is the disc that we see when we look up at the Milky Way, and that was first explored by Herschel's technique of star gauging. It is the disc that we notice most prominently in photographs of other galaxies. Yet there is more to our galaxy than the disc. There is also the halo, which comprises by far the largest part of the galaxy's volume, and possibly of its mass as well. Indeed, it is partly the very size of the halo that makes it relatively inconspicuous. Though as much as 80 percent of all the mass of our galaxy may lie in the halo, the stellar density there is much lower than in the galactic disc. That is not the only reason, however; for as we shall see below, the halo stars are also less luminous than those of the disc.

Dynamics of the Halo

We have seen that most of the stars in the disc revolve in nearly circular orbits about the galactic center. The stars in any one vicinity, therefore, will tend to have relatively low velocities with respect to each other—generally no more than 20 or 30 km/sec. In the neighborhood of the sun, however, about one star in 20 has a considerably greater velocity—perhaps as much as several hundred km/sec. This does not mean that they are really moving faster with respect to the galactic center—in fact, the opposite is generally true. Their high velocities with respect to the sun and most of the other stars of the disc are the result of their traveling in very different kinds of orbits. Unlike the stars of the disc, these **high-velocity stars** do not share a common, nearly circular motion about the galactic center, but move in randomly oriented, highly elongated orbits. Often, these orbits do not lie in the galactic plane, but are inclined at various angles to it, so that the stars pass through the plane only for relatively short periods on their way to regions above and below it. Such stars are halo stars, for it is in the halo that they spend most of their time; we find them in the disc only as brief visitors. (Brief, of course, must be interpreted on a stellar time scale. A high-velocity star in the neighborhood of the sun will probably remain our neighbor for many thousands of years.)

For many halo stars, even the notion of elongated elliptical orbits is something of an oversimplification. These stars bob up and down through the plane of the galaxy many times during each trip about the galactic center (Figure 8–8). Their motion represents a kind of "stitching" of the galactic plane, like that of a sewing needle through a piece of cloth. This pattern is the result of the gravitational field of the disc, which, not being spherical, cannot be taken as a point mass. As a star falls toward the disc, its velocity increases due to the disc's gravitational attraction. It attains its greatest velocity when passing through the disc, and

Figure 8-8
At left, the orbit of a typical halo star, which "stitches" through the galactic plane in response to the disc's gravitational attraction. At right, a "high-velocity" star. Its high velocity with respect to us is a result of its non-circular orbit. Most of the stars of the disc move in similar, nearly parallel orbits, and thus have low velocities with respect to each other. Because of the star's continuous interaction with the gravitational field of the disc, it does not travel in a closed orbital path at all. Such stars are generally members of Population II.

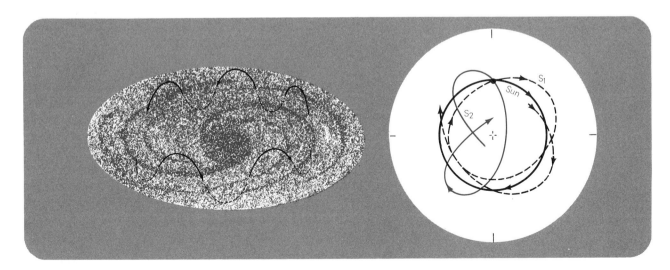

Relaxation Times

We have been speaking of the orbits of various sorts of stars in the galaxy, as we observe them today. But can we be sure that these orbits have not changed drastically in the billions of years since the galaxy was formed? Each star moves in the smooth, overall gravitational field of all the other stars of the galaxy. But close encounters among nearby stars can unpredictably alter these motions to some extent. The question is, to how great an extent?

The answer comes from the study of the **relaxation times** in various stellar environments. The relaxation time is the average time required for the orbit of a particular star to be significantly altered by its chance interactions with other stars. Such interactions tend to randomize the motions of the stars, so that they no longer reflect the conditions that prevailed when they were formed. Thus the effect of relaxation is to obliterate some of the evidence about the early history of the galaxy.

When the relaxation time is calculated for stars in different environments, it is found that stars in open clusters have the shortest "memories." Within 10^6 years, their motions have been scrambled by mutual perturbations. Thus in open clusters the orbit of one star can be calculated only by computing the orbits of all the other stars, and their detailed interactions with each other. In globular clusters, the relaxation time ranges from 10^8 to 10^{11} years, while near the center of the galaxy it is on the order of 10^{11} years. In the galactic plane, in neighborhoods like that of the sun, the figure is about 10^{12} years. The sun, in other words, could probably circle the galactic center for a trillion years before its orbit was appreciably disturbed by other stars.

The galaxy itself is estimated to be not much more than 10 billion years old. Thus we can be confident that the orbits of most stars do preserve information about the past history of the galaxy. In the following chapter, we will see how such evidence can be used to throw light on the question of how galaxies are formed.

slows down as it proceeds on away from it. We can see, therefore, that the passage through the plane by such a star will be fleeting; it will spend the vast majority of its time in the halo, far above or below the plane.

The Globular Clusters

It is not easy to observe those halo stars that do not happen to be passing through our neighborhood just now. As we have mentioned, halo stars tend to be dim, and they are widely scattered over a vast volume of space. It is much easier to observe them when they are bunched together in **globular clusters.** Over a hundred of these are known, and there may be others on the far side of the galaxy from us, hidden by the obscuring dust of the galactic plane. Like the halo stars, the globular clusters seem to travel in elongated orbits about the center of the galaxy. Their distribution in space is that of a flattened sphere, extending roughly as far as the galactic disc, but only about half this distance (60,000 light-years) in the direction perpendicular to the disc. This is essentially the shape of the halo itself; in fact, given the low observability of halo stars, it is the distribution of the globular clusters that we use to help define the boundaries of the halo.

As can be seen from the accompanying table, the globular clusters contrast markedly in almost every way with the open, or galactic, clusters of the plane. They contain many more stars, are symmetrical in shape, and are probably relatively stable structures—unlike open clusters, which rapidly disperse. Perhaps the most significant differences, however, have to do with age. We saw in Chapter 7 that the age of a cluster can be estimated from its Main Sequence turn-off point on the H-R diagram. When H-R diagrams of open and globular clusters are compared, we find that the Main Sequences of the former are often nearly complete. This means that even the massive, fast-evolving blue giants of spectral types O and B have yet to leave the Main Sequence—a sure indication that the cluster must be very young.

Globular clusters, by contrast, are invariably very old. The hot, luminous blue stars have long since completed their evolution, and only the cooler yellow and red stars remain on the Main Sequence. The most luminous stars in globular clusters are red giants—stars well advanced into old age. This conclusion is confirmed by another, closely related piece of evidence. The open clusters are rich in dust and gas, from which their stars formed, and may still be forming. In globular clusters, however, there is virtually no dust and gas remaining. Presumably it was all used up in star formation many eons ago.

Globular clusters are dense, spherical aggregations of old stars in the halo; open clusters are looser groups of young stars in the disc.

Table 8–1
Star Clusters

Type	Number of stars	Form	Distribution	Brightest stars	Age
Open Clusters	$10-10^3$	loose, irregular	Galactic plane, esp. spiral arms	Blue	Young ($< 10^8$ years)
Globular Clusters	10^4-10^6	tight, spherical	Galactic halo	Red	Old ($> 10^9$ years)

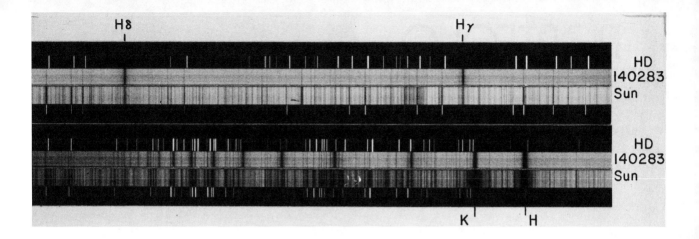

Figure 8-9

The spectrum of the sun, a Population I star, compared with that of a typical Population II star. About 1 percent of the sun's mass consists of elements other than hydrogen and helium, compared to .1 percent, or even .01 percent, for a Population II star. Notice the greater number and intensity of the absorption lines in the sun's spectrum.

The stars of Population II are old, and poor in heavy elements; Population I stars are young, and richer in heavy elements.

The Two Populations

Most of what we have said about the globular clusters is true of the halo in general. Unlike the galactic disc, the halo is completely free of dust and gas. There are no blue stars in the halo. The dominant color is red: luminous red giants, and faint red Main Sequence stars. All the facts, then, fall into a consistent pattern, and seem to point to a single conclusion: the galactic halo is populated exclusively by old stars. The argument is clinched by one final piece of evidence. We saw in Chapter 7 that stars synthesize elements heavier than helium (commonly referred to as "metals" in astronomical discussion, though this term is not strictly accurate in a chemical sense) late in their evolution. Elements up to iron may be created during the star's red giant phase; elements above iron, during supernova explosions. These heavier elements are dispersed into the interstellar medium when the star ages and dies—either through gradual mass loss from the star's outer layers, or the ejection of a planetary nebula, or the violence of a supernova.

Since these elements are thought not to have been present in the universe until created in stellar interiors, stars that formed a long time ago should be very poor in them. Stars that have formed more recently, from an interstellar medium already enriched with heavy elements from the deaths of earlier stars, should have them in greater abundance. A star's composition, in other words, should tell us something about its age.

The concentration of heavy elements in our sun is about 1 percent, and it is even higher in certain other stars of the galactic plane. But the farther we move from the plane, the lower (on the average) is the abundance of these elements. In the disc, a metal abundance of .1 percent might be typical. For stars in the halo and globular clusters, the figure is generally .01 percent or less. Though there are many gradations in the range of metal abundance, astronomers generally divide stars into two broad categories: **Population I** and **Population II**. For convenience,

ω Centauri (above left) in the southern sky is the brightest of globular clusters. Notice the elliptical shape.

M 13 in Hercules is the brightest of the northern globular clusters. In the photograph (below left) thousands of stars may be counted.

Open cluster in the southern sky. Called "The Butterfly", this cluster contains dozens of stars.

The double cluster in Perseus (above) contains hot, young stars. It is a naked eye object, but is not mentioned in Messier's list. Notice the relatively irregular shapes of the open clusters compared to the globular clusters.

we can regard Population I stars as those of the galactic plane—and especially of the spiral arms. All these stars are relatively young—some are virtually newborn. Population II stars are those found in the halo; all of them are very old. The population of the disc is intermediate in composition, and probably in age. In the nuclear bulge, there is apparently a mixture of Population I and Population II stars. As we shall see in the next chapter, the existence of the two stellar populations, and their prevalence in different regions of the galaxy, is an important piece of evidence in reconstructing the manner in which our galaxy, and galaxies in general, are formed.

Figure 8-10
A spiral galaxy photographed in blue light (left) and red light (right). The spiral arms, with their young blue stars of Population I, are clearly delineated in the photograph at left, while the older, redder stars of Population II can be seen concentrated toward the galactic center in the photograph at right.

Most people are accustomed to thinking of the space between the stars (if they think of it at all) as empty—a "vacuum." If we define a vacuum in terms of what we can achieve here on earth with our present technology, this is true. Yet by this standard, we would have to regard quite a bit of the space occupied by a red giant star as empty—hardly a reasonable conclusion. In fact, there is a good deal of matter in the galaxy that is not in the form of stars, but in the form of vast clouds of gas and dust. The fact that this material is very tenuous by earthly standards does not lessen its importance. We have already seen how the interstellar medium gives birth to stars. Now it is time to see what else we can learn about it and from it.

THE INTERSTELLAR MEDIUM

Interstellar Gas

Early telescopic observers discovered many objects that did not appear to be either stars or planets. These objects were usually described as "cloudy" or "hazy"; one observer likened them to patches of "luminous fluid." They were soon named **nebulas,** from the Latin word for cloud, but astronomers were unable to determine either their nature or their distances. In fact, they were first cataloged, not because of what they were, but because of what they were not. In 1781, the French astronomer Charles Messier published a catalog of fuzzy objects that were ***not*** comets; it was intended for the use of comet-hunters, who wished not to be distracted by the wrong quarry. Many well-known astronomical objects are still identified by their Messier numbers. The great nebula in the sword of Orion, for example, is M 42, while our neighbor, the great Andromeda galaxy, is M 31, and a prominent globular cluster in Hercules is M 13.

The prevailing theory about the nebulas was that they were exactly what they appeared to be: masses of glowing gas. In fact, that is what we mean when we use the term today. We can see from the examples just given that this guess was wrong for many of the objects on Messier's list; a majority of the Messier objects turned out to be globular clusters or external galaxies. Some of them did prove to be gaseous nebulas—but it took more than a century for that hypothesis to be confirmed.

The first evidence for the existence of gas between the stars was obtained spctrographically in the latter part of the nineteenth century. Astronomers found absorption lines in the spectra of several stars that seemed not to have been produced in the star's atmosphere. These lines had Doppler shifts indicating that the clouds producing them were moving at considerable velocities with respect to the stars in whose spectra they appeared. But where were these clouds of gas? It seemed quite possible that they might be confined to the regions of space near stars—perhaps shells of gas that had been ejected from the stars, similar to planetary nebulas or supernova remnants.

In 1904, however, this hypothesis was disproved by a discovery of the German astronomer Johannes Hartmann. Hartmann was studying the spectrum of the star Mintaka, in the belt of Orion. Mintaka is a spectroscopic binary (Chapter 5), so its spectral lines show periodic Doppler shifts. Superimposed on the star's spectrum, however, Hartmann found an absorption line of calcium that did not show any such displacements. This indicated that the calcium vapor did not share in the star's motion. The logical conclusion was that the gas producing this absorption was not associated with the star at all, but was situated in interstellar space, somewhere between the sun and Mintaka

Clouds of interstellar gas may produce absorption lines in the spectra of starlight.

Since 1904, absorption lines of many other elements have been found in the interstellar gas. The composition of the gas generally resem-

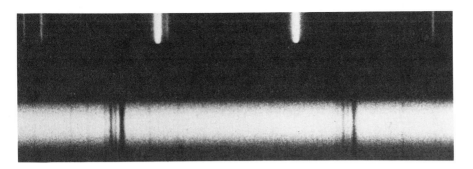

Figure 8-11
Sodium absorption lines produced by clouds of interstellar gas. The fact that the lines are multiple suggests that there are several clouds, each moving with a different radial velocity with respect to us.

bles that of the Population I stars; chiefly hydrogen, with small amounts of other elements that have presumably been released into the interstellar medium by supernova explosions. Some elements appear to be under-represented; it is thought, however, that these are to be found in the particles of interstellar dust, described below. Perhaps as much as 10 percent of the mass of the galaxy is in the form of gas. Because it is largely transparent to most wavelengths of electromagnetic radiation, however, this gas is not ordinarily very conspicuous. Much of our knowledge of it comes through observations at radio wavelengths, discussed later in this chapter.

EMISSION NEBULAS In some regions, however, the interstellar gas becomes highly visible. These beautiful masses of glowing gas are known as **emission nebulas.** We saw in Chapter 4 that a thin gas, when excited to incandescence, emits the same wavelengths that it absorbs. There are a number of ways in which the gas of the interstellar medium can be excited, but the most important is by the high-frequency radiation of very hot stars. Most of the interstellar gas is hydrogen, and hydrogen is easily ionized by any radiation with a wavelength shorter than 91.2 nm, in the ultraviolet region of the spectrum. (Recall that the violet boundary of the visible spectrum lies at about 400 nm.) The only stars that radiate enough energy in the ultraviolet portion of the spectrum to ionize an appreciable amount of hydrogen are the very massive, short-lived blue giants of spectral types O and B. These stars are often surrounded by large clouds of ionized hydrogen. Since ionized hydrogen is commonly designated H II, these clouds are called **H II Regions.**

As the hydrogen ions recapture electrons, and those electrons fall from higher to lower energy states, the hydrogen radiates its characteristic spectral lines. Much of this emission is in the lines of the Balmer series, particularly the red Balmer alpha line at 656.3 nm. For this reason, H II regions often appear red, as we can see from the plates in the Color Portfolio. The greens and blues that are also found can be attributed to the emission lines of other elements, chiefly oxygen and nitrogen.

The ultraviolet radiation of hot blue stars ionizes the surrounding interstellar gas, causing it to emit its characteristic spectral lines at visible wavelengths.

Very near a hot star, the radiation is so intense that the hydrogen gas remains always ionized. The intensity of the radiation, of course, diminishes inversely as the square of the distance from the star, and beyond a certain distance it is no longer strong enough to keep all the hydrogen ionized. As soon as a few neutral hydrogen atoms appear, they quickly soak up all the remaining ultraviolet radiation. The intensity of the radiation thus falls off very sharply beyond a certain distance from the exciting star. For this reason, H II regions tend to be quite sharply defined. Only between the two limiting distances, where there is a balance between ionization and recombination, do we observe the characteristic emissions of ionized hydrogen.

In general, clouds of excited, luminous gas are called emission nebulas. Not all emission nebulas are H II regions, however. Planetary nebulas, for example, also consist of luminous gas that has been excited by radiation from a nearby star. But they are not classified as H II regions because their substance is not truly part of the interstellar medium, but has been ejected by the star itself. Supernova remnants too consist of excited gas, but the excitation in this case is the result of heating caused by shock waves, as the expanding envelope plows through the interstellar material.

FORBIDDEN LINES   When astronomers first started to study the emission spectra of nebulas, they were very puzzled by the presence of lines that could not be identified, for they did not seem to be characteristic of any known element. For a time, it was thought that a new element had been discovered in the interstellar realm, and there was talk of naming it "nebulium." As it turned out, the name was not needed. The mysterious lines proved to be those of quite ordinary elements, chiefly ionized oxygen and nitrogen. The particular lines that were observed, however, were **forbidden lines,** never observed on earth. The reason for this is that the energy states that give rise to them are **metastable.**

We saw in Chapter 4 that most atomic energy states above the ground state have very short lifetimes. Atoms generally will not remain in an excited state for more than a millionth of a second before spontaneously emitting a photon and dropping to a lower energy state. Certain transitions, however, are more likely to occur than others. Some transitions are known as forbidden transitions — not because they are really prohibited, but because they are highly improbable compared to other possible transitions.

There are some energy states, known as metastable states, from which all possible downward transitions are forbidden. This does not mean that the metastable state is a dead end from which there is no escape — it simply means that the atom is unlikely to leave that energy state by radiation. At the densities normally encountered in the laboratory, collisions among atoms are very frequent. An atom in a metastable state will

Many of the emission lines produced by excited interstellar gas are forbidden lines, never seen under terrestrial conditions.

almost invariably be de-excited through collision with another atom before it has a chance to de-excite itself by emission of a photon.

The interstellar gas is so rarified, however, that collisions between atoms are few and far between. An atom in a metastable state is quite likely to remain in that state long enough—anywhere from a second to many years— to de-excite itself by radiating away a photon. Thus emission lines that are never seen under terrestrial conditions are often observed in the spectra of emission nebulas.

Interstellar Dust

The total mass of dust in the interstellar medium is thought to be only about $1/100$ that of the gas. But this dust makes its presence felt in several ways, owing to its ability to interact with electromagnetic radiation.

The interstellar gas consists of ions, atoms, and sometimes of molecules (discussed in the Feature). These particles are so small that electromagnetic waves, unless they are of just the proper frequency to be absorbed, pass right around them. Interstellar dust particles, by comparison, are thought to range in size from about 2 to 200 nm. This is of the same order as the wavelength of visible light. As a result, interstellar dust is not transparent to light. It tends either to absorb or to scatter it.

OBSCURATION   We mentioned earlier the role played by interstellar dust in obscuring distant regions in the galactic plane, including the center of our galaxy. Though some obscuration had been suspected earlier, it was not until 1930 that the magnitude of this effect came to be understood. The crucial work was carried out by Robert Trumpler, who was studying the size and brightness of star clusters. Trumpler found that clusters lying in or near the galactic plane were consistently fainter for their size than those lying far from it. This could mean that the clusters in the plane were more distant. But these clusters didn't appear any smaller, on the average. If they were really more distant, yet had the same apparent size, then their actual diameters would have to be much greater. There seemed no reason to believe that clusters near the galactic plane should always be larger than those lying in other parts of the galaxy. It made much more sense to assume that the light from these clusters was being absorbed on its way to us.

The distribution of dust is not uniform, so the amount of this interstellar obscuration varies considerably. In the plane of the galaxy, looking toward the galactic center, it is estimated that the light from a star loses about half its intensity for each 2000 ly it traverses. (This is in addition to the loss of intensity resulting from the inverse square law, and must be carefully allowed for by astronomers determining the distance of any object outside our immediate vicinity.) In some regions, dust is so thickly

Interstellar dust absorbs and scatters light from distant stars; dense clouds of dust produce reflection nebulas and dark nebulas

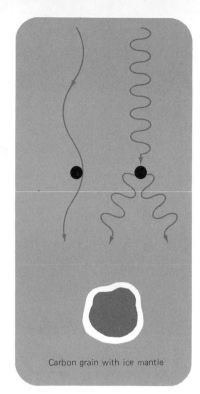

Figure 8-12
Left, scattering of light by tiny interstellar dust particles. Blue light is scattered more than red light, so that starlight becomes redder as it passes through the interstellar medium. At right, the hypothetical structure of the interstellar grains.

Because both absorption and scattering are greater for shorter wavelengths, starlight is reddened by its passage through the interstellar dust.

concentrated that virtually all the light from more distant stars is absorbed. We often see these dark nebulas silhouetted against bright emission nebulas or dense regions of the Milky Way. It was such a dark cloud that led Herschel to believe that he was looking at a hole in the heavens.

The amount of absorption is not the same for all colors, but varies inversely with wavelength: blue light, in other words, is more heavily absorbed than red light. This effect is discussed below.

SCATTERING   Dust particles of a wide range of sizes can cause absorption. In addition, the larger particles of dust may **scatter** starlight, dispersing it in random directions. Scattering is even more dependent on wavelength than absorption; short wavelengths are scattered far more than long ones. This is why, on earth, the sky appears blue and the sun yellow. Tiny particles in the atmosphere scatter the rays of sunlight. The shorter wavelengths of the visible spectrum, blue and violet, are scattered the most, creating the blue of the sky. The subtraction of this blue from the sun's light makes it look redder than it otherwise would. (The particles that make up a fog are somewhat larger than those that scatter sunlight, and scatter all visible wavelengths quite effectively. The wavelengths of infrared light, however, are long enough to penetrate fog, and infrared film is often used in aerial photography for this purpose.)

INTERSTELLAR REDDENING   As a result of the selective scattering and absorption of short wavelengths by interstellar dust, light reaching earth from a distant star is considerably redder than the light from an identical star in our neighborhood. Again, this effect must be taken into account by astronomers — in trying to estimate the temperature of a star from the color of its radiation by means of Wien's law, for example. Conversely, the distance to a star of known spectral type can often be estimated by the amount of reddening its light has undergone. Obviously a star of spectral type B that appears yellow to us has lost a great deal of its short-wavelength radiation to the interstellar dust, and therefore must be very far away.

REFLECTION NEBULAS   If enough dust is concentrated in one area, and enough starlight falls on it, the cloud of dust will itself become visible, as fog becomes visible in the beam of a car's headlight. We see such clouds of dust, illuminated by starlight, as reflection nebulas. Unlike emission nebulas, which absorb ultraviolet energy and re-emit it at visible wavelengths, reflection nebulas radiate no light of their own. The spectrum of a reflection nebula will be essentially that of the starlight illuminating it, not that of the elements present in the dust. But because the starlight is selectively absorbed and scattered by the dust, the light reaching us from the nebula will always be bluer than that of the original starlight.

PORTFOLIO

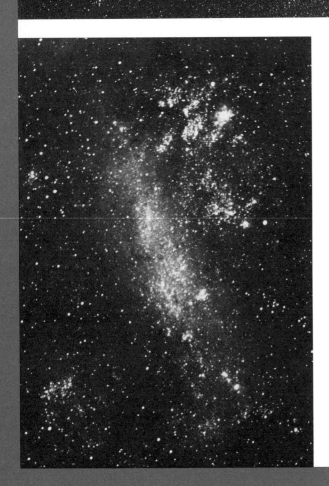

The "Trifid" and "Lagoon" nebulas in Sagittarius. These vast clouds of gas are seen against the background stars of the Milky Way.

At left, the Large Magellanic Cloud. In the upper right hand corner is the very bright "Tarantula" nebula.

Reflection nebula with unusual filamentary structure around Merope, a member of the Pleiades. (Opposite page, upper left.) The Pleiades are believed to be passing through this region of dust and gas.

Dark lanes of obscuring matter seen against starry background of the milky way. (Opposite page, upper right.)

The well known "Horsehead" nebula in Orion. In the picture at right notice the bright rim of the dark region.

Cosmic Rays

We have discussed a number of sorts of radiation that bombard the earth's surface, or at least the top of the atmosphere, from space. It may come as a surprise to learn that atomic and subatomic particles, traveling at velocities close to that of light, also reach us in a perpetual barrage from all directions. These high-energy particles are known as **cosmic rays**. Each second about 10^{18} of them enter our atmosphere: nearly one particle per second for each square centimeter of the earth's surface.

The composition of cosmic ray particles seems roughly to reflect that of the stars, and of the universe as a whole. Perhaps 83 percent are the nuclei of hydrogen atoms (protons); another 15 percent are helium nuclei (alpha particles). About 1 percent are electrons, while the remaining 1 percent consist of a wide variety of heavier elements. Because of their enormous velocities, the energy carried by cosmic ray particles is quite remarkable. Though a proton's rest mass is only 1.7×10^{-23} g, an energetic cosmic ray proton may strike the upper atmosphere with an energy equal to that of a brick dropped from a height of several feet.

Fortunately, however, most of the primary cosmic rays that arrive from space never penetrate to the earth's surface. They collide with the nuclei of atoms in the atmosphere, producing a variety of lighter nuclei and subatomic particles, known as secondary cosmic rays. These in turn may strike other nuclei; the result is a "shower" of less energetic secondary particles, many of which reach the earth's surface. Since the primary cosmic ray particles are charged, they tend to be deflected by the earth's magnetic field toward the poles. Thus the intensity of bombardment increases with latitude on the earth.

It is not known where cosmic rays are produced, or how they are accelerated to such high velocities. Since they arrive in equal numbers from all directions and at all times of the day and night, it is obvious that they do not originate in the solar system, or in our immediate neighborhood. At one time, it was thought that their origin was extragalactic, and that the particles were accelerated to their enormous velocities by strong magnetic fields of other galaxies, or even between the galaxies. It seems more likely, however, that they originate from some class of objects within our galaxy—perhaps supernovas. They are accelerated and trapped in spiral paths by the galaxy's magnetic field, which we are coming to realize may be far stronger and more important than previously believed. There is thus a vast and more or less, permanent reservoir of cosmic ray particles traveling in random paths throughout the galaxy. If this were not so, it would be very hard to account for the observed flux of cosmic rays reaching earth. At their great velocities, cosmic rays traveling in straight line paths could traverse the entire galaxy and escape into intergalactic space in no more than 100,000 years at most. Thus particles would constantly be leaving the galaxy, and it would be necessary to imagine an enormously more productive source for them than any so far proposed.

NATURE OF INTERSTELLAR GRAINS   We do not know for certain what sort of particles make up the interstellar dust. We can infer certain upper and lower limits for their probable size from the absorption and scattering effects they produce, but it is difficult to determine their composition, since we cannot capture them for study, and since they do not emit spectra of their own. One plausible theory holds that they are composed chiefly of carbon, in a form similar to the graphite used to lubricate locks. Studies with rocket-borne ultraviolet detectors reveal an absorption peak at about 216 nm, which is consistent with the carbon theory. So is the fact that the absorption seems to vary inversely with wavelength. However, reflection nebulas seem somewhat too bright to be composed of naked carbon particles. It has therefore been suggested that the carbon particles are covered with a thin layer of ice.

The formation of the particles is also not well understood. It can be shown that the interstellar medium is not dense enough for the particles to have condensed there. Presumably, then, they are produced within the relatively cool, outer layers of stars. It has been proposed that the particles condense in these regions, rather like snow crystals forming in the upper atmosphere, and are driven out into space by radiation pressure. This mechanism may also explain the infrared emissions sometimes detected from the vicinity of cool type M stars. The radiation is thought to originate in clouds of silicate dust particles

The tiny grains of interstellar dust may consist of carbon particles; they are produced in the outer layers of cool stars.

Radio Observations

As we saw in the first Interlude, radio astronomy was born in the 1930s. It did not accomplish much, however, until after World War II. The war brought about great advances in the technology of radar— the same technology that is used in radio astronomy. (In fact, the radar wavelengths are among those most widely and productively used by radio telescopes.) Another advance took place in part through a historical accident.

Jan Oort, a pioneer in studies of the galaxy's structure and dynamics, spent the war years in his native Holland, which was occupied by the Nazis. With little opportunity to practice observational astronomy, Oort and his students devoted their efforts to theoretical studies. One of the questions that occupied Oort was the potential of the infant technique of radio astronomy.

We have already seen how essential are the lines in optical spectra for determining radial velocities. Oort realized that unless lines of some sort could be found in radio spectra, the usefulness of radio astronomy would be extremely limited. He set his students the problem of finding suitable **radio lines**: radio wavelengths at which some component of the interstellar medium could be expected to absorb or emit an appreciable amount of energy.

Since radio wavelengths are very long, they correspond to very small energy transitions in the radiating atom. In the hydrogen atom, for example, radio lines are produced when the electron moves from energy level 107 to 106, or 106 to 105. The probabilities of these transitions are so low, however, that the lines produced are extremely weak. What was needed was a transition that happened frequently enough to produce a strong line that could be easily detected.

One of Oort's students, C. H. van de Hulst, solved the problem. Both the proton and the electron in the hydrogen atom have their own magnetic fields. The energy state of the atom differs — very slightly — depending on whether the two fields are pointing in the same direction or in opposite directions. When an atom changes from one alignment to the other, therefore, a photon should be emitted. The frequency of the radiation is 1420.4 MHz, corresponding to a wavelength of about 21 cm. This lies in the radio spectrum.

The 21-cm radiation from neutral atomic hydrogen, predicted by van de Hulst, could not be detected by the equipment available in 1944. By 1951, however, technology had improved sufficiently for this line to be observable. Since then, it has been a most useful tool for radio astronomers in mapping our galaxy, and other nearby ones as well. Before the discovery of the 21-cm line, the only way to explore the spiral structure of our galaxy was to study the distribution of young clusters, O and B stars, and H II regions. The presence of dust in the galactic plane seriously hindered such studies. Moreover, there was no way of mapping the many

Neutral hydrogen atoms in space emit a characteristic radio line at a wavelength of 21 cm.

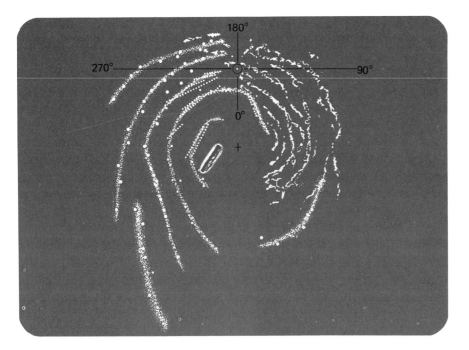

Figure 8-13
A radio map of our galaxy, based on the 21-cm radio emissions of neutral hydrogen atoms. The galactic center is indicated by the cross, the sun's position by a circle. The spiral arms can be seen much more clearly in this way than through visible observations.

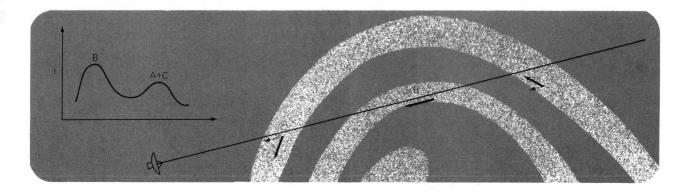

Figure 8-14
The lines in radio spectra show Doppler shifts, just as do lines in visible spectra. In this example, the gas in arms A and C have similar radial velocities, though they are moving in different directions. B has a different radial velocity, and thus a different Doppler shift. Such effects make radio mapping quite complex.

The radio emissions from neutral hydrogen atoms and from various molecules enable us to map the galaxy and see into the dense clouds where stars are forming.

areas of the galaxy where the interstellar gas was cool and un-ionized. The 21-cm radiation, which penetrates dust easily, solved both these problems at once.

In addition to delineating the structure of the spiral arms, radio studies have detected several peculiar phenomena in various other regions of the galaxy. Near the galactic nucleus, for example, clouds of hydrogen seem to be moving rapidly outward from the center of the galaxy. We do not know what has caused these movements of the interstellar gas; we do know, however, that the galactic center is probably the scene of intense activity now, which may have been even more violent in the past. Evidence for this can be found in the existence of high-velocity hydrogen clouds which appear to be falling in toward the galactic plane in a number of locations. It has been suggested that these clouds may represent material ejected in the past by some explosive event in the galactic nucleus. Having failed to attain escape velocity, they are now raining back down on the galaxy like ashes after a volcanic eruption. It is also possible that this material is intergalactic in origin, and has merely been captured by the galaxy's gravitational field. However, we have no other evidence for the existence of such intergalactic gas.

Useful as it is, the 21-cm line has its limitations. In very dense clouds of gas, most of the stellar ultraviolet radiation is effectively screened out. The hydrogen in such clouds remains cool enough to exist in the molecular form, H_2, rather than as individual atoms. Molecular hydrogen does not radiate at 21 cm, and so, until recently, such clouds have been quite impenetrable to our observations. The development of molecular radio astronomy, however, described in the Feature, has provided astronomers with a valuable new window on some of the most interesting regions of the galaxy.

Interstellar Chemistry

Optical astronomy has been very successful in dealing with objects such as the sun and other stars, which radiate most of their energy in the visible spectrum. But there are many objects that radiate strongly at radio wavelengths, and faintly or not at all at visual wavelengths. Today we study such objects with radio telescopes, and radio observation has become one of the most fruitful techniques of modern astronomy.

In optical astronomy, atoms are of supreme importance, because of the visible spectra they produce. But atoms produce few lines in the radio region of the spectrum. Molecules, on the other hand, have many closely-spaced energy levels. The transitions between these levels give rise to lines in the radio spectrum. Thus molecules really come into their own in radio astronomy, and astronomers have been able to undertake extensive studies of regions in our galaxy where molecules are abundant. Some of these regions, such as the galactic center, and the dense clouds that are the birthplaces of stars, have stubbornly resisted observation by optical means, but molecular radio astronomy is opening them to our understanding.

The study of interstellar molecules began in the 1930s, when absorption lines of interstellar CH, CN, and CH+ were identified in the optical spectra of distant stars. The next interstellar molecule to be found was the hydroxyl radical (OH) in 1963. By this time radio astronomy had become quite sophisticated, and it was not very difficult to detect the 18 cm radio radiation from OH. In fact, it appears in retrospect that many molecules could have been found even in the late 1950s, but no serious attempt had been made to observe them. Molecules were not expected to form under the conditions believed to exist in interstellar space, and ultraviolet radiation from stars was expected to destroy those molecules that did form.

It was in the face of this rather discouraging opinion of the majority that a team of physicists at Berkeley, using a 21-ft radio telescope, began to look for short wavelength radio waves from ammonia

(NH_3). In 1968, they reported detecting the 12.6 mm emissions of ammonia from the direction of the galactic center. This discovery did not fire much immediate enthusiasm for molecular studies amongst radio astronomers. After all, the ammonia had been found in the galactic center, and the galactic center was known to be a very peculiar place. But then water vapor (H_2O) at 1.35 cm was detected, not only in the galactic center, but also in two clouds quite far from it. This fairly well launched radio astronomy of molecules; since then, about 40 other molecules have been reported.

When water vapor was detected, it was found to be radiating thousands of times more strongly than expected. This was the result of a phenomenon known as **maser emission.** We have seen that a photon with the proper excitation energy can boost an atom to a higher energy state. Curiously, a photon with exactly the excitation energy can also induce an excited atom to **give up** a similar photon and become de-excited. The two photons, of equal energy, then travel in the same direction and in phase. Normally this **induced** or **stimulated** radiation is not significant because it is quickly absorbed by other atoms of the gas that are in a de-excited state. There are circumstances, however, in which there may be an abnormal number of atoms in an excited state, and

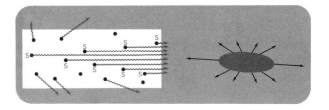

The maser effect. Each passage of a photon stimulates the emission of another photon with the same wavelength. Consequently, the resulting radiation is most intense in the direction in which the cloud is most extensive.

very few in a de-excited state. The stimulated radiation would then be free to travel and stimulate other excited atoms into radiating. This 'chain reaction' can build up into an avalanche of radiation. (The word maser comes from *m*icrowave *a*mplification by *s*timulated *e*mission of *r*adiation. Lasers work on the same principle, but at wavelengths that are in or near the visible spectrum.)

The maser effect builds up over a distance and is therefore affected very strongly by the geometry of the cloud. In an elongated cloud, the radiation can be thousands of times more intense along its length than in a direction perpendicular to it. Thus the intensity that one observes depends on where one is with respect to the cloud. This complication makes it very difficult to understand the size and density of the cloud from studying its radiation.

In March, 1969, observers at the National Radio Astronomy Observatory in Green Bank, West Virginia, detected the absorption line of formaldehyde (H_2CO) at 6.21 cm. This molecule was soon found to be quite widespread in the galaxy; more than two thirds of all galactic radio sources show H_2CO absorption. Delighted by this, the observers turned their radio telescopes to dark interstellar clouds that did not lie in front of radio sources. They hoped to see the 6.21 cm line in emission. But instead, they saw it once again in absorbtion. The question then was, what radiation could the formaldehyde in the clouds be absorbing? The only possible answer turned out to be the cosmic 3 K black-body radiation Chapter 10. This meant that the formaldehyde was effectively colder than 3 K—actually not more than 1.8 K. The only way this could happen is through a reversal of the maser effect, called the **inverse maser effect**.

Formaldehyde is a common chemical used for preserving animal parts and corpses. But conditions in interstellar space are so different from those in the laboratory that a 'new chemistry' had to be developed in order to understand how even such simple molecules behaved. This need for new studies was reinforced by the discovery in space of other molecules such as carbon monoxide (CO), a common urban pollutant; formic acid (HCOOH), which makes an ant's bite hurt; hydrocyanic acid (HCN), the lethal gas used in gas chambers; ethyl alcohol, (C_2H_5OH) which makes one drunk; and methyl alcohol (CH_3OH), which makes one blind.

Soon all talk of molecules not existing in space was forgotten. A new field, astrochemistry, emerged at the interface between astronomy and chemistry— just as, a hundred years earlier, astrophysics had emerged from the need to integrate astronomy and physics. There appeared a new type of investigator, equally at home in both chemistry and astronomy, and discussion of molecular structures became fashionable at observatories. This new discipline promises to help astronomers in resolving many difficult problems on the most interesting frontiers of astronomical research, such as star formation, galactic structure, cosmology, and the search for life in the universe. It has also returned some important dividends to the study of chemistry.

The two principal problems that confront the astrochemist are to explain how molecules form in interstellar space, and how they escape swift destruction by harsh ultraviolet radiation. The first problem arises because two colliding atoms have an excess of energy they must give up before they can unite to form a molecule. Normally they emit a photon, which carries away the excess energy. But under the conditions that prevail between the stars, this process is very improbable, and the atoms just bounce off each other.

The clue that leads to the solution of both problems is that molecules are found in dense, dusty clouds. The existence of such clouds had not been suspected because they are opaque to optical wavelengths. They also produce very little 21-cm radiation, because most of the hydrogen in the clouds is in the molecular form H_2. Atoms of various sorts stick to the surfaces of dust grains. Reactions nearly impossible for free atoms become easy when there are grains of dust around to carry away the excess energy. Atoms stuck to the grains interact, goaded by the mild ultraviolet radiation that seeps into the cloud. As the molecules form, they give up excess energy to the dust grain, and are dislodged from the surface as a result. The dust which enabled them to form in the first place now enables them to survive by shielding out most of the ultraviolet radiation. The simple molecules can then interact with other molecules to form more complex ones.

Observations show that there is a definite depth at which each particular kind of molecule is found within the cloud. At a greater depth the ultraviolet radiation is too weak to induce the formation of new molecules; at a shallower depth the ultraviolet is too harsh for the molecules formed to survive. It is quite common to see a layered structure in interstellar clouds. The toughest molecules, such as CO, are nearest the surface; the most delicate, such as H_2CO, are near the center. Thus it is quite clear why interstellar molecules had to await radio astronomy. If a certain region is open to optical observation, it is by that very fact exposed to destructive ultraviolet radiation. Most molecules can exist only within clouds that are quite opaque to visible light. This makes them uniquely useful when we wish to understand regions hidden to optical means.

Star formation occurs within vast complexes of dust and gas. Astronomers once hoped to understand the process by means of giant optical telescopes, but the opacity of the clouds frustrated these hopes. Radio astronomy, using molecular spectral lines, is on the verge of enabling us to understand these regions. Molecules appear to be present wherever star formation is suspected. In fact, by virtue of their ability to radiate away energy and cool the gas, they probably play a significant role in initiating and accelerating the collapse of a protostar. Their presence enables us to map such regions. Today, research based on the observation of molecule-laden clouds is beginning to clarify the relationship between star formation and the shock waves associated with the galaxy's spiral arms.

The structure of the inner regions of the galaxy is also being studied through molecular mapping. Maps of the galactic center have been made at 6 cm and at 2 cm using the lines of H_2CO. Maps have also been made at the 2.6 mm wavelength of CO. Cloud complexes estimated at millions of solar masses have been detected, and some observers have found evidence of a rotating, expanding ring of gas at the galactic center. Other finer details are being discovered, thanks to the greater resolving power of these shorter wavelengths compared to the 21-cm line of hydrogen.

One unexpected bonus of molecular radio astronomy has been the discovery of interstellar deuterium. Deuterium, or heavy hydrogen, is found in the molecule DCN. Interstellar deuterium is exciting because it cannot have come from stars. It 'burns' very easily in nuclear reactions, and would be consumed as soon as it was created inside a star. The deuterium must therefore have come from the primeval fireball from which the universe is thought to have originated (Chapter 10). Its presence is an important factor for cosmological theories, which must now explain how the deuterium came to be.

Also of cosmological interest is the discovery that many molecules have energy transitions that correspond to wavelengths around 1 mm. This is where the peak of the cosmic black-body radiation is expected to lie. It is a difficult part of the spectrum to observe, being too short for all but the most sophisticated radio techniques, and too long for most infrared detectors. Molecules show us a way out of this difficulty. They interact with such radiation and reradiate it at other, more accessible wavelengths. By observing these other wavelengths, it is possible to explore the nature of the cosmic background radiation. As we shall see in Chapter 10, the fate of many cosmological theories depends on whether or not the background radiation proves to have a definite black-body curve.

Not all the contributions of interstellar chemistry have been astronomical. Since the discovery of molecules in space by radio observations, there has been much progress in studying the radio-frequency spectra of molecules—a field that had been largely neglected before 1968. Radio astronomy has also

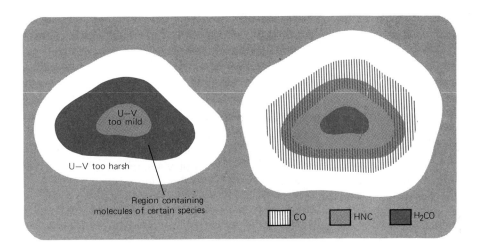

U—V
too mild

U—V too harsh

Region containing
molecules of certain species

▥ CO   ▨ HNC   ■ H₂CO

made direct contributions to chemistry. The discovery of interstellar HNC is a case in point. A few years ago, astrochemists were trying to identify the molecule that was producing a particular spectral line. The unknown molecule had been named X-ogen. HNC (a different arrangement of the atoms of HCN) had been suggested as a possible candidate. Chemists had suspected the existence of such a molecule, but had been unable to detect it in the laboratory, and had eventually given up the search.

When HNC was investigated theoretically, it was found that its emission frequency should be slightly higher than that of the X-ogen line. A search was made, and the HNC line was found near the predicted wavelength. Its exact frequency was then fed back into the theoretical calculations, permitting chemists to deduce a more precise model of the structure of HNC. (You may be wondering, in the meantime, what the mysterious X-ogen turned out to be. The evidence is not conclusive, but astrochemists now believe that it is HCO⁺.)

Radio astronomy is likely to be most useful to the chemist when he wants to study short-lived radicals. A radical may be thought of as a molecular fragment—a half-completed molecule yearning to be whole. Under normal laboratory conditions, radicals do not usually exist long enough for close study. But in interstellar clouds of low density they are so isolated from interactions with other molecules that they may have relatively long lifetimes. Optimistic astrochemists now envision the galaxy as a vast laboratory in which exotic types of molecules and chemical processes can be studied.

Among the frontier areas of astrochemical research at the moment is the search for amino acids in space. Amino acids are basic to life as we know it, and were probably among the precursors of life on earth. (Interlude IV). The presence of amino acids in the interstellar medium would be strong evidence for the possibility of life elsewhere in the universe.

Our galaxy is a system of about 100 billion stars. It consists of a disc about 30 kpc in diameter and 1 kpc thick, surrounded by a spheroidal halo flattened at the poles. In the thin central plane of the disc are the spiral arms, containing many hot, young, blue stars, open clusters, gas, and dust. The halo contains fainter, redder stars, and globular clusters—dense spherical aggregations with up to a million members. At the center of the disc is the spheroidal nuclear bulge, within which lies the small, mysterious galactic nucleus, the scene of great activity.

Observing the Milky Way with his telescope, Galileo discovered that it consists of myriad closely-spaced, faint stars. Herschel was the first to correctly deduce the shape of the galactic disc by counting stars in representative regions of the sky. But because of light-obscuring dust in the plane, he erroneously concluded that the sun was near the center of the system. In 1917, Shapley's mapping of globular clusters showed that the sun is actually closer to the edge—about 9 kpc from the center, we believe today.

Studies of galactic rotation show that near the center, the disc rotates as a solid body, with all the stars having similar periods. Farther out toward the periphery, the rotation of the stars conforms to Kepler's laws, with stars closer to the center having the shorter periods. The sun, moving at about 250km/sec, completes its journey around the galactic center in about 230 million years. The halo stars and globular clusters move in elongated elliptical orbits, randomly oriented, about the galactic center.

The spiral arms of the galaxy are density waves, where the concentration of the stars, dust, and gas is unusually high. The waves move about the galactic center at about half the speed of the galactic rotation, so that stars and interstellar material are constantly overtaking and passing through these regions. When this happens, the dust and gas are compressed, and the rate of star formation is increased, producing the hot, young blue giants that mark the spiral arms.

No such stars are found in the halo, nor is there any dust or gas in that region. The halo and globular cluster stars are chiefly faint red or yellow ones; the brightest are red giants. Thus we think that these stars were probably formed long ago, and that star formation has since ceased in the halo. Further evidence for this is found in the low abundance of elements heavier than helium in halo stars. They must have been formed before the interstellar medium was enriched with heavier elements from dying stars. The old stars of the halo are known as Population II stars; the young stars of the galactic plane constitute Population I.

The interstellar medium consists of gas and tiny particles of dust, ranging in size from a few nm to a few hundred. About 9 percent of the galaxy's mass may be gas, and 1 percent dust. Evidence for the existence of interstellar gas, which is transparent to most wavelengths of radiation, is provided by interstellar absorption lines in the spectra of stars. Where the gas is ionized by ultraviolet radiation from hot O or B stars, the atoms emit their characteristic wavelengths as they recapture electrons and drop to lower energy levels. Such clouds of glowing gas constitute emission nebulas. Most of the radiation comes from ionized hydrogen, so the emission areas around hot stars are often called H II regions. Some of the lines produced by emission nebulas are forbidden lines, which make their appearance only at the extremely low densities found in the interstellar medium.

The interstellar dust absorbs and scatters starlight passing through it. Both processes affect shorter wavelengths more than longer ones, so that light passing through the interstellar medium is not only dimmed, but also reddened. In some places, dust is so thick as to block out all light, creating a dark nebula. Clouds of dust may also reflect or scatter the light of nearby stars, producing a reflection nebula.

EXERCISES

1. Explain Herschel's method of estimating the size and shape of the galaxy. Why did his model of the galaxy differ from the one we accept today?
2. How did Shapley locate the center of our galaxy?
3. What evidence do we have that the galaxy rotates?
4. Why do the orbital velocities of stars in the galaxy first increase, and then decrease, as we move outwards from the center?
5. What sort of objects are found in the galaxy's spiral arms? How does the density wave theory explain their presence, and the stability of the arms?
6. What are high velocity stars? Are they aptly named? Why do we find few halo stars in the sun's vicinity?
7. How do globular clusters differ from open clusters? Why do we believe that they are among the oldest objects in the galaxy?
8. Why are Population II stars thought to have been formed earlier than Population I stars?
9. What can you say about the color of the halo, in general, in comparison with the disc? Explain your answer. Why are there no type O and B stars in the galactic halo?
10. What is a Messier object? Why do you think that Omega Centauri, the brightest globular cluster in the sky, is not a Messier object?
11. How did Hartmann prove the presence of interstellar gas?
12. Why did astronomers think they had discovered an element called nebulium? Why was this theory later discarded?
13. What are H II regions? Where are they found? Why do they have sharp boundaries?
14. Why do we think that interstellar obscuration is caused by dust? How can we estimate the sizes of dust particles?
15. Why is the clear sky blue? Why is fog white?
16. Why are reflection nebulas generally bluer than the stars whose light they reflect?
17. Why is the 21-cm line of hydrogen important to radio astronomers? How is it produced? What areas of the galaxy cannot be observed at this wavelength?

CHAPTER NINE

GALAXIES

What is the universe made of? When we are asked this question, we usually think first of the very small. "The universe is made of atoms," we are likely to answer, for atoms are often considered the fundamental, indivisible building blocks from which all substances are made. Even atoms, however, can be broken down into various subatomic particles, including the recently discovered and still mysterious quarks. Indeed, it seems probable that the nuclear physicists will soon be telling us that the universe is made of quarks.

But there is another way of approaching the question — to look for the *largest* recurring unit of the universe. A house, for example, is certainly composed of atoms — but if someone asked, most of us would be more likely to say that it consists of boards, or bricks. Similarly, if you asked an astronomer what the universe is made of, he would be very likely to answer, not "atoms," or "stars," but "galaxies," or even "clusters of galaxies."

Yet the existence of galaxies other than our own is a quite recent discovery. This is true despite the fact that three galaxies in addition to the Milky Way are actually visible to the naked eye. Two of these can only be seen from the southern hemisphere. The third, however—the great Andromeda galaxy—can be observed fairly easily on a dark night, if you know where to look. It appears to be a small, luminous patch, rather like a fuzzy star, or a tiny detached fragment of the Milky Way. Certainly it was seen by Greek astronomers, who recorded their observations of several similar objects. The Greeks, however, had no idea what they were.

Speculation and Debate

After the invention of the telescope, astronomers could observe these objects more closely, and speculation about their nature grew. In 1694, the Dutch astronomer Christian Huygens named them nebulas, from the Latin word for cloud. We know now that most of the nebulas seen by these early observers were really nebulas in our present sense of the term —that is, luminous clouds of interstellar dust and gas, of the sort described in the previous chapter. But one of the naked eye nebulas (the Andromeda galaxy, mentioned above), and several of those discovered telescopically, were of a very different nature.

That nature was soon guessed, but not easily proven. In the eighteenth century several people, including the German philosopher Emmanuel Kant, suggested that the nebulas were actually "island universes"—great stellar systems that rivalled our own. To a world that had barely recovered from the discovery that the sun was not the center of the universe, this idea proved too unsettling to be accepted. Moreover, it was only a hypothesis; there was little if any hard scientific evidence that could be marshalled to support it. At that time, the nature of even our own galaxy was not yet clearly understood. As a result, the debate over the true nature of the nebulas and the possible existence of external galaxies continued into the present century.

In the early 1900s, astronomers succeeded in obtaining the spectra of some of the brighter nebulas. Analysis of these spectra revealed two facts that tended to support Kant's island universe theory. First, the objects emitted continuous spectra crossed by dark absorption lines, similar to those of stars. This suggested that the nebulas were not patches of luminous gas, but groups of individual stars too distant to be resolved by telescopes. Second, measurement of the Doppler shift of the spectral lines indicated that the nebulas were speeding away from us at very high velocities. It seemed unlikely that any objects receding so fast could be part of our stellar system; their velocities should long ago have carried them far beyond its boundaries.

In addition to this evidence, there was the fact that several astronomers believed they had resolved individual novas in the brighter nebu-

Though the existence of external galaxies was suggested as early as 1750, convincing proof did not become available until 1924.

Figure 9-1
The Andromeda galaxy, M 31. This spiral galaxy is probably quite similar to our own, but it is somewhat larger and more massive. About 2¼ million light-years distant, it is a neighbor in our Local Group of galaxies. The bright patches nearby are dwarf elliptical satellite galaxies. (See also Color Portfolio I.)

Figure 9-2
A region of the Andromeda galaxy, showing its resolution into individual stars. Only the brightest of its several hundred billion stars can be resolved in this way.

las. If the luminosities of these stellar outbursts were assumed to be equal to those observed in our Milky Way, then the nebulas were at least a million light-years distant—far outside our galaxy. But other investigators denied these reported sightings. Moreover, some astronomers claimed to have measured the proper motions of stars within the nebulas, which would hardly have been possible if they were truly so distant.

The debate reached a climax of sorts on April 26, 1920. On that date, the two leading proponents of the opposing theories appeared before the National Academy of Sciences to defend their positions. H. D. Curtis of Lick Observatory supported the island universe hypothesis. His opponent, Harlow Shapley of Mount Wilson Observatory, contended that the nebulas were part of the Milky Way galaxy. We know today that Curtis was right, but at the time, Shapley's arguments seemed at least equally convincing to most astronomers. Neither could marshall the evidence for a decisive victory.

The issue was settled, however, just four years later by Edwin Hubble. Hubble studied three of the more prominent nebulas with the 100-inch Mount Wilson reflector—the largest instrument of its kind at the time. Within the nebulas he found several stars which he was able to identify as **Cepheid variables.** By means of these useful astronomical yardsticks (see Box), Hubble determined that the nebulas in question were at least 800,000 light-years away—much too far to be part of our own galaxy. They had to be separate galaxies.

Hubble discovered Cepheid variable stars in several external galaxies, enabling him to estimate their distances.

Had Shapley been right, the entire universe—consisting of our own galaxy, and perhaps the Magellanic Clouds—would have been no more than a few hundred thousand light-years in diameter. Hubble's discovery dramatically extended the size of the observable cosmos to include innumerable galaxies, whose distances from us must be measured in the millions and even billions of light-years.

Distribution of Galaxies

After Hubble had confirmed the existence of galaxies beyond the Milky Way, his next step was to study their distribution in space. Hubble realized that a survey of the entire observable universe with the 100-inch telescope would take many lifetimes. Instead, like Herschel before him, he turned to a sampling technique, photographing 1285 small regions distributed over most of the visible sky. Within these regions he found some 44,000 galaxies. From this sample, Hubble estimated that there might be as many as 100 million galaxies within reach of the 100-inch telescope. The introduction of the 200-inch instrument in 1952 increased the number of observable galaxies to nearly a billion.

Nearly a billion galaxies lie within the range of the largest telescopes.

At present, our instruments are limited chiefly by the brightness of the night sky, which is never completely black even under the most favorable conditions. The tiny amount of light reflected by the sky itself ri-

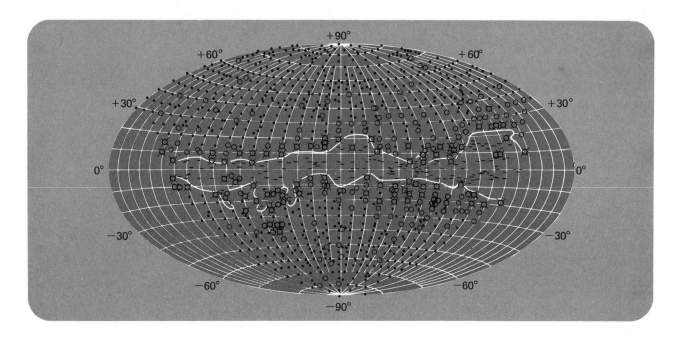

Figure 9-3
A map of visible external galaxies, made with the 100-inch Mount Wilson telescope. The dots do not represent individual galaxies, but regions of the sky photographed with the telescope. Solid dots represent areas where galaxies were densely distributed; open circles are areas where few galaxies are seen, and dashes, areas where no galaxies are observed. The empty region running through across the center of the diagram, which corresponds to the equator of our own galaxy, is the Zone of Avoidance. There is no reason to think that galaxies are really scarce in the direction of the Zone—it is merely that they are obscured by dist in the plane of our galaxy. The Zone of Avoidance coincides fairly well with the visible Milky Way.

vals the faint glimmer of the most distant galaxies, making the use of longer exposures unproductive. It seems likely that future observational tools and techniques—a large telescope in orbit, for example, or on the airless surface of the moon—will enable us to see still more galaxies at even greater distances. It has been estimated that the number of galaxies in the universe may equal or exceed the number of stars in our own galaxy. Thus we confront a universe of perhaps several hundred billion galaxies, each comprising, on the average, some 10 to 100 billion stars.

In Hubble's surveys, the highest concentration of galaxies seemed to lie in the region of the galactic poles. Near the Milky Way itself—that is, the disc of our own galaxy—few if any galaxies could be observed. This region Hubble named the **Zone of Avoidance.** He concluded, however, that there are probably as many galaxies in these parts of the sky as in any other. They are merely obscured from our view by the dense dust clouds of the Milky Way. In fact, when allowance for this effect is made, the galaxies appear to be distributed uniformly throughout the universe. On a large scale, in other words, the universe seems to be homogeneous (everywhere equal in density) and isotropic (the same in all directions). As we shall see in the following chapter, these observations are very important for the study of cosmology—the history of the universe as a whole.

CLUSTERS OF GALAXIES On a small scale, however—that is, when only a particular region of the universe is studied—the distribution of galaxies appears to be non-homogeneous. Instead the galaxies tend to occur in relatively compact groups known as **clusters** of galaxies. The

The Cepheid Yardstick

Cepheids are yellow supergiant stars whose luminosities vary in regular cycles. The periods of different Cepheids range from about a day to 100 days, but for any one star it is constant. In 1908, Henrietta Leavitt of the Harvard College Observatory discovered a remarkable property of Cepheid variables that enables us to use them as astronomical yardsticks. Studying a number of Cepheids in the Small Magellanic Cloud, a satellite galaxy of the Milky Way, she found a definite relationship between the period of each star and its brightness.

Since the Small Magellanic Cloud is so far away, all the stars in it can be regarded as about the same distance from us—just as we can consider all the people in Moscow to be about equally distant, though they may live in widely scattered parts of the city. Thus the relative brightnesses of the Cepheids in the Magellanic Cloud must indicate their true relative luminosities. What Leavitt had discovered, therefore, was the existence of a **period-luminosity law** for Cepheid variables.

Once the luminosity of an object is known, its distance can be found easily by means of the inverse square law. Though Cepheids are relatively scarce, they are very luminous—up to 10,000 times as luminous as the sun—and can therefore be seen for great distances. If the luminosity of a Cepheid could be determined merely by observing its period of variation, astronomers would have an extremely simple way of getting the distance of any Cepheid—and thus of any stellar system, such as a globular cluster or an external galaxy—in which Cepheids could be found. Cepheids, in other words, could serve as a yardstick for the measurement of distances beyond our own galaxy.

There was one serious obstacle to using the Cepheids in this way. In 1912, the distance of the Small Magellanic Cloud was not yet known. Neither was the distance of any other Cepheid, for no Cepheid happens to be near enough to have an accurately measurable parallax. Thus the period-luminosity law could only be stated in terms of *relative* luminosities—the actual luminosity corresponding to any particular period was unknown. The yardstick itself, in other words, was uncalibrated. However, if the distance—and thus the true luminosity—of even a single Cepheid were known, the exact period-luminosity relationship would be determined.

The first Cepheid distances were eventually determined by statistical methods; the results were not very precise, but they enabled Hubble to establish the existence of galaxies beyond our own. More recently, a dozen Cepheids have been found in clusters whose distances are known through other methods. This has enabled astronomers to refine the calibration of the Cepheid yardstick.

Despite these successes, an unsuspected flaw lay concealed in the apparently foolproof chain of reasoning that first established the scale of extragalactic distances. There are actually two kinds of Cepheids. The so-called ''classical'' Cepheids, like their prototype Delta Cephei, are Population I stars. Type-two Cepheids, by contrast, are Population II stars, and their period-luminosity rela-

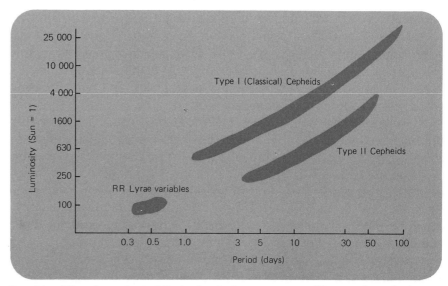

tionship differs from that of their classical counterparts (Figure, above). An astronomer can use either kind to calculate distances. But if he confuses the period-luminosity values of the two classes, his distance determinations will be off by a factor of 2 or more.

This is exactly what happened when Hubble and other investigators first applied the Cepheid yardstick to measure the distance of M 31 and other neighboring galaxies. At that time, no one knew that there were two different types of Cepheids. Astronomers applied the period-luminosity curve obtained for Type II Cepheids (the easiest to observe, because of their location in the sparsely populated halo) in our galaxy, to the Cepheids visible in neighboring galaxies — all of which were Population I stars. Type I Cepheids are about four times as luminous as Type II Cepheids with identical periods. Underestimating the luminosity of the Cepheids in M 31 by a factor of 4 led investigators to underestimate their distance by a factor of 2 (the inverse square law).

The error was only discovered in 1952, when the 200-inch Mt. Palomar telescope went into operation. Using the 100-inch telescope, astronomers had detected Cepheids in M 31, but not RR Lyrae variables, which are considerably less luminous. If the distances then accepted for the nearby galaxies were correct, it should have been easy to find RR Lyrae stars with the 200-inch instrument. Yet none could be observed. There seemed no reason to suppose that a class of star generously represented in our own galaxy should be unaccountably absent in other, similar systems. The only likely explanation was that the neighboring galaxies were more distant than had been believed. This was the clue that revealed the error in calibrating the Cepheid yardstick, and led to its correction — thereby doubling, at one stroke, the scale of the cosmos.

fact that clusters are so numerous, and that a majority of galaxies are found in clusters, has led some astronomers to feel that clusters of galaxies, rather than individual galaxies or stars, may be the fundamental unit of the universe.

Our own galaxy belongs to a fairly modest cluster about 3 million light-years in diameter, commonly known as the Local Group. The two largest members are both spiral galaxies—our own Milky Way galaxy, and the great Andromeda galaxy, M 31. They lie at opposite ends of the group, about 2¼ million light-years apart. Most of the other members of the cluster are considerably smaller and less massive, and seem concentrated near one or the other dominant galaxies.

It seems likely that, in addition to the 20 or so known galaxies in this cluster, there are others, obscured by interstellar dust, still awaiting discovery. The objects designated Maffei 1 and 2, for example, may well be members of the Local Group. They lie in the Zone of Avoidance, however, and thus cannot be seen with optical telescopes; we know of their existence only through observations at infrared and radio wavelengths, which are not scattered by interstellar dust. (If these systems are included in the Local Group, we must regard it as a rather large cluster than was previously thought, for they are nearly 9 million light-years distant—about four times as far from us as M 31. Maffei 1, a giant elliptical galaxy, would also replace M 31 as the most massive member of the group.) And only recently a dwarf galaxy, not yet named, was found at a distance of about 55 thousand light-years—very close to the edge of our own galaxy, in fact.

Further out in the universe there are other, larger clusters, some of them containing thousands of members. About 70 million light-years from us, for example, is the giant Virgo group, about 7 million light-years

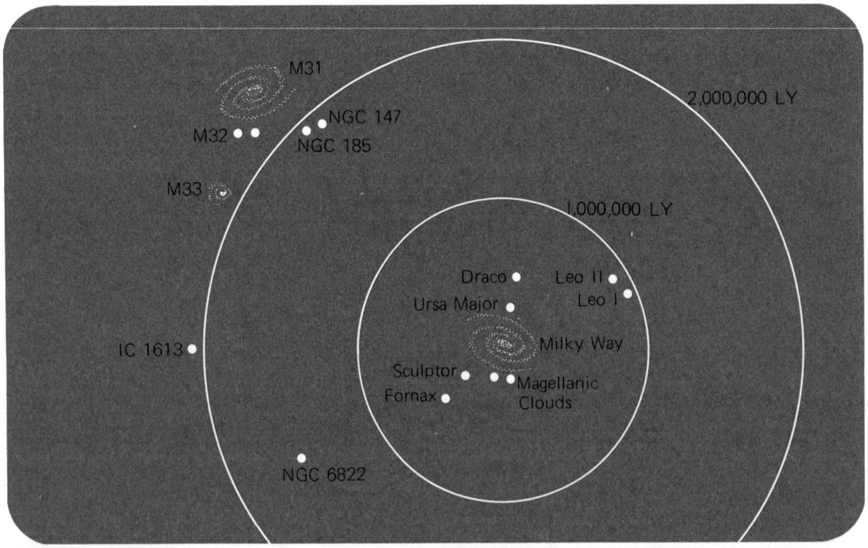

Galaxies occur in clusters which may have thousands of members.

Our galaxy is part of a small cluster, several million light-years across, known as the Local Group.

Figure 9-4
A map of our Local Group. Most of the members are dwarf elliptical galaxies, considerably smaller than our galaxy. The group is dominated by two large spirals—ours and the Andromeda galaxy.

PORTFOLIO

The Coma cluster contains hundreds of galaxies, mostly ellipticals. Many of them look like stars in the halftone above.

The irregular cluster in Hercules contains various galaxy types. Notice the pairs of interacting galaxies.

The Distance Pyramid

Cepheids are among the most important of astronomical distance indicators, for they enable us to leap the gap from our own galaxy to others. But the Cepheid yardstick is only one of a series that astronomers have devised in their attempt to measure the distances to the stars and galaxies.

It is not just curiosity as to the scale of the universe that makes astronomers devote so much of their time to distance measurements. As we pointed out in Chapter 2, unless we know the distance of an astronomical object, it is very hard to learn anything else about it—its size, for example, or the amount of energy it is radiating. Conversely, knowing something about the nature of an object often gives us an important clue to its distance. Thus distance measurement and physical understanding are mutually interdependent at every stage of astronomical inquiry.

The methods used for determining astronomical distances can be ranked in a hierarchy and arranged schematically in the form of a pyramid. Each method is useful only for certain types of objects, and within a particular range of distances. Moreover, the accuracy of each method depends upon the results obtained using the methods below it in the hierarchy. Thus each serves as a kind of stepping stone to the next.

The Cepheid method, for example, can only be used because a handful of Cepheids can be observed in clusters whose distances have been determined. The distances to these clusters are obtained by fitting their Main Sequences to that of the Hyades cluster (Chapter 6). But this is only possible because the distance of the Hyades has been obtained through application of the moving-cluster method (described briefly in Appendix 0). And the moving-cluster method in turn is double-checked and confirmed by parallax.

When we use the Cepheid yardstick to find the distances of other galaxies, therefore, the results we obtain can be no better than those obtained through the methods below it in the pyramid. Similarly, when the Cepheid yardstick was doubled in 1952, this change affected several other methods of distance determination, which had been calibrated on the basis of Cepheid-derived distances. The entire structure of inference thus is no more reliable than its weakest element. Fortunately, though, there is more than one method available for certain ranges of distances, and one technique can sometimes be sued to check the results of another.

As one ascends the pyramid, however, the percentage of error for each technique increases. This is because the methods used are indirect ones. Moreover, the uncertainties at each step of the pyramid add up. When Cepheids are used as distance indicators, the probable error is about 20 percent. Further up the pyramid, the uncertainty of the results becomes even greater. Our estimates for the most remote objects are constantly being revised, and may still be in error by a factor of 2 or more.

Women astronomers at the Harvard College Observatory early in this century. Most of them were employed to perform routine calculations, which before the advent of computers were elaborate and tedious. Henrietta Leavitt, who discovered the period-luminosity relationship for Cepheid variables, giving us our distance yardstick to the galaxies, is in front of the window to the left of the doorway. Annie Cannon, who helped establish the spectral sequence (Chapter 4), is 4 persons farther to the right.

At the very base of the distance pyramid lies our knowledge of the astronomical unit — the distance between the earth and the sun. This distance can be determined with extreme accuracy by applying the laws of mechanics, which govern the periods and sizes of the planetary orbits (Chapter 5), and by measuring the distances to the planets of the solar system with radar beams. A knowledge of the astronomical unit is important because the AU gives us our baseline for the measurement of stellar parallax.

Parallax is the most direct means of measuring distances, but it is only reliable out to a distance of about 100 light-years. Beyond that range, it must be supplemented by other geometrical methods, based on statistical analysis of the proper motions and radial velocities of groups of stars. The most useful of these is the moving-cluster method, described in the Appendix. By means of it, astronomers can find the distance of the Hyades and other open clusters, which lie at the edge of, or just beyond, the effective range of parallax measurement.

The next step upward can be made in a number of closely related ways, all of them involving the H-R diagram and the relationship between temperature and luminosity of Main Sequence stars. From the color of a star, or from the analysis of its spectrum, we can determine its temperature (Chapter 4). Knowing its temperature enables us to fix its approximate place on the Main Sequence band, and thereby find its luminosity. This method was described in Chapter 6. When the temperature determination is made from the star's spectrum, it is known as the technique of spectroscopic parallax. It is most useful for Main Sequence stars, though it can also be employed in a cruder way for giants and dwarfs.

The H-R diagram can be used in this way to find the distances of individual stars; the method really comes into its own, however, when applied to clusters. This is the method of cluster fitting, also described in Chapter 6. The technique involves matching the Main Sequence of one cluster to that of another cluster whose distance is already known. Hence the importance of the moving-cluster method in determining the distance of nearby clusters such as the Hyades. Because of the large numbers of stars involved, the accuracy of cluster fitting is greater than can be obtained when the distance of only a single star is measured by spectroscopic parallax.

The techniques described so far are applicable only within the confines of our own galaxy. Moreover, they are effective chiefly for measuring the distances to the Population I stars and clusters of the galactic disc. They work less well for Population II stars. Spectroscopic parallax and cluster fitting require that the Main Sequence on the H-R diagram be clearly marked. But as we pointed out in Chapter 7, the Main Sequence for Population II stars does not coincide with that for Population I stars. Population II stars are not found in the open clusters whose distances are gotten by the moving cluster method, nor are very many of them close enough for their distances to be found by parallax measurement. Thus the Main Sequence line for Population II stars is harder to fix. This makes it more difficult to determine the distances of halo stars and globular clusters. Fortunately another independent method is available that does not depend on the H-R diagram: the use of RR Lyrae stars as standard candles, described in the previous chapter.

RR lyrae stars are extremely useful for determining distances within our own galaxy—especially those of the globular clusters. However, they are not luminous enough to be seen in any but the nearest galaxies of the Local Group. Cepheids, by contrast, are bright enough to be visible out to about 20 million light-years. They can thus be used as yardsticks to a considerable number of external galaxies, beyond the confines of the Local Group.

The use of both Cepheids and RR Lyrae stars as yardsticks depend on the same principle: finding objects whose luminosity is known, and determining their distance by applying the inverse square law. The key is of course discovering classes of objects whose true luminosities can be easily established by some observable characteristic. In the case of Cepheids and RR Lyrae stars, this recognizable characteristic is the period of variation. To supplement and extend the Cepheid yardstick, however, somewhat more indirect methods must be used. All of them involve comparisons between objects in our own galaxy and apparently similar ones in other galaxies. They thus depend on our assumption that the same sorts of objects are found, and the same physical laws obeyed, in all parts of the universe. This is a difficult notion to prove, but an even more difficult one to do without.

Table 9–1   The Distance Pyramid

Method	Objects used	Effective distance	Method	Objects used	Effective distance
Radar reflection	Planets	Determination of AU	Period-luminosity	RR Lyrae & Cepheid	
Celestial mechanics	Planetary orbits	and distances within solar system	relationship	variable stars	to nearby galaxies
			Comparison with	Brightest stars,	
Parallax	Nearby stars	to 100 ly	brightness of	novas, globular	
Moving cluster	Open clusters	100–500 ly	local objects	clusters	~40 million ly
Spectroscopic parallax	Individual stars		Comparison of size	H II regions	~50 million ly
Main Sequence fitting	Star clusters		Comparison of brightness	Supernovas, Galaxies in clusters	~2 billion ly
		to edge of galaxy	Red shift	Distant galaxies and QSO's	to 10 billion ly

Once the distances to remote parts of our own galaxy, and to other nearby galaxies, are determined by the Cepheid method, many such comparisons become possible. Galaxies contain many classes of easily recognizable objects, the luminosities of which appear to vary little from one galaxy to the next. One such comparison involves novas, which at their peak luminosity are some 16 times brighter than Cepheids, and can thus be seen at 4 times the distance. By comparing the apparent peak brightness of a nova in a distant galaxy with a similar one in a nearby galaxy whose distance is known, the distance to the more remote galaxy can be easily calculated.

The same method can be applied using the brightest supergiant stars, which are thought to be about equally luminous in all galaxies. The brightest globular clusters in nearby galaxies also seen to be about equal in luminosity, and can thus be used as distance candles. This method is made possible only because we know the distances (and thus the luminosities) of globular clusters within our own galaxy by means of RR Lyrae stars. All three of these classes of objects—novas, bright supergiants, and globular clusters—have about the same luminosity, and can be used out to about 30–40 million light-years.

The calibration of objects lying at even greater distances depends upon the same principle of comparing the unknown with the known. For example, investigators have measured the sizes of H II regions in our own galaxy, and in neighboring galaxies whose distances are known. These regions of luminous, ionized hydrogen that envelope extremely hot stars are often quite large—as much as 30 to 40 million light-years in diameter. If we assume that the largest H II regions in more remote galaxies are of comparable size, we can then determine their distance simply by measuring their apparent angular diameters. This method is useful for distances as great as about 50 million light-years.

Distances beyond 50 million light-years can only be roughly guaged by using still more luminous objects: supernovas, and entire galaxies. Some supernovas, for example, are actually brighter than many galaxies, and can be seen at distances of hundreds of millions of light-years. If a supernova in a distant galaxy is truly as luminous as the ones we observe closer to home, than the distance of the galaxy can be calculated using the inverse square law. We can also use the galaxies themselves as the basis for such comparisons. Like stars, galaxies are found in clusters. Also like stars, the brightest galaxies in any cluster seem to be about equal in luminosity. By comparing such distant, luminous galaxies with bright galaxies in our own vicinity, the distances to entire clusters of galaxies can be estimated. This method can be used for galaxies lying up to 2 billion light-years away.

At this point, we are already quite close to the limits of our observations. Beyond 2 billion light-years, only the most luminous objects—very bright galaxies and QSO's—can be seen, and our only way of judging distances is the cosmological red-shift, described in the following chapter. This yardstick can also be used for less remote objects, but becomes supremely important near the edge of the observable universe, for it is our key to our understanding of the large-scale structure of the cosmos. It rests, however, on the distances determined by the methods we have described above.

in diameter. The furthest observable clusters — faint smudges on a photographic plate, in which the brighter individual galaxies can barely be distinguished — are billions of light-years away.

Even clusters of galaxies, however, may not be the largest observable units of our universe. Some astronomers believe that the known clusters of galaxies fall into some 16 **superclusters,** ranging in diameter from about 60 to 300 million light-years. It has been suggested that our own Local Group may be part of a supercluster dominated by the Virgo cluster. The existence of superclusters, if confirmed may tell us much about the origin and early history of the universe.

Types of Galaxies

Hubble's other important contribution to the study of galaxies was to devise a system of classification based on their form. He divided galaxies into three principal categories: spiral, elliptical, and irregular. Of the large, bright galaxies that tend to dominate our observational samples, a substantial majority are spirals — perhaps as many as three out of four. But as we saw in Chapter 6, in connection with stars, a sampling composed of objects that seem brightest to us is not necessarily representative. Spirals tend to be inherently large and luminous, and thus highly observable, whereas many ellipticals are less luminous. When this factor is taken into account, it seems likely that spirals comprise no more than 50 percent of all galaxies.

Spiral galaxies range from 20,000 to 100,000 light-years in diameter. Our own Milky Way galaxy is typical of the type. Like our galaxy, spirals have a well-defined nucleus and a large spheroidal halo composed of Population II stars. The disc of the galaxy, consisting chiefly of Population I stars, is divided into spiral arms which give the galaxy the appearance of a rotating pinwheel. The arms are prominent because they contain many bright, young stars, along with heavy concentrations of the gas and dust from which they form. Some spirals exhibit a bar-shaped structure extending through the nucleus. These barred spirals comprise about 30 percent of all spiral galaxies. Both normal and barred spirals are further classified on the basis of their form, as shown in the accompanying portfolio.

Elliptical galaxies account for about 25 percent of all observed galaxies — though, as we mentioned above, they are probably really much more numerous than that figure would suggest. Like spirals, elliptical galaxies are classified on the basis of their structure into categories designated E0 to E7. E0 galaxies are spherical, or nearly so, while higher numbers represent increasing degrees of flattening, with E7 found in a wide range of sizes, from giant systems like the radio galaxy M 87, with an estimated mass of 10^{13} suns, to dwarf ellipticals which seem little more than oversized globular clusters.

Even when they are large, elliptical galaxies tend to be less luminous

Hubble first classified the galaxies as spirals, ellipticals, and irregulars; with refinements, these categories are still in use.

Elliptical galaxies, composed of older Population II stars, tend to be fainter and redder than spirals galaxies of similar mass.

than spirals. This is because they are composed of old Population II stars. This means that the highly luminous, short-lived stars have already completed their evolution and ended their careers as energy producers. Only the less massive, less luminous stars remain — red and yellow stars from the lower portion of the Main Sequence. Elliptical galaxies contain very little gas or dust — an additional indication that the formation of new stars has long since stopped.

Finally there are the irregulars — generally small galaxies which lack a clearly symmetrical form, and thus cannot be fitted into any of the above categories. Irregular galaxies, which comprise only about 3 percent of observed galaxies, seem to consist largely of young stars. Small irregular systems are often joined by gravitational forces to large spiral galaxies. Our own Milky Way galaxy has two such satellites, known as the Magellanic Clouds, easily visible to the naked eye from the southern hemisphere. These systems, which are about 180,000 lightyears distant, are rich in hot, bright blue stars, clusters, and nebulas.

THE FORMATION OF GALAXIES

The birth of galaxies is a subject not yet fully understood. We mentioned earlier that our knowledge of star formation is somewhat limited by lack of observational evidence — it is not easy to see stars when they are forming. This is even truer of galaxies. Star formation is apparently still going on in our vicinity, but it seems very unlikely that any galaxies are now forming in our part of the universe — or anywhere else, so far as we know. The creation of galaxies seems to belong to a past epoch — perhaps as much as 15 billion years ago. This conclusion is based on the fact that all galaxies contain at least some old stars — we see none that consists solely of young stars.

It is true that in the astronomical realm the past is not completely closed to us, for whenever we look through our telescopes we look backward in time. Observing an object a billion light-years away, for example, we are seeing light that started on its journey toward us a billion years ago. This means that we are seeing the object, not as it is "now," but as it appeared a billion years earlier. When we look at distant parts of the universe, we are also seeing a much younger universe, and the farther into space we look, the farther into the past we see.

Galaxy formation seems to have taken place in a past epoch, for we see no young galaxies.

If we wish to observe young galaxies, then, we must seek them, not nearby, but at the very edge of the observable universe, 12 or 15 billion light-years away. So far, however, we have not observed any object farther from us than about 9 billion light-years. There is no reason to suppose that infant galaxies would be bright enough to be visible at such distances, or that we could see enough detail to be useful even if they were. Thus the study of how galaxies come into being has been largely theoretical, and is likely to remain so. Nevertheless, some very convincing models of galaxy formation have been developed in the past few years.

PORTFOLIO

M 81, above, is a beautiful example of a spiral galaxy. It is about 10 million lt.y. away. Notice the spiral structure of the dust lanes. The spiral arms are studded with bright, hot stars. Notice the faint, parallel streaks running diagonally, just to the right of the central bulge. They are believed to be the result of the explosion that took place in the galaxy M82 which lies beyond the lower right corner of the photo.

In the left hand column of the opposite page we have three spiral galaxies that exemplify the types Sa, Sb and Sc. The tightly wound spiral arms of the Sa galaxies become increasingly less so as we proceed to Sb and to Sc. The arms are also less regular and more dusty. Star formation has virtually stopped in the Sa spirals, but is still going on in the other types; notice the bright young stars in the spiral arms of the Sc galaxy.

In the right hand column we have three barred spirals of types SBab, SBb and SBc. Upto a third of all the spirals may be barred.

A

B

C

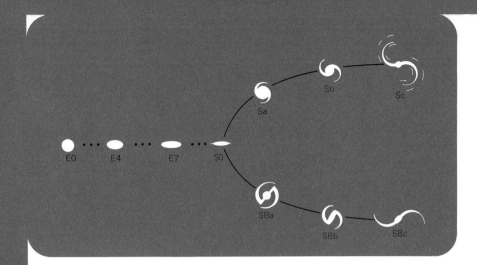

E0   E4   E7   S0

Sa   Sb   Sc

SBa   SBb   SBc

Hubble's famous diagram classifies the galaxies according to their shapes. No evolutionary significance is implied by this arrangement; at the present moment none has been clearly established. NGC 205 is one of two elliptical companions of M 31, the galaxy in Andromeda. It is rather small, with two dust patches near the center. This photograph was taken with the Mt. Palomar telescope in red light.

The Message of Galactic Forms

When Hubble first classified the observable galaxies according to their forms, it was thought that the various types might represent different stages of evolution. Some astronomers proposed that all galaxies begin as ellipticals, and then evolve into progressively more open spiral shapes. Others proposed just the opposite evolutionary scheme, with open spirals as young galaxies which would eventually evolve into ellipticals as their arms became more and more tightly wound about their nuclei.

Today, however, the evidence weighs heavily against both of these scenarios. Neither spirals nor ellipticals can be considered young galaxies, for very old stars are found in both types. Instead, it seems likely that the form of each galaxy is more or less fixed at birth—ellipticals remain ellipticals, and spirals remain spirals. But what factors determine these forms? Let us look first at spiral galaxies, such as our own Milky Way, and see what their structure can tell us about their possible origin.

It is easy to fall into the habit of thinking that our galaxy is essentially a disc, as Herschel originally supposed. Indeed, it is because of the density of stars in the galactic disc that we see the Milky Way, from which our galaxy derives its name. This impression is reinforced by photographs of other spiral galaxies similar to our own—the great Andromeda spiral, for example, (Color Portfolio), or NGC 4565 (Figure 8–6). In such pictures the disc of the galaxy is conspicuous because of its rich concentration of bright young Population I stars, gas, and dust.

As we saw in the previous chapter, however, the disc of a spiral galaxy is surrounded by a large, spheroidal halo of relatively faint stars, studded with globular clusters. In fact, it is now thought that most of the mass of our galaxy—perhaps as much as 80 percent—may lie in the halo, rather than the disc. Thus it is rather misleading to think of spiral galaxies as flat discs. Their overall form is spheroidal—similar, in other words, to elliptical galaxies. So we have already narrowed the apparent gap between the principal types. The question, then, is not "Why do galaxies come in two very different shapes?" but "Why do some galaxies have a disc of young stars, gas, and dust in their equatorial planes (spirals), while others don't (ellipticals)?"

When we take into account the shape of the halo, not just the galactic disc, we find that spiral and elliptical galaxies are not so dissimilar in form.

A clue to the answer is provided by the nature of the two stellar populations found in spiral galaxies. The stars of the halo and globular clusters all belong to Population II. Most of them are yellow and red stars from the lower end of the Main Sequence. The hotter, more massive stars have already completed their evolution, and since there is no longer any gas or dust in the halo, there are no young stars forming to take their place. All the halo stars are relatively poor in elements heavier than helium, suggesting that they were formed long ago, before the interstellar material was enriched with metals from dying stars.

The halo stars, then, are old, and all about equally old. They seem to have formed very early in the history of the galaxy—perhaps even before the galaxy took its present shape. Thus they tell us something about the galaxy when it was young. Though there is no gas or dust in the halo now, there must have been at one time in order for the halo stars to have come into being. Thus the galaxy must once have been a roughly spherical cloud of gas, at least as large as the present halo.

Condensation of Galaxies

This suggests that galaxies, like stars, began as large, diffuse clouds of gas, gradually condensing under the influence of their own gravitational forces. When the material of such a protogalaxy became dense enough, it began to break up into clouds of about 100,000 solar masses. Within these clouds, star formation began, and each eventually became a globular cluster. Over the eons, some of these clusters gradually broke up. Their scattered remnants comprise the halo stars.

This scenario accounts for the Population II stars of the halo and the globular clusters, but it does not explain the existence of the disc. Indeed, if all the material of the protogalaxy had condensed into stars at this stage, there would be no disc. Even while star formation was going on, however, the collapse of the protogalactic cloud was continuing. The way in which such a collapse can take place depends on the angular momentum of the cloud. If the original large, diffuse cloud is spinning, the conservation of angular momentum requires that it must spin faster and faster as its radius decreases. It can be shown (see Box) that a rapidly rotating cloud will not collapse symmetrically, like a balloon losing air, but will flatten out. That is, it will tend to collapse into a disc, but the disc itself will not contract beyond a certain radius.

A rotating protogalactic cloud will tend to flatten as it contracts.

It is important to note, though, that this is true only for clouds of gas, not for stars. When such clouds fall into the equatorial plane of the galaxy, their collisions stop them, and they remain concentrated in the disc. Clouds falling into the galactic plane from one direction collide with those falling from the other direction, and their kinetic energy is converted into heat, which is radiated away. The gas settles into a thin disc, whose density is not great enough for star formation until it is compressed by density waves (Chapter 8). Stars and globular clusters, however, pass easily through each other and through the disc without appreciable interaction, and continue on. As a result, their orbits about the galactic center are not confined to the plane of the galaxy, but extend far above and below it, into the halo.

We can now put these results together and sketch, in broad outline at least, the way in which different types of galaxies are formed.

ELLIPTICAL GALAXIES In elliptical galaxies, star formation must have taken place relatively quickly. While the original cloud of gas was still collapsing, virtually all of its material had broken up into smaller clouds, and those clouds had condensed into individual stars. The result

The Collapse of a Galactic Cloud

The fact that a rotating cloud of gas will tend to flatten out as it collapses is another consequence of the conservation of angular momentum. Imagine a large, diffuse cloud of gas, spinning slowly on a vertical axis. Let us consider first what will happen to a region of gas lying in the equatorial plane of the cloud, near its periphery. The gas of this region, which we can call A, is acted upon by the gravitational attraction between it and the cloud as a whole, which will be directed toward the center of the cloud. In response, the gas at R will move inward, closer to the center.

Since no external forces are at work, the angular momentum of the cloud remains unchanged throughout its collapse. This means that as it becomes smaller, it must spin faster and faster. The faster it spins, the greater the tendency for the material of the cloud to fly off into space, like marbles rolling off a whirling merry-go-round. To keep the cloud intact, an ever greater centripetal force is needed. This force is supplied by gravity.

The gravitational force, acting inversely as the square of the distance, increases rapidly as the cloud contracts. It can be shown, however, that the centripetal force required to hold the material of the cloud in a circular path increases even more rapidly—inversely as the cube of the distance from the cloud's center. As the material of region A falls inward, therefore, it will eventually reach a point at which the gravitational force on it is just sufficient to keep it in a circular path, It cannot fall any further in toward the center; to do so would require more centripetal force than gravity could supply.

This suggests that the rotation of a protogalactic cloud will limit the extent of its collapse in the equatorial plane. But what about those parts of the cloud above and below the equatorial plane—say at point B in our diagram? The same argument can be used to show that there is a limit to how far this material can move toward the axis of rotation. There is no limit, however, to how far it can fall down (or up) parallel to the axis of rotation—that is, towards the equatorial plane. Gravitational attraction will cause more and more of the material in the protogalactic cloud to fall into the plane. The cloud, originally roughly spherical, will in time come to resemble a disc.

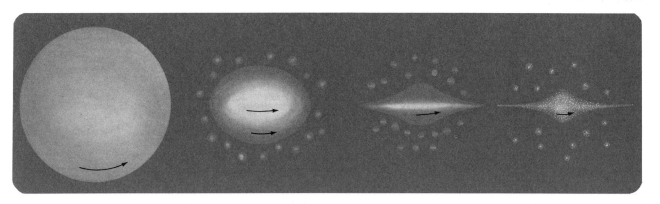

of this process would be, at first, a galaxy consisting of a large number of globular clusters. Because of their proximity to each other, these clusters would soon lose their separate identities through gravitational interaction. Since all the dust and gas would have been used up in the initial phase of star formation, there would be no young stars in such a galaxy, and no new stars created. The entire galaxy would consist of Population II stars.

Presumably, protogalaxies with little initial angular momentum would undergo little or no flattening, and evolve into E0 (spherical) galaxies. Protogalaxies with greater angular momentum would flatten more markedly before star formation had gotten well under way, and would develop into E6 or E7 (spheroidal) systems.

SPIRAL GALAXIES   In spiral galaxies, however, star formation was for some reason slower, or the collapse of the galactic cloud faster. Which of these was the case, and why, we do not know for certain, though it is thought that here again the angular momentum of the protogalaxy is likely to have been a factor. Whatever the reason, the result was that a considerable part of the original cloud collapsed into the equatorial plane of the galaxy **before** it had had a chance to condense into stars. From this material the younger Population I stars of the disc and spiral arms were formed, and continue to form to the present day. In the halo, no gas or dust remain — all this material was used up in the formation of the globular clusters, or fell into the disc. Thus no young stars are to be found in this region.

In spiral galaxies, some dust and gas collapsed into the central plane before it had time to condense into stars.

To summarize, then, nearly all galaxies are essentially spheroid in shape. The key difference between elliptical and spiral galaxies is the presence, in the latter, of a flat disc of gas, dust, and young stars. The reason for this difference is probably to be found in the relative rates of galacitc condensation and star formation in the two types of systems.

One final point remains to be mentioned. We have said that Population II stars are very poor in elements beyond helium. But the original clouds from which the galaxies are formed are thought to have contained virtually **no** elements heavier than helium. Where then did the Population II stars get even the small amounts of metals that they have? One plausible explanation is that Population II stars, old as they are, do not represent the first generation of stars formed in the universe. Before they came into existence, an even earlier generation of stars must have lived and died, enriching the interstellar material with heavy elements created in their interiors. We see none of these ancestral stars today — that is, no stars composed **solely** of hydrogen and helium. These first stars must therefore have been extremely short-lived. From this we know that they were very massive. The existence of these stars may have some bearing on the collapse of the galactic cloud described earlier. When the formation of stars was discussed in Chapter 7, we saw that as a cloud of gas collapses, its temperature and pressure increase. Eventually the pressure is high enough to prevent further collapse until the cloud can cool down. Protostars do this by radiating away energy, and the same would presumably be true of the larger galactic cloud. It happens that, because of their many, closely-spaced energy levels, the atoms of metals are much more efficient radiators of energy than hydrogen and helium. The presence of metals, therefore, ought to help a protogalactic cloud in its collapse. The precise significance of this, however, is not yet certain.

ACTIVE GALAXIES

Because galaxies are so large, it takes a long time for any event on a galactic scale to unfold. Consequently, it is tempting to think of galaxies as extremely stable structures. But the discoveries of the past two decades have shown this idea to be quite misleading. It has become more and more evident that a great deal of activity—often quite violent—is continually occurring in the nuclei of many galaxies, including our own.

Our knowledge of these events has come largely through observations made outside the visible spectrum. We know that the vast majority of stars, radiating essentially as black bodies, emit most of their electromagnetic energy chiefly in the visible and near infrared. Many galaxies, however, have been found to be extraordinarily bright over a much wider range of wavelengths, including the radio, far infrared, ultraviolet, and X-ray parts of the spectrum—sometimes in several of these regions. Much of the energy radiated by active galaxies, therefore, must be **nonthermal,** due to some process other than the black body radiation of their constituent stars. These outpourings of electromagnetic energy are evidence of a variety of cataclysmic events taking place within the nuclei of galaxies. Some of these events are not yet fully understood, and are being studied with great interest.

Many galaxies are sources of strong non-thermal radiation in various parts of the spectrum.

Radio Galaxies

All normal galaxies, including our own, emit some radio waves. The output of radio energy from the Milky Way galaxy is estimated to be about 10^{38} ergs/sec. This seems at first a large amount of energy—25,000 times that radiated by our sun at all wavelengths combined. Yet compared to the total amount of energy produced by all the galaxy's hundred billion stars, it is hardly a drop in the bucket. Some galaxies, however, emit as much as 10^{45} ergs/sec at radio wavelengths—a million times the radio output of an ordinary galaxy. Such **radio galaxies** actually generate more radio energy than they do visible light.

The study of radio galaxies was at first hampered by the difficulty in locating them. The development of radio astronomy led to the detection of many strong radio sources, but finding their optical counterparts was often not easy. The resolution of radio telescopes is much coarser than that of optical instruments. Essentially, a radio astronomer could only point to a relatively large region of the sky and say, "somewhere in this area there is an object that emits radio waves." In that area, however, a photograph at visible wavelengths might reveal dozens of different objects—stars, supernova remnants, galaxies—any one of which might conceivably be the source of the emissions detected by the radio telescope.

Several techniques have been developed to overcome this limitation. One involves the use of two or more widely separated radio receivers. The resolving power of the combined array is then nearly that of a single instrument as large as the distance between the components. Another ingenious method makes use of the moon to help pinpoint the location of radio sources. Radio astronomers record the exact moment when the edge of the moon occults (that is, blocks off) the radio signals from a particular source, or the moment when occulatation ends. The position of the moon in the heavens can be determined with extreme precision. By recording a series of such occulatations, it is possible to determine not only the location of a radio source, but its exact size and shape as well.

When a great many radio galaxies were found and photographed by optical astronomers, a few looked indistinguishable from ordinary galaxies. The majority, though, turned out to be quite peculiar in appearance. Some seem to have double nuclei, as if two galaxies were in the process of collision. Others are surrounded by large gaseous envelopes. Still others exhibit bright filaments of gas jetting from their nuclei, suggesting that a violent explosion might be in progress.

Cygnus A, for example—the first radio galaxy to be discovered—is about 555 million light-years from earth. Despite its great distance, however, it is one of the strongest radio sources: it emits more energy at radio wavelengths then all the stars of our galaxy radiate at *all* wavelengths combined. Cygnus A appears to have a double nucleus, suggesting that it may actually be two galaxies colliding. Another possibility is that a huge explosion has occurred within the galaxy. As a result a dense lane of obscuring dust now crosses its nucleus, making it appear double.

Since the discovery of Cygnus A, thousands of other radio galaxies have been located. In the Virgo cluster, for example, is the giant elliptical galaxy, M 87 (Figure 9–8)—one of the nearest of known active galaxies. M 87 is not only a strong radio source, but also emits an enormous amount of energy as X-rays. Shooting forth from its nucleus there is a bright blue jet of gas some 3000 light-years long. A fainter jet in the opposite direction has also been detected. Still another curious radio source is Centaurus A (Figure 9–7), an elliptical galaxy about 15 million light-years distant. This object is unusual for the dark band of dust that stretches across its equator. It too may be a pair of galaxies in collision, or, more likely, a galaxy that has experienced an explosion.

A radio map—an image built up by charting the signals emitted from various regions of the galaxy—reveals some interesting facts about the

Figure 9-5

A radio interferometer. In the first drawing, all the radio waves from a distant object arrive in phase—that is, their crests arrive simultaneously at both antennas. The resulting signals reinforce each other, creating a stronger signal. But as the earth turns, and the signals strike the antennas at a slightly different angle, they go out of phase—a crest reaches one receiver at the save time that a trough reaches the other. The resulting signals cancel each other out. Using such a system, it is possible to pinpoint the location of a radio source with much greater accuracy than could be achieved by either antenna separately.

Synchrotron radiation, emitted by many active galaxies, is produced by energetic electrons traveling in magnetic fields.

Figure 9-6

Synchrotron radiation is emitted when electrons traveling at velocities near that of light are accelerated in spiral paths around lines of magnetic force. The radiation is always emitted in the direction in which the electron is moving at that moment. The resulting radiation is polarized— that is, all the electromagentic waves vibrate in the same plane, rather than at random angles, as is the case with ordinary radiation such as we receive from stars.

location of radio sources in galaxies. In some cases, radio signals are emitted by relatively small regions within the galaxy. Many galaxies, on the other hand, seem to possess double radio sources, one on either side of the visible galaxy. Typically, each radio-emitting area measures about 300,000 light-years across, and the center of each is separated from the visible galaxy by some half-million light-years. Closer study has shown that in most of these cases, there is a fainter radio-emitting region at the center of the optical galaxy, so that the source is really triple—one within the galactic nucleus, and two others flanking it.

What is responsible for the radio emissions of these unusual galaxies? A clue is provided by the radiation itself, which has a distinctive spectrum, quite unlike that of black body radiation. The spectrum is much flatter; it does not exhibit the distinct peak characteristic of thermal radiation. Moreover, the strength of the radiation increases at longer wavelengths, and there is a sharp cut-off at the high frequency end.

Such a spectrum is characteristic of **synchrotron radiation,** so called because it was first seen experimentally in a particle-accelerating device known as a synchrotron. Synchrotron radiation is produced when electrons traveling near the speed of light are accelerated in a magnetic field. The electrons, moving in helical or circular paths around the lines of magnetic force, give up some of their kinetic energy in the form of electromagnetic radiation. A distinguishing feature of such radiation is that, unlike black-body radiation, it does not depend on the temperature, but rather on the kinetic energy of the electrons and the strength of the magnetic field. It is also highly polarized—that is, the electromagnetic waves vibrate in only a single plane (Figure 9–6). A high degree of polarization is in fact observed in both the radio emissions and visible light of radio galaxies.

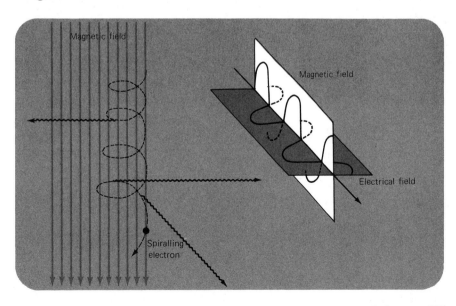

Figure 9-7
Centaurus A, a strong radio source in the southern sky. The dark band may be a dust lane, created by some tremendous explosion within the galaxy.

Figure 9-8
M 87, another radio galaxy. At left, a radio map of the object, superimposed on an optical photograph in which it is possible to make out the jet of material being ejected from the galactic center. In the longer exposure photo at right, the jet is difficult to discern, but the many globular clusters surrounding this elliptical galaxy can easily be seen.

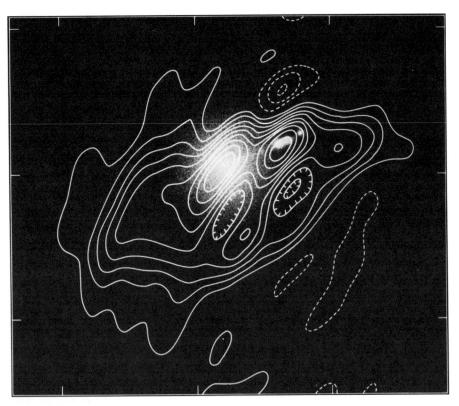

The amount of energy required to produce synchrotron radiation—and thus the powerful radio emissions of these galaxies—is immense. It has been estimated to be as much as 10^{60} ergs. The source of this energy is still being debated. Some investigators have suggested that it may be produced by the collision of two galaxies. Other proposed mechanisms include the gravitational collapse of a galactic nucleus, a chain reaction of supernova explosions among closely spaced stars near the galactic center, or the presence of a massive black hole within the galaxy. Some of these theories are discussed more fully in the following sections.

Galaxies in Collision

In the past quarter century, astronomers have photographed many pairs of galaxies that appear to be colliding, or to have collided quite recently. Such collisions are much more likely than collisions between individual stars. Our sun, for example, has a diameter of about 1.4×10^6 km. Its distance from its nearest neighbor (typical of stellar distances in the galactic disc), is about 4×10^{13} km, or almost 30 million times its diameter. The Milky Way Galaxy, by contrast, which measures some 100,000 light-years across, is separated from its nearest sizeable galactic neighbor, M 31 (the Andromeda galaxy), by about 2.3 million light-years—only 23 times its own diameter. Comparatively speaking, therefore, space is far more thickly populated with galaxies than a typical galaxy is with stars. Thus galactic collisions may be relatively common events.

Since galaxies are much closer together, for their size, than individual stars, collisions of galaxies are not extraordinarily rare.

So great are the spaces between the stars in a galaxy that even during a collision between two galaxies, the probability of close encounters between individual stars remains very small. The galaxies pass through each other like two swarms of bees, without contact. The clouds of interstellar dust and gas of the two systems, though, are quite likely to intermingle during such a collision. The results of this process are still being debated. Some scientists theorize that the interaction of the clouds may produce many local regions of relatively high density. These dense pockets may later develop into stars. Others believe that the collision might have the opposite effect, sweeping the dust and gas out of one or even both of the galaxies, and thus halting the formation of new stars in these systems.

The collision of two galaxies could theoretically generate enough energy to account for the observed synchrotron radiation from radio galaxies. However, the kinetic energy involved in the collision would have to be converted into electromagnetic radiation with an extraordinarily high efficiency. No one has yet suggested a satisfactory explanation of how this could take place. Moreover, investigators have found thousands of radio sources, and only a few of them appear to be galaxies in collision. It seems, then, that while some colliding galaxies may be emitting radio waves through a mechanism that is not yet understood, not all radio galaxies can be accounted for in this way.

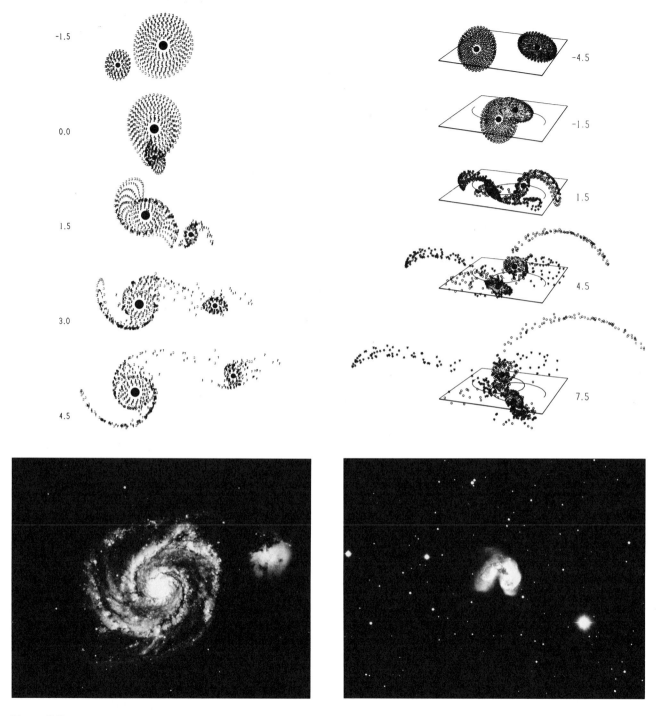

Figure 9-9
Computer simulations of interacting galaxies. These studies
may shed light on the creation of galaxies or pairs of galaxies
such as those in the photographs. At left, the spiral galaxy
M 51 and its companion. At right, two peculiar spiral galaxies
that appear to be colliding. (See also Color Portfolio I.)

Exploding Galaxies

More recently, some astronomers have proposed that explosions within the nuclei of radio galaxies are responsible for the enormous outpouring of radio energy. This hypothesis received support when observers discovered radio galaxies that were obviously experiencing violent upheavals. An example of such a galaxy is M 87, mentioned previously. It appears that the bright blue jet of hydrogen projecting from the center of this system is the result of an ongoing explosion in the nucleus.

Another dramatic example of an exploding galaxy is M 82 in Ursa Major, about 12 million light-years from us. Photographs of this peculiar object show jets of hydrogen spewing forth from the galactic center. These jets are moving at high speeds and extend some 1200 light-years. The quantity of matter being ejected is considerable — the equivalent of several thousand average-sized stars. It has been estimated that the explosion has been going on for some 1½ to 2 million years. M 82 is also emitting huge quantities of infrared energy.

A number of galaxies seem to be undergoing violent explosive events.

Are the explosions in the galactic nuclei the source of the powerful radio signals emitted by radio galaxies? Quite possibly. Supporters of this hypothesis point to the fact that radio sources associated with most radio galaxies are located on either side of the visible galaxy. These sources, they suggest, are actually hot gas clouds whose magnetic fields have captured the streams of high-speed particles hurled from either side of the exploding nucleus. As they become trapped by the cloud's magnetic fields, the particles spiral around the lines of the field at nearly the speed of light, emitting radio waves through the synchrotron.

Seyfert Galaxies

A possible link between normal and exploding galaxies may be represented by **Seyfert galaxies.** Discovered in 1943 by the American astronomer Carl Seyfert, these unusual spiral galaxies have small but very brilliant nuclei that are evidently the scene of violent activity. The spectra of Seyferts often reveal broad emission lines, indicating the presence of hot gas clouds in rapid, turbulent motion. In some cases, this gas is apparently being ejected from the galactic nucleus at nearly 3000 km/sec.

Seyfert galaxies also exhibit a variety of other peculiar characteristics. Some, for example, vary in their visible luminosity over a period of months. Others resemble radio galaxies in their powerful radio emissions, but the intensity of the radio signals, like that of their light, varies markedly with time. Some Seyferts emit huge quantities of infrared radiation; one is also known to be an X-ray source. Virtually all radiate a great deal of ultraviolet radiation, which in some cases appears to be polarized. Despite the fact that they are much smaller and less massive than the average galaxy, Seyferts emit 100 times more energy at all wavelengths than most galaxies.

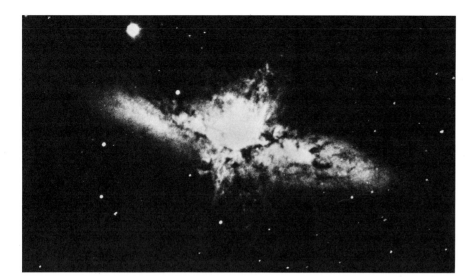

Figure 9-10
Two photographs of the exploding galaxy M 82 (see also Color Portfolio I). The top picture was made in red light, the bottom in blue light. Notice how the ejected material is more prominent in the red photo. The red color may be due to hydrogen Balmer alpha emissions, or heavy concentrations of dust — quite possibly both.

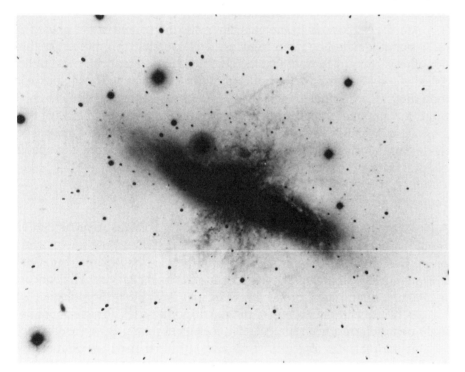

The fact that the brightness of Seyfert nuclei can vary in a period of only a few months places an upper limit on their size — they cannot be more than a few light-months across. To see why this is so, imagine what would happen if such an object, 1 light-year in diameter, suddenly flared up through some unknown process, and all parts of it instantaneously doubled in brightness. If the diameter of the object is 1 light-year, then light from the side farthest from earth will take a full year longer to reach us than light from the nearer side. We would see, not a sudden doubling

The fact that Seyfert galaxies and QSO's can fluctuate in brightness in short periods of time indicates that the light-emitting region must be small.

Figure 9-11
An exploding Seyfert galaxy photographed in the light of the hydrogen alpha line. Since Seyferts are considered to be a possible link between "normal" galaxies on the one hand, and such mysterious objects as exploding galaxies and QSO's on the other, this peculiar galaxy is the object of much study. Notice how far the filaments of gas extend.

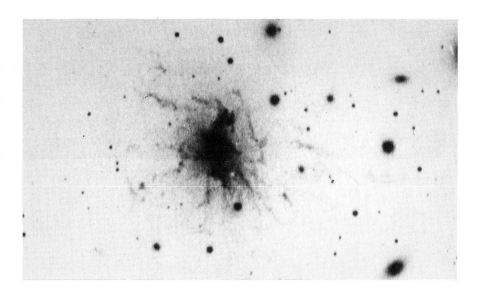

Figure 9-12
A more typical Seyfert galaxy. Notice the very bright nucleus. The outer regions of the galaxy cannot be easily seen in this photo. A barred spiral galaxy is at the lower right.

in the brightness of the object, but a gradual increase in brightness as light from successively more distant parts of the object reached us. It would be a full year before the change in brightness was complete. To change in brightness within half that period of time, the object would have to be correspondingly smaller—no more than 6 light-months across. In fact, though, it would not even be possible for all parts of an object to brighten simultaneously. Since nothing can travel faster than light, no physical process that might cause such a change in luminosity—an explosion, for

Galaxies 287

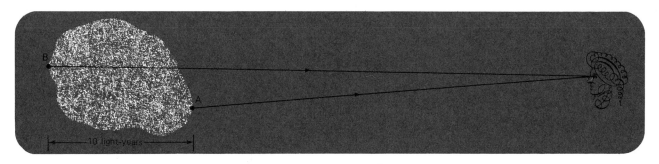

10 light-years

example—could propagate through the object faster than the speed of light. If such an explosion started at the center of our hypothetical object, it would take at least 6 months to reach its edge. This only reinforces our conclusion, however, that the size of an object sets a limit on how rapidly its observed brightness can change.

Seyfert galaxies comprise about 1 or 2 percent of all spiral galaxies. Some astronomers feel that Seyferts do not really represent a special **class** of galaxy at all. Instead, they propose that all spiral galaxies may go through Seyfert phases during their existence—perhaps many times. If this is true, then to account for the observed population of Seyferts, we must assume that a typical galaxy spends as much as 2 percent of its life as a Seyfert.

Quasars and QSO's

In 1963, astronomers succeeded in locating the optical counterpart of a particularly strong source of radio emissions. The object was listed in the catalog of radio sources as 3C 273. On photographs, it turned out to be a bright, blue, star-like body. In fact, it had been seen many times over the years, and had always been recorded as a star. This odd object proved to share many features with the active nuclei of Seyfert galaxies. Its luminosity varied over short periods of time, indicating that it was extraordinarily small for its huge output of energy. Moreover, its spectrum showed the broad emission lines characteristic of hot turbulent gases.

What was most unusual about 3C 273, however, was the pattern of its spectral lines. They seemed to correspond to no known elements. When the explanation was found, it turned out to be a very simple one, but its implications were far-reaching. Common elements such as hydrogen, iron, and magnesium were the source of the mysterious spectral lines; they had not been recognized because of their enormous Doppler displacements. The spectral lines of 3C 273 had all been shifted toward the red end of the spectrum by almost 16 percent. This meant that the object was receding from us at about 44,000 km/sec—nearly $\frac{1}{6}$ the velocity of light.

As we shall see in the following chapter, there is a relationship between the distance of extragalactic objects and the speed with which they

Figure 9-13
The time necessary for an object to change in brightness sets an upper limit on its size. Light from the far end of this object will reach the observer 10 years after light from the nearer edge. If all parts of the object became more luminous at once, this change would take 10 years in the eyes of the observer.

are receding from us. If 3C 273 is really receding at the velocity suggested by its spectral shift, it must be some 2 billion light-years distant. Yet 2C 273 is a relatively bright object. Tobe so bright at that distance, it must be enormously luminous — some 40 times as luminous as a typical galaxy. It seems impossible that any star could produce such an outpouring of energy. Yet judging by the rapid fluctuations of its light intensity, 3C 273 is far too small to be a galaxy.

Because of its strong radio emissions, 3C 273 and several similar objects discovered shortly thereafter were first designated "quasi-stellar radio sources," or quasars for short. Since then, however, hundreds of additional objects have been found with almost identical properties, but that are not radio sources. For this reason, astronomers now refer to all star-like objects with large red shifts as **"quasistellar objects,"** or **QSO**'s.

These mysterious objects are the most distant that astronomers have yet observed. The red shift of the remotest QSO discovered thus far, 4C 255, indicates that it is receding from us at over 90 percent the speed of light, and is thus some 9 billion light-years away. If this is true, then we are seeing 4C 255 as it was 9 billion years ago — not very long after the birth of the universe, according to modern cosmological theories. However, a great deal of controversy still surrounds the observed red-shifts of QSO's. To understand the issues involved, it is necessary first to know something of our ideas about the history of the universe as a whole — that is, about cosmology. This is the subject of the next chapter. Then we will return to the QSO problem in the following Interlude.

SUMMARY

As early as the eighteenth century, there had been speculation that some of the "nebulas" seen by telescopic observers were not clouds of luminous gas within our own galaxy, but distant stellar systems rivalling the Milky Way. The existence of external galaxies, however, was not proven until 1924, when Hubble discovered Cepheid variable stars in the Andromeda nebula, and used them to prove that it was far too distant to be a part of our galaxy. Cepheids are stars whose luminosities can be found directly from their periods of variation. When the luminosity of a Cepheid is detmined in this way, the inverse square law can be applied to give its distance. Cepheids thus serve as yardsticks for determining the distance of any cluster or galaxy in which they can be found.

Many millions of galaxies can be photographed with the largest telescopes, and their total number may run into the hundreds of billions. They appear to be distributed symmetrically in space about us. Because of dust in the disc of our galaxy, however, we can see very few external galaxies through the plane of the Milky Way. This region is known as the Zone of Avoidance. The distribution of galaxies is not smooth and uniform, for most galaxies occur in clusters, with anywhere from a handful of members to several thousand. The Milky Way galaxy is part of a cluster known as the Local Group. It contains about 20 members, the most prominent being M31, the Andromeda galaxy, about 2¼ million light-years distant. The existence of a still larger unit—the supercluster, or cluster of clusters—has been suggested but not proven.

Galaxies are classified according to their form. About ¾ of the bright, easily observable galaxies are spirals, like our Milky Way. These systems have prominent discs of young, luminous, hot, blue stars, along with much gas and dust, organized into coiling, pinwheel-shaped arms. Some spirals seem to have bar-shaped structures running through their nuclei; these are known as barred spirals. Elliptical galaxies range in form from spherical to flattened, lens-like shapes. Because they consist of older, Population II stars—less massive, and therefore cooler and less luminous than those in the discs of spiral galaxies—elliptical galaxies are generally dimmer and redder than spirals. A few galaxies are classed as irregular.

Galaxies are thought to have formed from the gravitational collapse of vast clouds of gas. If such a cloud is originally rotating, it will spin faster as it collapses, thereby tending to flatten out. In elliptical galaxies, star formation is thought to have finished before the completion of this collapse. In spiral galaxies, however, some of the original cloud collapsed into the galactic plane before condensing into stars. This material later formed the Population I stars of the disc and spiral arms. Star formation is still going on in these regions. In elliptical galaxies, all the dust and gas has already been used up, and no new stars are forming.

Many galaxies emit large amounts of non-thermal electromagnetic radiation in the radio, infrared, ultraviolet, and X-ray regions of the spectrum. This radiation tells us much about the violent activity characteristic of many galactic nuclei. One source of such non-thermal radiation is the synchrotron mechanism, whereby fast-moving electrons accelerating in magnetic fields produce a spectrum quite unlike that of essentially black-body radiators such as stars. This is probably the source of the radiation from many radio galaxies. Most radio galaxies are peculiar in their visual appearance as well. Some appear to be pairs of galaxies in collision. Others seem to be experiencing violent explosions in their nuclei, with large quantities of matter being ejected.

A link between normal and exploding galaxies may be provided by Seyfert galaxies—spirals with unusually small, bright nuclei that emit large quantities of energy in the ultraviolet, and often at radio and infrared wavelengths as well. The luminosities of Seyfert nuclei vary over short periods of time, indicating that the radiation-emitting region is very small. It is thought that all spirals may undergo periods of Seyfert activity repeatedly during their lives. Quasi-stellar objects, or QSO's (formerly called Quasars) also have a lot in common with Seyfert galaxies: small size and fluctuating but very intense luminosity in many regions of the spectrum. The spectra of QSO's, however, exhibit enormous red shifts, indicating that they are receding from us at great velocities. They should therefore be extraordinarily distant—the farthest known at this time is some 9 billion light-years away. To be so bright at such great distances, QSO's must be astonishingly luminous. We are not yet sure that QSO's are really as distant as their red shifts seem to indicate; or, if they are, how so much energy can be produced by such small objects.

1. The term *nebula* was often used for two very different kinds of objects during the eighteenth and nineteenth centuries. What were the objects, and why were they confused?

2. What was the Shapley-Curtis debate about? Why was it so difficult to prove that there are galaxies other than our own?

3. How did Hubble establish the existence of external galaxies?

4. How are Cepheid variables used to measure distances? What are the probable sources of error in this method?

5. Describe some of the methods used in determining extragalactic distances. Why do these techniques become increasingly inaccurate when we measure distances to the more remote galaxies?

6. What is the Zone of Avoidance? How can we detect the presence of other galaxies obscured by the dust in our galaxy?

7. What is the Local Group? How large is it, and how many members does it include? Which are the dominant galaxies in the group?

8. Why do some astronomers believe that even clusters of galaxies are not the largest structures we can detect?

9. Describe the various galactic types. How does a spiral differ from an elliptical galaxy?

10. If an Sa galaxy and an E3 galaxy have the same mass, which will be more luminous? Why? What can you predict about the colors of these galaxies?

11. How many galaxies can you see with the naked eye? Where can they be found, and what do they look like? What types are they?

12. What do the two stellar populations tell us about the formation of galaxies? Describe briefly how spiral and elliptical galaxies are thought to have formed.

13. Why do we think there may have been a generation of stars formed before those of Population II that we see today?

14. What is a radio galaxy? Does our galaxy emit radio waves? Is it a radio galaxy?

15. What is synchrotron radiation? Why is it called nonthermal radiation? How do we recognize it?

16. Why is it difficult to find the visible source of radio emissions even after it has been located with a radio telescope? How can this difficulty be overcome?

17. What are Seyfert galaxies? Why are they considered interesting objects for study?

18. What are QSO's? Why has this term replaced the earlier term "quasars?" What do QSO's look like on photographs? If you were asked to survey the skies in a search for QSO's, how would you go about it?

19. How does the rapid fluctuations in luminosity of QSO's indicate that their energy must be emitted from a small region?

20. A QSO is found to change in brightness over a period of a month. What is the maximum size of the light-emitting region, in km?

COSMOLOGY

Suppose you were asked to give a definitive description of a guinea pig--its habits, appearance, life history, reactions. You might begin by studying the one you keep as a pet; then you would study others that belong to friends and neighbors. To check that your observations and conclusions apply to all guinea pigs, and not just the ones in your immediate vicinity, you might decide to visit a pet store and observe all the different kinds they have in stock. Soon you would have formulated a general description of some specific guinea pigs, and you would have reason to believe that the description would fit any other guinea pigs you might encounter.

Astronomers can apply somewhat the same method to formulating a general description of a star. They begin by carefully studying the nearest star, our sun, and then go on to look at other stars in our neighborhood. They can check their conclusions by analyzing the light from distant stars. By comparing observations of many different stars, astronomers have found out what stars are made of, how they produce energy, and how they evolve.

With improved methods of obtaining and analyzing data, astronomers have begun to make comparative observations even of galaxies. But there the usefulness of the comparative method seems to end. We cannot study the universe by observing this one and then comparing it with others. By definition, there is but one universe. Since the universe is all there is, no comparison, no repetition of observations under the same conditions, is possible. Technological breakthroughs may provide us with better ways to study distant parts of the universe, but the fundamental problem will remain. No matter how well we can see our universe, it is very hard for us ever to be sure that our conclusions about it are valid, with no other universes to check those conclusions against.

It is possible that the ultimate questions about our universe will have to be answered by speculation or logic, rather than empiricism. Yet some of the methods of science do apply to even so difficult a problem. Astronomers presently try to answer questions about the universe by building models of its evolution. The models must incorporate all the existing observational evidence, and explain it. Newly discovered data then serves as a kind of test of the model. If the new data fits into the predictions of the model, the model can stand. If the data cannot be explained or predicted by the model, then the model needs correcting.

Since the universe includes everything, it is unique; we cannot test our theories by comparing it with others.

DESCRIBING THE UNIVERSE

Most of the terms commonly used to describe the universe are better descriptions of human feelings about the universe than they are of its physical reality. We say it is "unimaginably vast," or "limitless," or sometimes "as old as time itself." Such statements may possibly be true, but they are not very useful for scientific reasoning. Astronomers would prefer descriptions that are more concrete, more quantitative.

The Shape of the Universe

How do you measure the universe? We tend to think of it as a problem involving the distance between two unknown points, like that of measuring the distance from the earth to some star, or from our galaxy to the next. It seems as if all we need is a long enough measuring stick—a unit bigger than light-years or megaparsecs—and a telescope good enough to see the distant points at the "ends" of the universe.

The problem with this imaginary solution is that it is unconsciously based on the assumption that we are measuring flat surfaces. In our daily lives, as well as in most scientific studies, we take for granted the ancient concepts of Euclidian geometry. In this geometry, the sum of the angles of a triangle is 180°, parallel lines never meet, and the area of a circle is πr^2. This geometry seems "logical" to us because it is so common.

In the nineteenth century, a suspicion gradually arose that Euclidian geometry was not the only possible geometry. Some other set of assumptions might prove to offer a better description of the cosmos. From these suspicions two kinds of non-Euclidian geometries emerged: hyperbolic and elliptical geometry.

Elliptical geometry is based not on flat space, as in the case with Euclidian geometry, but on curved space. It is demonstrated on the surface of a sphere (space with a positive curvature). On the surface of a sphere, the shortest distance between two points, or **geodesic,** is a great circle, which divides the surface in half. In elliptical geometry, parallel lines are impossible; any two large circles drawn on the face of a sphere will meet in two opposite points. Triangles have angle sums greater than 180°. For example, the sum of the angles of the triangle drawn on the earth by connecting the equator with the 0° meridian and the 90° meridian of longitude is 270°, not 180°. It is also true that the surface of a sphere has a fixed, finite area, but no boundaries. A voyager starting from any point on a sphere, and traveling along a great circle, will eventually return to his starting point.

Hyperbolic geometry is based on the concept of space as negatively curved. It is usually demonstrated on a saddle-shaped surface. Under the axions of hyperbolic geometry, the sum of the angles of a triangle is always less than 180°, and more than one parallel to a given line may be drawn through a point not on that line.

It is very difficult to visualize the results of non-Euclidian geometries in space of more than two dimensions, or the results of any geometry in space of more than three dimensions. Yet it is clear that we need not be restricted to Euclidian geometry in our explanation of the cosmos. For example, if we think of space as having the properties of a spherical, rather than a flat surface, we can see how it could be both finite and boundless. Imagine the universe to be like the *surface* of a basketball. The ball has a fixed, limited size, and yet to an ant crawling on its surface, there are no bounds or limits to his travels. In fact, such a universe is quite consistent with Einstein's general theory of relativity. According to this theory, space and time cannot be considered independently, but must be treated as a single entity, called **space–time.** The presence of mass alters the geometry of space–time, causing it to assume a positive curvature. Light rays, which always travel along geodesics, can be used to measure the curvature. If the density of matter in the universe is great enough, the universe will automatically be closed—that is, it will have a finite size. But although Einstein's equations are thought to be correct, they are too general. There are in infinite number of possible universes in which they would be true, so they do not tell us why our one universe is the way it is.

We are familiar with the geometry of flat space, but on a cosmic scale, space itself may be significantly curved.

Is the universe open or closed? Does it have space bounds? Time bounds? No one yet knows a scientific answer to these questions. In fact, we are not even perfectly sure that they are the right questions.

The Uniformity of the Universe

The models of the universe currently considered valid by astronomers are all based on several important assumptions. We begin by assuming that the physical laws governing matter and energy are the same in other parts of the universe as they are on our planet. This assumption seems reasonable, because after all, we have already identified fundamental particles that make up all the matter we have been able to test. More importantly, it is the only assumption we know how to make: who could imagine what another set of physical laws might be like?

Models of the universe are based on another assumption of uniformity. It is assumed that matter is more or less evenly distributed throughout the universe. Presumably the universe is not very dense in one place and very empty in another. (Of course, we are talking about very large regions of space—large compared even with superclusters of galaxies.) A third assumption of model builders is that time and space have the same properties throughout the universe. Thus it does not matter where a galaxy is, or in what direction it is traveling; the same phenomena are presumed to occur in the same way.

THE COSMOLOGICAL PRINCIPLE One way of summing up the assumption that the universe is uniform throughout is the **cosmological principle.** It states that all observers everywhere should see the universe in essentially the same form. The way we on earth see the universe should be no different from the way the universe appears from a planet in a distant galaxy.

This principle introduces a number of interesting problems. For example, if it is strictly true, there can be no edge to the universe, since an observer on the edge would see many celestial objects on one side and empty space on the other—and that would differ from the view from the center of the universe. There are various ways of overcoming this problem. One is to assume that the universe is open, or infinite. Another is to assume that the basic geometry of the cosmos is curved rather than flat, and closes on itself. As we have seen, in such a universe, there is no edge, and each observer will always consider himself to be at the center.

The Expansion of the Universe

One other feature central to current models of the universe is that it is believed to be in the process of expansion. The evidence for this expansion comes primarily from the radial velocities of distant galaxies. About 1912, it was observed that many distant galaxies had large Doppler red shifts, indicating a high velocity of recession.

At that early date, observations were confined to galaxies no farther away than 65 million ly. But by 1930, Edwin Hubble, working at the

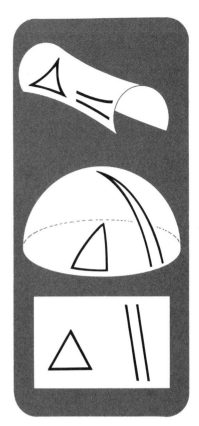

Figure 10-1
The familiar geometry in which parallel lines never meet and the sum of the angles of a triangle is 180° is possible only if space is flat. But space can also be curved. If the curvature is positive, as on the surface of a sphere, parallel lines are impossible, and the sum of the angles of a triangle always exceeds 180°. If the curvature is negative, as on the surface of a saddle, many lines can be drawn parallel to a given line through an outside point, and the sum of the angles of a triangle is always less than 180°. Since the curvature of space involves four dimensions, rather than just the two shown in the figure, it is impossible to visualize.

Mount Wilson Observatory, had accumulated information about velocity and distance for a large number of galaxies. The velocity he obtained by measuring the galaxy's red shift; the distance he determined through observation of the brightness of Cepheid variables within the galaxy. Although this method caused him to underestimate all the distances, the data was consistent enough for him to take the next step. Hubble concluded that, with a few exceptions in our Local Group, all galaxies were receding from us. Moreover, the farther away they were, the faster they were receding. In mathematical terms, what Hubble found was that, for each galaxy,

$$\text{velocity} \propto \text{distance} \qquad \text{or} \qquad v = Hr$$

where H is a number now known as the Hubble constant. This formulation is known as **Hubble's law,** or the law of red shifts.

THE HUBBLE CONSTANT In 1930, Hubble announced a calculation for the value of the constant in his equation. He put it at 150 km/sec million light-years. Since then, his measurements of both distance and velocity have been repeatedly refined. Each recalibration has resulted in a decrease in the Hubble constant. Today, the accepted value, as measured by Allan Sandage, is about 17 km/sec per million light-years.

IMPLICATIONS OF HUBBLE'S LAW Clearly, Hubble's law implies a continual expansion of the universe, rather as if it were a balloon that was being inflated. Initially, there were attempts to explain away the evidence of the red shift; for example, it was postulated that the shift means not greater velocity, but a loss of energy by photons of light that

All distant galaxies exhibit red shifts, indicating that they are receding from us; the velocity of recession for each galaxy is proportional to its distance.

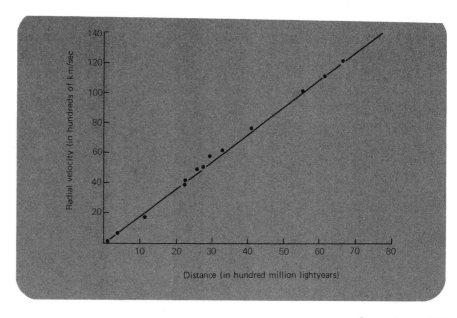

Figure 10-2
Hubble's law tells us that all galaxies are receding from us at velocities proportional to their distances. But this does not necessarily mean that we are at the center of the expansion. If the entire universe is expanding, like the surface of a balloon being inflated, then all points will be receding from all other points at velocities proportional to their separation. An observer at any point will think that all the galaxies are fleeing him.

have traveled great distances. The trouble with this "tired light" theory is that it is an entirely new phenonemon that violates the accepted laws of physics. That doesn't mean it couldn't occur, but scientific arguments based on special cases of exceptions to the laws are generally considered to be weaker than those that explain things in accordance with these laws.

Shortly after the 1912 observations of the galactic red shift, Einstein published his theory of general relativity. He suggested a model of the universe based on a four-dimensional space-time curvature. Once theoreticians became aware of the observations that indicated the universe was expanding, they were able to show that Einstein's equations could be made entirely consistent with the observed expansion simply by adjusting some constants in the solution. Unfortunately, as we mentioned above, these equations can also be made consistent with other models which are not expanding, so they do not explain the expansion.

If all the other galaxies are rushing away from us, are we the center of the expansion? Does Hubble's law also imply that we are at the center of the universe? In 1930, such a conclusion was unacceptable. The earth-centered view of the universe had long ago passed out of fashion; moreover, the notion violates the cosmological principle, which says that the universe looks the same to all observers.

Since the universe is expanding uniformly, an observer at any point will think he is located at the center of expansion.

To see why the expansion of the universe away from us in all directions does not necessarily mean we are at the center, it helps to return to the analogy of the balloon. Suppose you take your little cousins Eddie and Bill to the park one spring morning and buy them each a Mickey Mouse ballon. As the balloonman blows it up, the features of Mickey's face will gradually move apart. If you pick a point on the uninflated ballon—say the left eye—and try to watch how it changes in relation to the other features, you will find that they all move away from it. The right eye moves away, the ears move away even faster. No matter which feature you pick as your initial starting point, however, it will always seem that the *other* features are rushing away from it. And the farther away another feature was to begin with, the faster it will appear to rush away.

According to the currently accepted model of the universe, it is expanding in just the same way as the balloon (though with the added complication of four dimensions to imagine instead of two). Thus an observer at *any* point in the universe will always find that the distant galaxies seem to be running away from him, as if he were at the center of the universe. This explanation has the virtue of being consistent with the cosmological principle.

THE AGE OF THE UNIVERSE   Hubble's law tells us how to calculate the velocity at which galaxies are moving away from us; thus we can predict their positions in one billion years, or 10 billion, or 100 billion. Presumably we could also apply the same logic to past time. If we project backwards rather than forward, we would find the galaxies moving closer

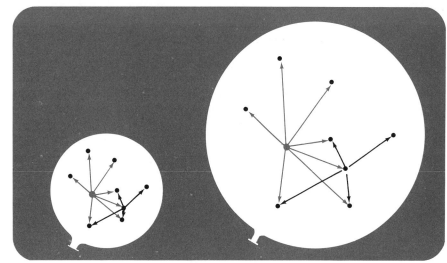

Figure 10-3
A modern diagram showing Hubble's findings. The more distant the galaxy, the greater its velocity of recession. The best current calculation of Hubble's constant is 17 km/sec per million light-years.

and closer together, until eventually they arrived at the stage when the universe was in its original condition—whatever that was. The time it takes the galaxies to move backward to that original moment, when they occupied a very small volume of space, is thus the age of the universe. It is possible to calculate this age from Hubble's law. It turns out to be simply the reciprocal of the Hubble constant, H; that is, $1/H$. This works out to roughly 18 billion years. Of course, that figure presumes that the velocity has been the same throughout the history of the universe, and many astronomers disagree. They believe it likely that the velocity has decreased over time, through the braking force of gravitational attraction. They therefore think the universe is younger—current estimates range from 11 to 16 billion years old.

COSMOLOGICAL MODELS

None of the things we can observe about the universe can answer one central question: Is the universe we know the same as the universe that has always been and always will be? Or do we see just one particular stage in its evolution? Within the last ten years, most astronomers have come to agree that the universe changes. Their evidence is the way recently available data fits into models based on these two assumptions. On the whole, the evidence seems to fit better into the models that assume change than those that assume no change.

Steady-State Models

The cosmological principle assumes that all observers should see the universe the same way, no matter where in the universe they are. In the 1940s, some astronomers suggested that the principle should apply not only in space but in time as well. They called this idea the **perfect cosmological principle;** it states that all observers should see the universe the

Against the Fall of Night?

Many people do not find it particularly uncomfortable to think of an infinite universe filled with stars. It even has a theological appeal to those who wish to see it as a manifestation of an infinitely powerful God. The German astronomer Heinrich Olbers (1758–1840) thought of just this sort of universe. He assumed that the universe was infinite, unchanging, and uniformly populated with stars. (We know nowadays that stars are not uniformly distributed in space, but the argument is unchanged if we replace 'stars' with 'galaxies' or 'clusters of galaxies'.) We can mentally divide such a universe into spherical shells of uniform thickness, centered on the earth. The amount of light reaching earth from each shell depends on the volume of the shell and its distance from us. According to the inverse square law, the light from a shell *decreases* as the square of its radius, since that is its distance from us. But the volume of the shell *increases* as the square of the radius. Thus the light contributed by each shell is the same, regardless of its size. In the infinite universe there are an infinite number of these shells. Therefore an infinite amount of light should be reaching the earth. Why then is the sky dark at night?

Olbers' Paradox, as this is called, is resolved when we recall that the universe is expanding. The light from a shell at a great distance suffers a proportionately greater red shift. This involves a reduction in the energy of photons reaching us from the shell. Consequently the distant shells contribute less light than Olbers expected them to.

It has been shown that Olber's paradox cannot be resolved merely by postulating a finite but boundless universe. Thus the fact that the night sky is dark is a supporting argument for the expansion of the universe!

same way no matter where or *when* they observe it. According to the perfect cosmological principle, the universe should have looked roughly the same 10 billion years ago as it will 10 billion years from now. It should be in a **steady state.** In other words, the universe would be infinitely old; it would have no birth date, and would never end.

One problem with this model arose immediately. How can the universe be in a steady state when Hubble's law seems to show that it is expanding? An expanding universe would gradually get less and less dense; it would therfore look different to observers at different times in its history. This objection was answered by postulating the **continuous creation** of matter. In this view, as the universe expands, it is slowly filled with new matter. In order to maintain the present density of the universe, one new hydrogen atom would have to be created in each cubic meter of space every ten billion years. Though this seems a small amount, it is by no means negligible. Converted into its energy equivalent, it corresponds to about 2 percent of the electromagnetic energy contributed by stars to the same volume of space. Critics of the theory, naturally, asked how this creation could occur; its supporters replied that they did not know, but that it was certainly no harder to imagine than the creation of all the matter in the universe in a single instant.

Philosophically, the steady-state model is very attractive. It pictures the universe as literally endless in space and time—just what one imagines a universe ought to be. It does not suggest that individual stars, or even galaxies, live for ever—merely that the number of particular kinds of stars and galaxies have been and will continue to be the same for all time. It also avoids the difficult questions for which science has no answer whatsoever: What made the universe begin? What was there before it existed? At the time it was proposed, the steady-state model had scientific as well as philosophical appeal. Since the value of the Hubble constant was thought to be much much larger than it is now computed, that made the age of the universe, or $1/H$ much smaller. Thus the age of the universe was put at about 2 billion years, whereas the sun was thought to be several times that old. The steady-state model neatly avoids this problem, since it assumes that the universe was never "born." You cannot run the movie of the expanding universe backward to find the date of its birth; all you would find if you looked back 10 or 20 billion years would be the same universe, expanding at the same rate.

> *The steady–state theory is based on the assumption that the universe is uniform in time as well as space.*

When the value of the Hubble constant was adjusted downward, the calculated age of the universe increased proportionally. The discrepancy between the calculated age of the universe and the age of the objects it contains vanished. Thus the steady-state model became slightly unnecessary. Moreover, observational evidence began to appear that did not fit the model. For example, the model allows for the birth, aging, and death of any specific object in the universe, but predicts that any *class* of object will always be observed with about the same frequency. Yet there is unmistakable evidence that there were once many more QSO's in the universe than there are now. Another lack of fit between observational data and the model is the absence of evidence that the galaxies differ in age. If some of them were recently created to fill in the emptiness left by the expansion of the universe, then in any given region of the universe, some galaxies should be either older or younger than others. But, on the contrary, all neighboring galaxies seem roughly the same age.

> *The fact that the universe appears to have been very different billions of years ago undermines the steady state theory.*

Today the steady-state model has few supporters. Trying to fit all the evidence now available into this model requires too many contortions. It has thus been superceded by evolutionary models of the universe.

Evolutionary Models

Evolutionary models of the universe assume that it has a life history—that it will look different to observers separated in time (although the same is not true of observers separated in space). In the evolutionary model, the expanding universe is simply getting less and less dense, and any local region of space less markedly curved.

THE BIG-BANG MODEL   The best-known, and probably the most widely accepted, model of the universe is the big-bang model. According to this model, the universe began as a hot (perhaps as much as 10^{12} K), dense ball of radiation (that is, photons) and subatomic particles. The theory makes no attempt to explain the origin or creation of this primeval fireball; nor is it possible, within the framework of the theory, to ask what existed before it. Wherever it came from, the fireball represents the beginning of the universe.

As soon as it came into existence, the fireball began to expand. Within less than 2 minutes, it had expanded enough for the temperature to drop to 10^9 K. This temperature was low enough to permit some of the fundamental particles to combine and form nuclei. Within two hours, it had cooled to 10^8 K, and ionized atoms of hydrogen, helium, and deuterium had been formed. For the next 100,000 years or so, the universe existed in this state: a hot ionized gas filling all space uniformly. The ionized gas was opaque to radiation—that is, radiation was constantly passing from atom to atom, being absorbed and re-emitted.

Eventually, the continued expansion caused the temperature of the gas to drop to 3000 K. That was cool enough to permit the formation of

Figure 10-4
Left, a static universe. Center, an expanding universe. As the galaxies recede from each other, the average density of matter in the universe decreases. At right, a steady state universe. As the galaxies fly apart, new matter is created in interstellar space, and new galaxies form. The density of matter in the universe thus remains constant.

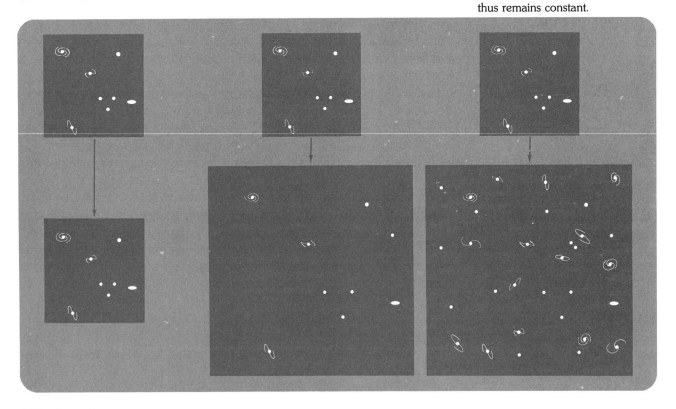

neutral atoms, as nuclei began to capture free electrons and hold them in stable configurations. At that point, the gas ceased to be opaque to radiation. The universe became transparent, and radiation was free to travel almost unchecked. The next stage was the formation of galaxies and stars, with the eventual creation of new and heavier elements within the early stars. This occurred through the condensation of the cosmic gas, as described in Chapters 7 and 9, and probably took about a billion years.

COSMIC BLACK-BODY RADIATION What evidence supports this model of the universe in preference to others that have been suggested? In the late 1940s, when the merits of steady-state versus evolutionary models of the universe were being hotly debated, George Gamow and his colleagues suggested one kind of evidence to look for. At the time the cosmic gas became transparent, it no longer hindered the passage of radiation, and the radiation that existed at that time was released to travel freely through the universe. Some of that radiation should still be present. But where could we look for it? We know that the radiation was liberated a long time ago. In astronomy, if we wish to look at things that happened a long time ago, we must look for them very far away. The cosmic black-body radiation, therefore, should seem to come from the very periphery of the visible universe. (In fact, it would **define** the limits of the visible universe. Trying to look any farther would mean looking back into time before the universe became transparent, and nothing could be seen!)

Since the "past" of the universe is all around us, in every direction, the cosmic background radiation should seem to come equally from all directions. However, due to the expansion of the universe, the wavelength of the radiation should have been stretched out — that is, red-shifted — by a factor of about 1000. Thus the radiation, originally that of a black body at about 3000 K, would now appear as long-wavelength radio waves, such as would be emitted by a black body at a much lower temperature: about 3 K. The technology of that time, however, could not detect any such radiation.

The breakthrough came quite by accident. Two scientists at the Bell Telephone Laboratories in New Jersey, Penzias and Wilson, were testing a satellite antenna. They found they were getting a kind of static — a weak signal at 7.35 cm wavelength that seemed to come from everywhere at once. In fact, as later testing with a specially built radiation detector confirmed, they were picking up the radiation from the early gaseous stage of the universe. The radiation is just such as might be expected to emanate from a black body with a temperature of 2.7 K.

It was this evidence that tipped the balance against the steady-state model and for the evolutionary big bang model. Since the radiation shows that the universe once had a very hot phase, it contradicts the perfect cosmological principle on which the steady-state model is based. The evidence is strongly in favor of an evolving universe. The main thrust of research today is the search for evidence that will provide additional details about that evolutionary process, and to reduce our uncertainty about the step-by-step scenario for the birth of the universe.

The Future of the Universe

Once we have rejected the steady-state model, which says that the universe will always be as we see it now, we are left with the question of the future of the universe. If it is evolving, what will it evolve into? The big bang model tells us how the universe was born, but it says nothing about how — or whether — it will die.

The answers to this question are still rather tentative. They depend on the selection of a geometric model of the universe. If the universe is open, rather than closed on itself, it will continue to expand forever. If it is closed, however, like the spherical surface described earlier, there are several further possibilities. It may continue to expand forever, though the rate of expansion may diminish with time due to the retarding force of gravity. If the density of matter in the universe is great enough, however, the force of gravity may be sufficient to stop the expansion eventually. The universe would then begin collapsing, somewhat like a ball that has been thrown into the air and, failing to attain escape velocity, falls back under the influence of gravity.

The result of such a collapse is not known with any certainty, but many people think that the universe would eventually return to its original state — a hot, dense fireball. Possibly, then, the process could begin all over again. This theory, known as the **pulsating** or **oscillating universe** model, can be considered intermediate between the big bang and steady state model. Although it includes change, the change would be an endlessly repetitive cycle. It is possible to show, however, that absolutely no information can survive through this hot, dense state, which thus serves to wipe out the past forever. If the universe has passed through many such big bangs, we have no way of ever knowing it.

Figure 10-5
For very remote galaxies, plotting distance against velocity may not give the straight line that Hubble's law implies. In theory, the nature of the deviation can tell us the shape of the universe and its future. In the figure at left, curve A represents the relation between velocity and distance that we would expect in a positively curved, closed, pulsating universe. Curves B and C represent open universes, with flat and negatively curved space, respectively. Curve D is what we would expect in a steady state universe. Since the velocities and distances of remote galaxies are hard to measure independently, the available data is not conclusive. The figure at right depicts the way the size of various universes would change with time. Curve A represents a closed, pulsating universe, curve B an open universe that will expand forever. Curve C is the hypothetical case of an empty universe. Notice that even if the universe is open, the force of gravitation slows its expansion.

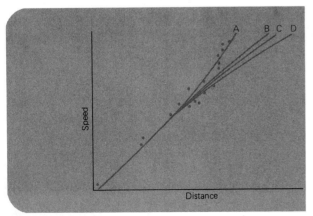

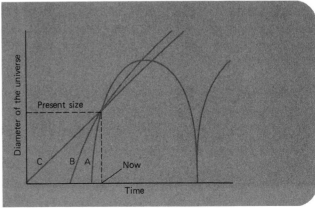

Where Has All The Mass Gone?

William Kingdon Clifford died young, but he left to the world the thought that gravitation is geometry. Half a century later, in 1915, Einstein confirmed his idea. One of the consequences of Einstein's theory of gravitation is that the shape of the universe depends on the amount of mass in it. Specifically, whether the universe is infinite, or whether it is so strongly curved that it closes in on itself, depends on its density. It has been calculated that a density of 10^{-29} g/cm^3 would be sufficient to close the universe. This doesn't seem like very much, when we recall that the density of water is about 1 g/cm^3 — a hundred-billion-billion-billion times greater. A density of 10^{-29} g/cm corresponds to a single atom of hydrogen in a volume of six cubic feet. Yet if we calculate the observable density of the universe by adding up the masses of the galaxies in a given volume, it turns out to be roughly a hundred times too low. This is an excellent reminder of how empty the universe really is, but it has distressed some astronomers, who believe that the universe is in fact closed. If they are right, though, where is the missing mass? No one knows for sure.

The problem is reproduced on a smaller scale, but more troublesomely, within clusters of galaxies. Most galaxies in clusters have velocities so high that the cluster would appear to be in danger of flying apart. The known mass of the galaxies in each cluster does not seem to be enough to hold them together. Why, then, have the clusters not all dispersed? No easy solution to the problem appears in sight. Perhaps we are underestimating the masses of the galaxies. Perhaps the whole cluster is filled with invisible but massive objects — black dwarfs, for example, or black holes. No one knows.

Still nearer home, we find the same problem of missing mass within our own galaxy. We have seen that halo stars oscillate up and down through the disc of the galaxy. We can determine their average velocity and the average height that they reach above the plane of the galaxy, before they fall back. From this information we can determine the mass of the disc itself. The calculation is simple, very much like trying to determine the mass of a planet by throwing up a baseball into the air. Let us say you can throw a baseball to a certain height on earth. On another planet, of the same size but much more massive, you would not be able to throw it nearly as high. The mass of the disc of the galaxy estimated in this way is called the **Oort Limit**. Now, if we add up all the *observed* matter in our galaxy it adds up to about half the Oort Limit. It is not definitely known where the missing mass is. Some say it is in the form of small, hard-to-observe stars. Others say that it might be in the equally elusive molecular hydrogen. But again, no one is certain.

It is not clear, though, whether there is enough matter to close the universe; nor is it clear, if so, whether there is enough to reverse its expansion. The problems of finding out are formidable. It is very hard to make observations that might shed any light on these issues. The ideal way to find out if the expansion of the universe is slowing down, and how rapidly, would be to determine whether Hubble's constant, which gives the relationship between distance and velocity, has varied with time. But to do this involves measuring the velocities (that is, the red shifts) and distances of galaxies in the past—in other words, galaxies that are very far away. This is extremely difficult to do, since for such remote objects there is no clear way of measuring distances. In fact, we usually **assume** Hubble's constant to be the same everywhere, and use it to calculate distances. In other words, we do not have any independent check on the Hubble relationship for distant objects.

The majority of astronomers today believe that the universe may be open. If so, the process of expansion will continue to its inevitable conclusion: the heat death of the universe. The radiation will continue to lose energy through the red shift. All the matter in the universe will be locked in dead stars or black holes. No heat transfer will be able to take place. In this state of deathly equilibrium the universe will remain forever.

If the density of matter in the universe is great enough, gravitational forces may halt the expansion of the universe and cause it to collapse.

SUMMARY

Since the universe is all there is, there can be only one universe. We cannot test our theories about it by comparing it to others. Nor can we observe its evolution over a period of time, although we see the more distant regions at an earlier period in their history than those in our vicinity. We cannot meaningfully ask what preceded the universe, or where it came from. Moreover, to talk about the universe at all, we must assume that the laws of nature are the same everywhere, and always have been—which is almost impossible to prove. Despite these limitations, however, we can still formulate models of the universe and its probable history, trying to make them fit our observations as well as possible.

The question of the size of the universe is closely connected to that of its shape. We know that space itself need not be flat, like that described by the axioms of Euclidean geometry. The universe may be open and infinite—this would be the case if the geometry of space were flat or hyperbolic. But space may be positively curved, and if the curvature is great enough, the universe may be closed on itself, like the surface of a sphere. Such a universe would be finite in size, though unbounded. Einstein's theory of gravitation implies that the shape of the universe depends on the density of matter in it. A great enough density should produce a closed, spherical universe. But we cannot be certain whether the density of the universe is really sufficient for this, or even whether Einstein's model is correct.

When we talk about the universe, we generally assume that it is essentially uniform throughout. This implies the cosmological principle—that an observer in any part of the universe will see essentially the same sort of universe. The basic empirical fact about our universe is that all the galaxies seem to be receding from us. The more distant the galaxy, the greater its Doppler red shift, and thus its velocity of recession. This relationship is known as Hubble's law. It is explained by assuming that the universe is expanding, as the surface of a balloon expands when the balloon is inflated. It is not necessary to think of ourselves as at the center of the expansion, for all points in space are receding from all other points, and an observer anywhere would see the same thing. Hubble's law permits us to calculate the age of the universe by tracing the flight of the galaxies backward in time. This method gives an estimate of 11 to 16 billion years.

Two chief models of the universe have been proposed. The steady-state model is based on the perfect cosmological principle—that the universe is not only the same everywhere, but always has been and will be much as it is now. This implies that as the galaxies fly apart, matter is continuously created in interstellar space, from which new stars and galaxies form. This is no more mysterious than assuming the creation of all the matter in the universe at one time. But current observational evidence does not support the steady-state theory.

Today, evolutionary models of the universe, such as the big-bang model, are favored. This model proposes that the universe began as an enormously hot, dense ball of photons and subatomic particles. As it expanded, it began to cool. In time, hydrogen atoms, and some helium as well, formed, and eventually began to condense into galaxies and stars. The most convincing evidence for this model is the discovery of a uniform cosmic background radiation in the radio spectrum, such as might be emitted from a black body at a temperature of 3 K. This is thought to be the radiation left over from the primeval fireball, lengthened in wavelength by a factor of 1000 by the expansion of the universe.

Currently, speculation centers on the issue of whether the universe will expand forever, or whether the braking force of gravitation will eventually halt the expansion and cause the galaxies to fall back together. If this happens, the result may be another big bang, with the entire cycle starting over again. Such a cycle is postulated by the oscillating or pulsating universe model. If the universe continues to expand forever, though, it will slowly grow colder and emptier, until it reaches a state total thermal equilibrium, or heat death.

EXERCISES

1. How does cosmology differ from other branches of astronomy? Why do we resort to so many principles and assumptions in this field?
2. Explain how Hubble's relationship, $v = Hr$, is established.
3. Does Hubble's law imply that our galaxy is in a special position in the universe, from which all other galaxies are fleeing?
4. Show that the general expansion of the universe follows from Hubble's law.
5. The red shift of a galaxy is .017. What is its velocity of recession? What is its distance, in millions of light-years? What is the distance of a galaxy whose red shift is .34?
6. What is the difference between the cosmological principle and the perfect cosmological principle?
7. What are the attractive features of the steady-state theory?
8. Explain why the big-bang theory is now generally preferred to the steady-state theory.
9. What is the origin of the cosmic black body radiation? Why does it have the character of radiation from an object at 3 K?

Brightest known QSO, 3C273.

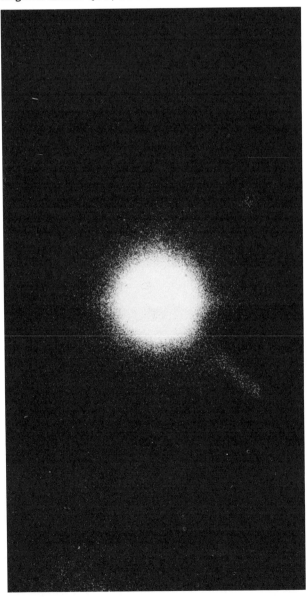

"On that night, I went home in disbelief. I told my wife, 'Something absolutely incredible has happened. . . .'" The words are those of Martin Schmidt, the Dutch-born astronomer now working in America. The date was February 7, 1963. The event was Schmidt's discovery that the spectrum of a peculiar object, known in the catalogs as 3C 273, had been shifted toward the red end of the spectrum by an astonishingly large amount. That discovery marked the beginning of a major astronomical puzzle that still intrigues investigators today. The large red shift of 3C 273 is the starting point for a simple chain of reasoning that leads us to a host of perplexing questions. A large red shift means a high velocity of recession. Evidently 3C 273 is speeding away from us at some 47,000 km/sec. Hubble's law tells us that an object with such a velocity of recession must be very distant—more than 2.5 billion light-years. But if it is that distant, how can it be so bright? Not by being large, certainly; the fluctuations in its light output prove that it cannot be more than a few light-months across. What is the source of its energy? We do not know. Does its red shift really indicate a high velocity of recession, or could it have some other cause? And even if it is receding so rapidly, does this mean that it is necessarily as distant as Hubble's law would suggest? Both of these questions have been asked, but not answered to everyone's satisfaction. Perhaps the best summary of the situation is to be found in the name given to the class of objects to which 3C 273 belongs: quasi-stellar objects. That is, star-like objects—objects that look a little like stars, but are not stars. In other words, we do not know what QSO's are.

The QSO story really begins three years before Schmidt's startling discovery. In 1960, radio astronomers located the optical counterpart of a strong radio source known as 3C 48. (the designation 3C refers to the third catalog of radio objects compiled at Cambridge University, England.) The small size of the object suggested that it was a star. Puzzled by the broad emission lines in its optical spectrum, however, observers did not know what to make of it. Then, in 1962, the exact position of another strong radio source, 3C 273, was pinpointed by means of lunar occultation, described in Chapter 9. At the time, there were not many radio telescopes in operation, and few radio sources had been identified in this way; astronomers were particularly eager to take advantage of the moon's passage in front of 3C 273. The work was done at Parkes, Australia, with the 210-foot radio telescope, and was considered so important that several extraordinary measures were taken to assure its success. Part of the mounting of the Parkes instrument—several tons of steel—were sawed away so that it could be pointed toward the position of 3C 273 in the sky. Tractors in the vicinity were fitted with special devices in their ignition systems, so that the electrical discharges would not create radio interference during the observations. When the data was finally obtained, it was considered so precious that duplicate copies were prepared and flown out of the country in two planes—in case one crashed!

The spectra of QSO's are also quite peculiar. Typically, they consist of three components: a continuous spectrum on which are superimposed both emission and absorption lines. Most of the continuum radiation is not that of a black body, but seems characteristic instead of synchrotron radiation. In addition, some QSO's, such as 3C 273, also radiate huge amounts of what appears to be thermal radiation at infrared wavelengths. Some theorists have suggested that the infrared emissions are produced by clouds of dust particles in or near the QSO, which are heated by its radiation.

The emission lines in QSO spectra enable investigators to determine their red shifts. These bright lines also tell us that somewhere within the object there are masses of low-density gas that are being excited by highly energetic radiation. The absorption lines in these spectra often show red shifts slightly lower than those of the emission lines, suggesting that clouds of cooler gas are being ejected from the hot center of the QSO.

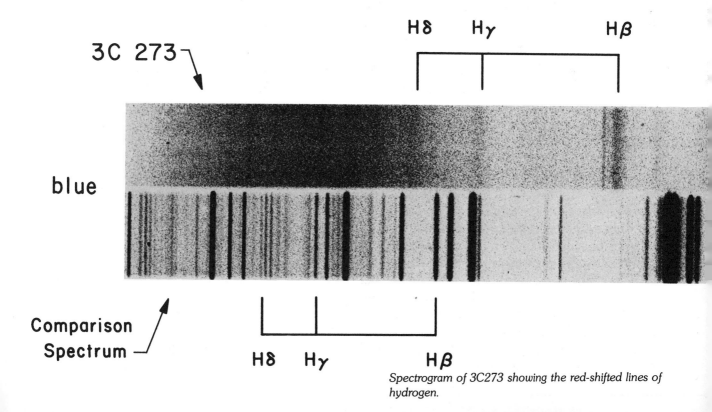

Spectrogram of 3C273 showing the red-shifted lines of hydrogen.

This spectral information has enabled investigators to construct a theoretical model of the typical QSO. At its center, there must be some energy source about which virtually nothing is known, save that it ejects huge quantities of high-energy electrons. The electrons are trapped in the magnetic fields of the gas clouds surrounding the source, and produce synchrotron radiation as they are accelerated in spiral paths around the lines of magnetic force. Around the center of the QSO are regions of hot, ionized gas which produce the spectral emission lines. In some photographs of QSO's, there are hints of ray-like patterns, suggesting filaments of an expanding cloud of hot gas, while in the case of 3C 273, a jet of apparently ejected material is clearly visible.

Surrounding the hot central regions are shells of somewhat cooler gas, expanding outward from the center at speeds approaching 150,000 km/sec. It is these clouds that produce the absorption lines in QSO spectra. Some spectra exhibit as many as six sets of absorption lines, each with a different red shift, indicating the presence of six separate clouds moving at different velocities. Whether the outer shells of gas are integral parts of the central energy source is not known at this time. The distance of these clouds from the core is also unknown.

What is the nature of the QSO power source? The atomic fusion process responsible for the energy of the sun and other stars could not generate such prodigious quantities of energy from a small mass, and a large enough mass would not be stable—it would be dispersed by the pressure of its own radiation. Several alternate sources have therefore been proposed, but none of the theories has so far proven entirely satisfactory. One hypothesis asserts that a QSO is a galaxy with a tightly packed nucleus containing billions of stars. The density of stars in this super-nucleus would be extraordinarily high—perhaps a hundred thousand times that found in the disc of our galaxy. Under these crowded conditions, collisions between stars are quite frequent, occurring with an average frequency of at least one per year. Such collisions would greatly accelerate the evolution of the stars in-

volved. Many supernova explosions might result, and the energy released by these explosions might in turn help to trigger others. Since supernovas not only release vast amounts of radiation, but also produce large numbers of high-energy electrons, this theory is very attractive.

Location of the optical counterpart of 3C 273 made it possible for astronomers to record its spectrum, and led to Schmidt's discovery of its huge red shift. Since then, many other QSO's have been found—all with large red shifts. The most distant known to date has a red shift of more than 3.5, indicating a velocity of recession greater than 91 percent the speed of light, and a distance of nearly 10 billion light-years. It may therefore be the most distant object of any sort we have yet observed. If so, it provides us with our farthest glimpse into the past of the universe.

Besides their star-like appearance and their large red shifts, QSO's have several other distinguishing characteristics. For one thing, they are extremely blue, indicating that much of their radiation lies at the shorter visible wavelengths. Indeed, it is by their blueness that QSO's generally first attract the attention of astronomers, after which their identity can be confirmed by measurement of their red shifts. Most QSO's are strong radio sources, but not all—that is why the term QSO has replaced the earlier designation Quasar. Both the radio and visible output of QSO's vary irregularly. Since its discovery, the optical brightness of 3C 273 has waxed and waned in no apparent pattern. Between 1963 and 1966, for example, it grew steadily brighter. By 1967, its brightness had diminished by about 50 percent, but then began to rise again. Since 1967, it has continued to fluctuate, gaining and losing brightness each year. About 20 percent of all known QSO's fluctuate even more sharply. One of them, 3C 446, has been observed to vary in brightness by as much as 200 percent in a single day.

The most remarkable characteristic of QSO's, however, is their energy output. Though even the brightest QSO's are about a hundred times too faint to be seen with the naked eye, they are nevertheless astonishingly brilliant when their enormous distances are taken into account. An amateur astronomer, using a modest 4-inch telescope, can pick out 3C 273, which is about 2 billion light-years away. By contrast, one would need a 60-inch telescope to see an average elliptical galaxy at the same distance. The luminosity of QSO's becomes even more amazing when we consider their small size. Many measure no more than a few light-months across, yet generate more energy than several hundred large galaxies, each containing billions of stars. Indeed, about once a year, 3C 130 produces some 10^{52} ergs in a single short burst—the equivalent of all the energy that the sun would generate if it shone for 100 billion years.

There may already be some observational support for the supernova hypothesis. Astronomers have recently become quite interested in a strange class of objects called BL Lacertids. The prototype of this class, BL Lacertae, was long thought to be a star. Closer observation, however, has shown this not to be the case. The first red shift of a BL Lacertid has only recently been obtained. It indicates that the object is over a billion light-years distant, and has a luminosity comparable to that of a bright elliptical galaxy.

If Seyfert galaxies are intermediate between QSO's and normal galaxies, BL Lacertids seem to be intermediate between QSO's and Seyfert galaxies. They have extraordinarily bright, compact nuclei that emit intense, polarized, highly variable radiation at visible, radio, and infrared wavelengths. All of these are hallmarks of a QSO. But the nuclei of BL Lacertids are surrounded by faint, fuzzy halos whose spectra contain absorption lines. This suggests that BL Lacertids may be extremely compact galaxies with high central stellar densities. If this is so, then it is reasonable to think that QSO's may represent merely a more extreme example of the same phenomenon. Despite this supportive evidence, however, we still have no explanation of how or why so many stars came to be concentrated in such a small volume in the first place, or why we do not see such objects nearer to us in space and time.

Many competing theories have been advanced. One holds that a QSO is a giant, spinning object that stores an enormous amount of energy in its rotation—a kind of super-pulsar, in effect. Another is that QSO energy is the result of the mutual annihilation of matter and antimatter. Each variety of elementary subatomic particle has, in theory at least, an antiparticle. These anti-particles have very short lifetimes in our universe. Whenever a particle and an antiparticle meet, both are entirely converted into energy. This process is thus 100 percent efficient, compared with the .7 percent efficiency obtained when matter is converted into energy by hydrogen fusion. If antimatter—matter composed of antiparticles—were somehow being created in our universe, or trickling in from an adjoining antimatter universe, it would provide an almost limitless source of energy. The trouble with this imaginative theory is that we have no evidence for the existence of antimatter—let alone an entire universe of it; nor is it clear where, in space or time, such a universe might be, or how it might leak into ours.

Some astronomers feel there is no need to construct elaborate theories of QSO energy sources. The answer to the QSO puzzle is right in front of our eyes: the red shift measurements are all wrong, and the QSO's are actually much closer to us than we think. If the QSO's are really local phenomena, their supposed astonishing luminosity vanishes immediately. In effect, this theory shifts the nature of the problem: it is no longer the QSO energy problem, but the QSO red-shift problem.

Some astronomers have proposed that the QSO red shifts are due not to the Doppler effect but to gravity. As we saw in Chapter 7, light leaving a strong gravitational field will lose energy, and its wavelengths will be lengthened—in other words, it will be red-shifted. This effect has been observed in the radiation leaving massive stars such as white dwarfs. It would take an object of staggeringly high density, however, to produce red shifts on the order of those observed in most QSO's. The trouble is that no one has yet proposed a plausible QSO model that would account for such densities. In

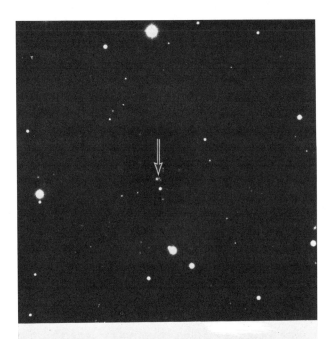

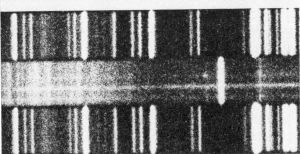

The QSO 3C295 in Bootes. This QSO has a larger red-shift than 3C273. The spectrum is characterized by emission lines.

by some cataclysmic event. We know that the galactic nucleus is indeed the scene of violent activity. But so are the nuclei of other galaxies. It is hard to see why such objects should not also be ejected from other nearby galaxies, in which case some of them should be headed toward us. However, no QSO's with blue spectral shifts have been found, indicating that none of them are moving in our direction. Moreover, a terrific amount of energy would be required to fling numerous QSO's from the galactic center. No one has yet explained how such energy could be generated.

A third non-cosmological interpretation of the QSO red shifts is a kind of catch-all for various theories. Its premise is that the huge red shifts that we detect are not governed by any known physical laws. This gives one the license to invent new physical laws to suit the observed phenomena. There is always a chance that this may yet prove to be necessary. Naturally, though, scientists are reluctant to consider this possibility until all others have been ruled out.

Perhaps the most serious case against the reliability of the red shifts as distance indicators has been presented by Halton Arp of the Hale Observatories in California. Using the 200-inch telescope, he photographed a number of galaxies and QSO's that seem to be lying very close to one another. Some of the galaxies and their QSO companions even appear to be connected by umbilical-like filaments of glowing gas. The best known and best studied example of such a relationship is the pair comprising the spiral galaxy NGC 4319 and the QSO Markarian 205. Photos of these objects show what appears to be a bridge of material linking the two. But the red shifts of the galaxy and the QSO differ widely; that of Markarian 205 has been measured at .07, compared with a value of .006 from NGC 4319. If both objects are really about equally distant from us, the QSO red shift cannot be considered an accurate distance indicator. This, say Arp and his colleagues, is strong evidence that the QSO's are local rather than cosmological phenomena.

fact, we know that the emission lines found in QSO spectra could come only from tenuous gas. Furthermore, if QSO's were really relatively near-by objects, and possessed gravitational fields of such magnitude, their effects should be detectable within our own galaxy.

Another theory holds that the QSO red shifts are really Doppler shifts, but are not cosmological. The QSO's, in other words, are really moving away from us at enormous velocities, but these velocities are not the result of the expansion of the universe, and so do not indicate that the QSO's are very distant. Rather, they are nearby objects that have been hurled from the nucleus of our own galaxy

A number of investigators, however, question

Arp's conclusions. Some, for example, hold that there is no bridge between the two objects; and indeed, such a structure does not show up on the photographs taken by other astronomers. It has been claimed that the bridge is in reality an artifact of the photographic process, created as the edges of the images of the two objects blurred together on the film. Some have proposed that the bridge is actually an arm of the spiral galaxy. Still others maintain that it is really only an optical illusion, created by the presence in the background of another galaxy, seeming to link the two objects.

The anti-cosmological faction, however, holds that these alternatives are smoke screens thrown up by those who do not wish to admit the evidence of their own eyes. Furthermore, say supporters of the local theory, there is other evidence which supports their hypothesis. Photographs of a group of apparently neighboring galaxies show that the red shifts of one of the group are discordant. The group is known as Stephan's Quintet, after M. Stephan, the French astronomer who discovered them. Four galaxies in the group have red shifts that are identical, or nearly so, whereas the fifth galaxy has a far lower red shift.

Again, there are two ways of viewing the situation. The anti-cosmological faction asks: "Since the five galaxies all seem to lie at the same distance from us, why should the red shift of one be different from the others? If the measurements are correct, then there must be something wrong with the red shift as a distance gauge." The cosmological faction, on the other hand, feels that Stephan's Quintet is not a quintet at all, but a quartet. The galaxy with the small red shift is not a member of the group, but a foreground object, much closer to us.

Supporters of the cosmological interpretation of the red shifts have sought additional evidence to bolster their argument. One logical way to do this would be to find a QSO in the company of a galaxy having the same red shift. This is exactly what has been done. Investigators have located three QSO's, each lying near a different cluster of galaxies. In all three cases, the QSO red shift proved to be similar to that of the neighboring galaxies.

This, say the cosmological theorists, is another indication that the QSO red shifts are accurate indicators of their distances, and that the QSO's are truly very distant objects.

Although each side has produced evidence in support of its claims, most astronomers now support the view that the QSO red shifts are indeed cosmological. This means that the QSO's are the most distant objects in the visible universe, and afford us our farthest look into its past. Needless to say, the QSO question is an important one for astronomy. If the QSO's are truly as distant as their red shifts imply, then the universe must have been very different billions of years ago than it is today. This is a strong blow against the steady-state theory. Nevertheless, the enormous energy production of these objects is still unexplained, as is their origin. We do not know why they should have existed in the early eons of the cosmos, but not today. If, on the other hand, the anti-cosmological theorists are right, and the QSO's are local phenomena, the energy problem is solved, but a much more serious one has taken its place. The downfall of the red shift as a distance yardstick would mean that a great deal of the most impressive theoretical work of recent years, especially in cosmology, would have to be abandoned. Clearly, most astronomers would be reluctant to take such a radical step. They feel that the missing pieces of the QSO puzzle are there in the natural world, waiting to be discovered. It is just a matter of time, and more study, before they are fitted into place.

INTERLUDE

THE SUN

From our earthly vantage point, the sun seems to be the brightest star in the universe. It is also our most important star, for without its light and warmth there would be no life on earth. To the astronomer, the sun is important for another reason as well. It is his main link to the billions of other stars in the universe, for it is the only star near enough for detailed and accurate analysis. Using the knowledge obtained from such study, we can test theories about stars in general. Much that we know about stars, we have learned from the sun.

But as large and important as it appears to us, our brilliant sun, by astronomical standards, is a very ordinary star. When classified according to size and temperature, it is a yellow Main Sequence star of spectral type G2. It stands roughly in the middle of the Main Sequence — between the dimmest and the most luminous, the smallest and the largest, the coolest and the hottest. Because of its proximity, it appears to be ten billion times as bright as the brightest star. But if it could be plucked from its place in our solar system and placed beside one of the giants or supergiants, we would immediately see what a modest star it is. Beside Rigel, which is roughly 15,000 times as luminous, our brilliant sun would appear feeble indeed. If placed next to the mammoth Antares, which is large enough to hold 36 million suns, our star would be barely noticeable.

What makes the sun a star, rather than a planet? Like other stars, the sun is a sphere of gas held together by gravity, prevented from collapse by gas pressure, luminous because of its high temperature, its store of energy constantly replenished by nuclear fusion taking place in its interior. The study of the sun as a star begins with an examination of its internal structure and physical conditions. Unfortunately it is impossible for us to investigate the sun's interior directly with our present technology, and it is hard to imagine a technology which would ever make such investigation possible. The temperature of the solar core, for example, is 16 million degrees K, millions of degrees hotter than the temperature at which all known substances vaporize. Indeed, we cannot even see into the heart of the sun because the gases of its outer envelope are too opaque.

However, there are many things we can say about the sun. This knowledge is obtained both from observation, and from the application of physical principles. For example, from observation we know its diameter, luminosity, and gravitational field, as well as the chemical composition and temperature of the surface layers. And theoretical physics gives us a fairly good idea of the processes responsible for its energy production and output, as well as its internal temperature, pressure, and density. All of these data, and more, can be related to one another in the form of mathematical equations to predict the physical conditions of any given point of the solar interior. By solving these equations with a computer, an astrophysicist is able to construct an overall model of the sun as a star. Before the computer, calculating a solar model could take up to a year. With modern high-speed computers, however, one can construct such a model in a matter of a few minutes. Refinements in the physics of stellar interiors are always being made, and improved models are constantly being produced.

The temperature at the sun's center is about 15 million K.

A Solar Model

The model shown in Figure 11–1, based on recent data, assumes that the original composition of the sun was about 74 percent hydrogen and 25 percent helium by weight. How are the variables in this model related? Let us first consider pressure, which has been derived from theoretical calculations. As we saw in Chapter 5, the sun is held together by gravitational forces and prevented from collapsing by the outward pressure of its internal gases.

In theoretical stellar models, scientists conventionally divide the interior into a series of imaginary concentric shells. In our model, the shells are labelled "fraction of mass"; this figure denotes the percentage of the sun's mass that is included within each shell. Each shell must have a pressure that is great enough to withstand the weight of all the shells above it. Thus the pressure increases with distance toward the center of the sun.

Figure 11-1

A solar model. We can see from this model that the temperature rises sharply from a few thousand K at the photosphere to a million K at a relatively shallow depth of the interior. Thereafter, it rises more slowly and smoothly to about 15 million K at the center. The density, on the other hand, rises slowly at first, and then very rapidly as we approach the core—90 percent of the sun's mass is concentrated in about 10 percent of its volume at the center. The energy production rises even more steeply, with half the energy coming from a region having only about 1/1000 the sun's volume, and virtually none produced outside the central 1½ percent of it volume. (In interpreting the radii on the diagram, recall that the volume of a sphere is proportional to the cube of its radius.)

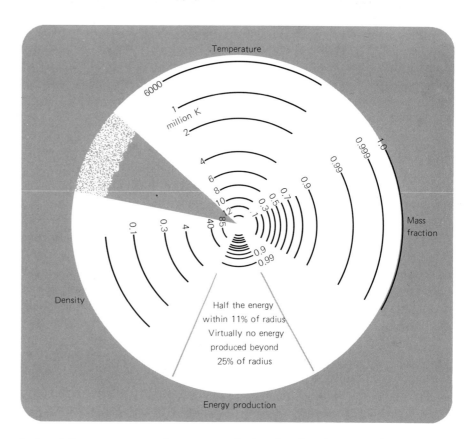

The density of matter at the center of the sun is 7 times greater than that of the densest metal on earth.

Indeed, the pressure at the core is calculated to be about one billion times the air pressure at the surface of the earth, which is about 14.7 lbs/in². It is the combined pressure of the shells which counterbalances the force of gravity and prevents the sun from collapsing upon itself.

This tremendous pressure suggests that the temperature at the core is about 16 million degrees. How hot is this? A small quantity of material from the core—an amount no bigger than the dot over this *i*—would radiate enough heat in every direction to roast to death anyone standing within 100 miles of it. Temperature gradually drops with distance from the center; at the surface, it is a relatively cool 6000 K.

Another factor related to pressure is density. Like pressure and temperature, density increases with distance toward the core. The model tells us that the density at the center of the sun is about 158 g/cm³—or about 7 times the density of osmium, the densest substance on earth. This is possible because the atoms in this region, having been stripped of their electrons, can be packed tightly together.

Another variable is the fraction of hydrogen that is present in each region of the solar interior. The amount of hydrogen decreases as we move toward the center. This is so because hydrogen is consumed in that region. The regions toward the surface, by contrast, contain 74 percent hydrogen—the amount the sun had on the day it was born. As we saw in Chapter 6, the core is a nuclear furnace in which hydrogen nuclei are continuously being fused into helium nuclei. Roughly 655 million tons of hydrogen are converted into 650 million tons of helium every second. About five million tons a second are converted into radiation energy, chiefly in the form of X-rays.

The age of the sun is estimated to be about 5 billion years. As we saw in Chapter 6, this figure has been derived from the age of the earth. Our model has been adjusted to fit this age. The model tells us that the sun has thus far exhausted about 50 percent of the hydrogen available for thermonuclear reactions. It should therefore live at least another five billion years. At that time, all the core hydrogen will have been consumed, and the core will contract. The outer envelope will expand and the sun will enter the red giant stage. Then the remaining helium nuclei in the core will fuse to form carbon nuclei. Needless to say, the sun will become very luminous and the earth will be scorched when these events occur. After orbiting for a while inside the sun, our planet will eventually evaporate.

In 4.6 billion years, the sun had consumed only half its original store of hydrogen fuel.

Thermonuclear reactions at the core generate an enormous amount of energy: 4×10^{33} ergs per second. This energy travels toward the surface initially in the form of X-rays. From the core to about three-quarters of the way to the surface, the X-rays travel upward by means of radiation transfer. This region may be called the **zone of radiation.** At the uppermost limits of this zone—some 150,000 km beneath the surface—conditions become right for the convective transfer of heat energy. Hot gas rises and gives its heat to the cooler upper regions. After it cools, the gas descends again to the solar interior, where it is heated and begins the cycle over again. This region is known as the **zone of convection.** Here, the temperature drops to about 150,000 K and the density decreases to about 0.6 g/cm³—less than the density of water.

Because the X-rays produced in the solar furnace must zigzag their way through thousands of kilometers of gas, being absorbed and reradiated every few centimeters, they require about one million years to reach the surface. In the process, they are repeatedly reradiated at ever longer wavelengths. Hence by the time they reach the surface, most of the photons are in the visible light range. It is these that we see emanating from the solar disk, and which make up what we call sunlight.

X-rays from the solar interior are absorbed and re-emitted at ever longer wavelengths on their million year journey to the surface; eventually they leave the photosphere as photons of visible light

Largely because of the earth's distance from the sun, barely one two-billionth of the energy emitted by the sun is intercepted by the earth's surface. And about 35 percent of this fraction is reflected back into space by the earth's protective atmosphere. Yet by human standards, even this amount is enormous. For example, the amount of energy reaching the earth in slightly less than twenty minutes is sufficient to meet our civilization's power requirements for an entire year.

THE QUIET SUN

Some of the data in our model of the solar interior comes from direct observation of the sun's outer envelope. A great deal of activity can be observed in the sun's outer regions. Some of these phenomena are relatively stable, taking place over vast areas for long periods. They are aspects of the **quiet sun.** At certain times, however, the sun becomes sporadically active, turbulent with phenomena that are relatively short-lived. These phenomena are often referred to as aspects of the **active sun.**

The phenomena of the quiet and the active sun both belong to the outer layers of the sun — its atmosphere. The atmosphere consists of the **photosphere, chromosphere,** and the **corona.** These three layers are not sharply distinct; they are convenient divisions which in fact merge with one another. Some astronomers prefer to think of the photosphere as the surface of the sun, and the other two layers as its atmosphere. But since the sun is a gas, it has no real surface. Therefore, when we refer to the surface of the sun, we are only speaking metaphorically about the photosphere. It is from this region that most of the photons leave on the last leg of their journey out of the sun.

The Photosphere

The photosphere is the visible layer of the sun, the bright disk we actually see. The bottom reaches of this 500 km-thick layer are so opaque that they prevent most of the photons from escaping directly from the sun's interior. The opacity of this region is due to the presence of a small number of negative hydrogen ions. A negative hydrogen ion is a hydrogen atom that has temporarily picked up an extra electron. The extra electron is loosely bound to the atom, and so can easily be detached by the absorption of a photon. Any photon in the visible part of the spectrum has enough energy to detach the electron. Thus the H^- ions serve as radiation sponges, absorbing photons throughout the visible range, and rendering the photospheric gases extremely opaque.

The presence of a few negative hydrogen ions, which absorb a wide range of wavelengths, makes the photosphere opaque.

The Photosphere at a Glance

Thickness	500 km
Temperature	6800 K (bottom) to 4500 K (top)
Density	2×10^{-7} g/cm³ to 3×10^{-8} g/cm³
Chief permanent features	granulation, limb darkening; source of continuous spectrum.

The Neutrino Hunt

How correct is our model of the sun, and how complete is our knowledge of its energy processes? Most astronomers would probably agree that our current theories account for the observed structure and behavior of the sun quite well. But an experiment is presently being conducted in the depths of an abandoned gold mine that casts some doubt on our ideas about solar energy production. The test, known as the solar neutrino experiment, is designed to catch and count some of the neutrinos emitted from the sun's core.

Neutrinos are subatomic particles created in the proton-proton reaction that is thought to account for most of the sun's energy output. If our current theories are correct, trillions upon trillions of neutrinos should be produced each second within the sun. Neutrinos have neither mass nor electric charge, and their ability to interact with matter is very weak. Indeed, they pass through matter almost as if it weren't there. A neutrino, which travels at the speed of light, reaches the solar surface within 2 seconds of its creation, though a photon, absorbed and reradiated trillions of times during its passage, requires over a million years for the same journey. After leaving the sun, neutrinos that happen to be heading our way reach the earth in about 500 seconds, pass through it (and anything else that might lie in their paths) and continue on unimpeded into interstellar space. Indeed, even as you are reading these words, trillions of neutrinos are passing through your body.

The value of neutrinos, therefore, is that they are messengers from the sun's core. They tell us what was happening within the sun just minutes ago, while photons can only bring us news that is already a million years old. The designers of the neutrino experiment have attempted to record the passage of some of these particles, in hopes of confirming our solar model. This is naturally not easy, since neutrinos interact with matter so infrequently. It has been calculated that a neutrino could pass through many parsecs of solid steel with only a very small chance of being stopped by an atom of the metal. Nevertheless, a relatively uncomplicated and apparently foolproof experimental arrangement was devised to count solar neutrinos—not every one, but a fraction that could be precisely calculated. The principle was simple—to place a large quantity of chlorine atoms together in one place. Every so often, one of the immense number of neutrinos passing through the chlorine should react with a chlorine atom to create an atom of radioactive argon. The number of argon atoms produced could then be tallied with an atomic counter.

This is exactly what has been done. The neutrino trap is a 100,000-gallon tank of the cleaning fluid perchlorethylene, each molecule of which contains four chlorine atoms. The tank is buried in a South Dakota gold mine, a mile below the earth's surface. This ensures that the neutrino detector will be well shielded from cosmic rays, which could have the same effect as neutrinos and thus confuse the experimental results. The tests have been in progress for a number of years. So far, the results have been surprising and, to many scientists, unsettling.

Theoretical calculations led the experimenters to expect that at least one atom of argon should be produced per day, on the average. But the number actually found has been consistently and dramatically lower. In order to detect any possible flaws in the experiment, the design of the apparatus, the equipment itself, the physical principles involved, and the theoretical calculations, have all been exhaustively reexamined. No errors have been found, and the neutrinos have still not appeared in the predicted numbers.

Does this mean that our solar model is incorrect? No one is sure yet. By and large, theoreticians have explained the discrepancy by claiming that there must be something wrong with the experiment, while experimental physicists have claimed that something must be wrong with our theories. Astronomers have generally argued that our understanding of neutrinos is at fault—they must possess some property unanticipated by physicists. Perhaps, for example, they are unstable, and decay into other particles. Physicists have answered that it is our imperfect knowledge of the sun that is to blame. One proposed explanation suggests that the internal structure of the sun itself undergoes cyclical changes every few million years or so. Expansion of the core produces a decrease in temperature and density, with a resulting drop in the proton-proton reaction rate that supplies the sun's energy. The production of solar neutrinos would decline proportionally.

If these conditions exist in the sun right now, they would account for the puzzling results of the neutrino experiment. Such a cyclical decrease in solar energy might also explain the appearance of ice ages, which have gripped great parts of the earth several times in the past. Whatever the answer, though, the neutrino problem is a serious one for astronomers and other physical scientists. It has led to a closer examination of present theories that attempt to account for the properties and evolution not only of the sun, but of stars in general.

When viewed with the naked eye, the appearance of the photosphere can be deceptive. For example, it seems to have a smooth, solid surface. However, spectroscopic analysis reveals that its surface is gaseous. Furthermore, it is mottled, possessing structures, known as **granules,** that look like grains of rice. Ranging in diameter from 300 to 1000 km, each granule is actually the top of a cell of extremely hot gas that is rising from the interior at the rate of about 2 km/sec. When it reaches the surface, the gas spreads, radiates its heat out into space, then descends to the solar depths. A granule typically lasts about five to ten minutes, before cooling and reentering the interior.

The granules comprise even larger gas cells known, appropriately enough, as supergranules, which are larger than the entire earth. In these structures, which reach diameters of 30,000 km or more, gases move from center to edge at the rate of about 0.5 km/sec. In addition, gas also rises at the center of the supergranule and descends at the edge in five-minute cycles. Supergranules typically last about 24 hours before descending to the interior.

SPECTRUM AND COMPOSITION OF THE SUN Ruminating on the state of astronomical knowledge in the 1830s, the French philosopher Auguste Comte (1798–1857) was moved to lament about the heavenly bodies: "We understand the possibilities of determining their shapes, their distances, their sizes, and motions, whereas never, by any means, will we be able to study their chemical composition." Had Comte lived two years longer, until 1859, he would have seen himself contradicted. In that year, the German physicist Gustav Kirchhoff discovered a way to determine the chemical composition of the sun by analyzing its spectrum.

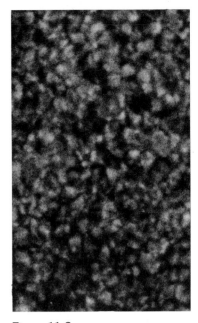

Figure 11-2
Solar granulation. The bright regions are hot gases rising from the interior; the dark regions, slightly cooler material falling back.

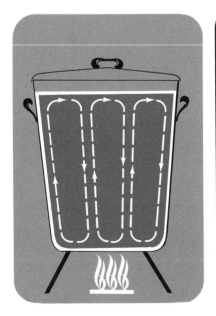

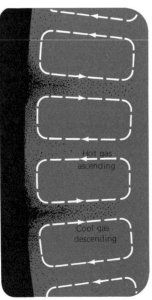

Hot gas ascending

Cool gas descending

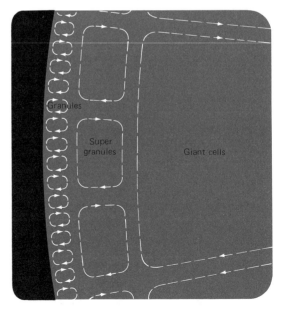

Granules

Super granules

Giant cells

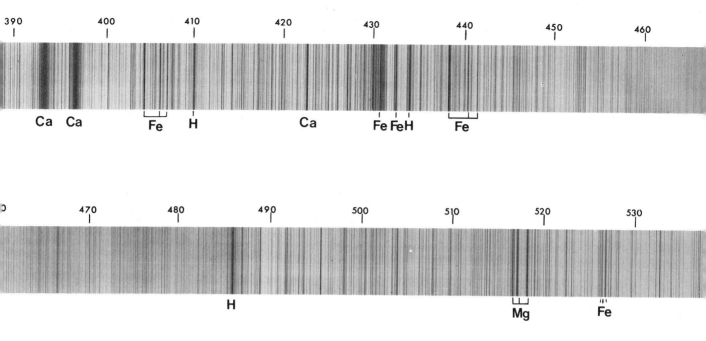

Figure 11-4
A portion of the solar spectrum from 390 nm (just beyond the violet boundary of the visible range) to 540 nm (in the yellow region). Thousands of absorption lines—each representing a particular atomic transition in a particular element—have been mapped. Some of the more prominent ones have been identified here. Among them are three of the Balmer lines of hydrogen (H) at 410, 434, and 486 nm. Fe stands for iron, Ca for calcium, and Mg for magnesium. The two calcium lines are the strongest in the entire spectrum. This does not indicate, however, that calcium is the most abundant element in the sun (see Chapter 4). These absorption lines are produced in the chromosphere and upper photosphere.

Figure 11-3
The process of convection, thought to account for the granulation and super-granulation visible on the solar surface.

Since that time, astronomers have extensively studied the radiation emitted by the photosphere and other atmospheric regions of the sun. They have found that the sun radiates not only in the visible part of the spectrum, but also in the infrared, ultraviolet, radio, and X-ray regions. The visible spectrum consists of a continuum of colors from violet to red. This visible continuum is crossed by thousands of dark lines, known as Fraunhofer lines, after the German physicist, Joseph Fraunhofer who discovered them. The Fraunhofer lines indicate that light of certain wavelengths is missing; the missing wavelengths are characteristic of the selective absorption by various elements. The presence of the dark lines, then, indicates the presence of these elements in the sun. Fraunhofer counted nearly 800 lines and mapped some 574 of them. Since, then, modern spectroscopic techniques have revealed more than 30,000 lines in the solar spectrum.

The sun's photosphere produces a continuous emission spectrum. This is to be expected because, as we saw in Chapter 4, such a spectrum is characteristic of a hot, dense gas. But the upper reaches of the photosphere also give rise to the dark-line absorption spectrum. How can a single region, presumably with the same chemical properties, produce two totally different spectra? The photosphere emits a continuous spectrum characteristic of its temperature. Light from this region then passes through the more tenuous outer layers of the photosphere and the chromosphere. Because of the lower density, atoms in these layers absorb their characteristic wavelengths, creating the dark lines.

By matching the Fraunhofer lines with those produced by various elements in terrestrial laboratories, scientists have been able to determine the chemical composition of the sun. Approximately 62 of the 92 elements known to occur on earth have been identified from Fraunhofer lines in the solar spectrum. It is believed that all 92 elements are probably present in the sun, but they have not yet been identified. Analysis of the Fraunhofer lines has shown that hydrogen is by far the most abundant of the solar elements: roughly 90 percent of all atoms in the photosphere are hydrogen. About ten percent of the photospheric atoms are helium. Other elements are present in much smaller amounts, the most abundant being oxygen, carbon, neon, nitrogen, and iron.

The Chromosphere

At its upper reaches, the photosphere merges into the extremely active chromosphere, or "sphere of color." The region takes its name from the bright red light of the hydrogen alpha emission line, which dominates the spectrum of the chromosphere.

Unlike the photosphere, the chromosphere has very few negative hydrogen ions to render it opaque. Because of this, and because of the intensely bright light emitted by the photosphere, the chromosphere is difficult to observe without special instruments. Indeed, up to the present century, this region could be observed only during a total solar eclipse when the moon occulted, or masked, the bright light of the photosphere. Thus, seventeenth-century scientists, viewing eclipses, wrote of a crimson streak that appeared at the edge of the moon just before and after totality. They were not certain whether the streak was produced by the sun or the moon.

Modern observations have revealed the nature of the mysterious crimson streak. During a total solar eclipse, the moon passes briefly between the earth and the sun. The moment before the sun's disc is totally eclipsed by the moon, a thin crescent of bright red light is visible at the sun's limb for a few seconds. This is the chromosphere. It disappears at totality when the moon occults it as well. The crescent then reappears,

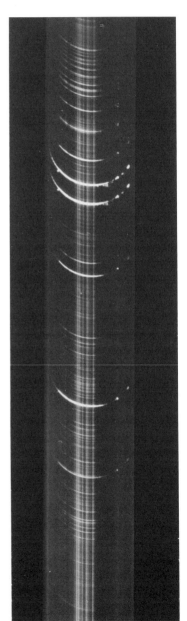

Figure 11-5
A flash spectrum. This is actually the emission spectrum of the chromosphere, visible only briefly during a solar eclipse, when the much brighter photosphere is covered by the moon. The flash spectrum is made without a slit; the arcs are images of the sun's narrow limb in the light of different wavelengths.

The Chromosphere at a Glance

Thickness	8000 km
Temperature	4500 K (bottom) to 100,000 K (top)
Density	10^{-8} g/cm^3 to 10^{-11} g/cm^3
Chief permanent features	spicules, flash spectrum; creation of absorption lines.

again for a few seconds, as the moon begins its transit away from the sun. When the sun's disk is not occluded, the light of the disk completely overwhelms the fainter light of the chromosphere. Because it is visible for such a brief period, this phenomenon is known as the **flash spectrum** of the chromosphere.

The flash spectrum is an emission spectrum that replaces the dark line absorption spectrum of the photosphere. Since this type of spectrum is produced only under conditions of high temperature and low density and pressure, the flash spectrum tells us that these conditions prevail in the chromosphere. The chromosphere, in other words, is hotter and more rarified than the photosphere. Studies have shown that the temperature of the chromosphere increases with distance from the photosphere. Density, by contrast, decreases toward the outer limits of the layer.

In our simplified three-layer model of the solar atmosphere, the chromosphere (along with the upper levels of the photosphere) is the region in which the dark Fraunhofer absorption lines are produced. Energy from the continuous spectrum of the photosphere is absorbed by atoms and almost immediately reradiated in random directions, so that only a small amount reaches us. But when the flood of radiation from the photosphere is blocked off during an eclipse, we can see the fainter emission lines from the chromosphere: the flash spectrum.

Outstanding features of the chromosphere are **spicules** (from the Latin word for "javelin"). These are numerous jets of very bright gas that shoot thousands of kilometers above the sun's disk. They are thought to be related to convective processes that occur in the granules of the photosphere. As many as 100,000 spicules can appear at any one time, each lasting five to fifteen minutes. Their blade-like shapes prompted one nineteenth century astronomer to liken the face of the chromosphere to a burning prairie.

Figure 11-6
Spicules, jets of materials at the surface of the photosphere, photographed in the red light of hydrogen alpha at 656.3 nm.

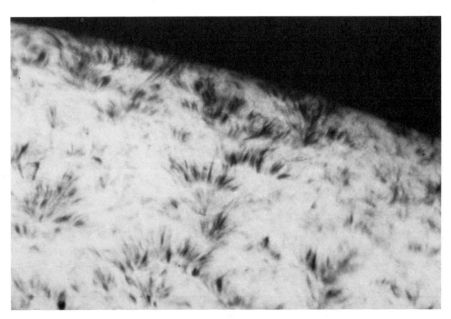

THE SPECTROHELIOGRAPH   The chromosphere and other solar regions can be studied at any time by the use of special instruments, such as the **spectroheliograph.** This device, which is really a modified version of the spectrograph, enables the observer to screen out all wavelengths of the solar spectrum except a single narrow desired wavelength band. Thus it is possible to observe only the radiation representing a particular atomic transition—that is, a single spectral line.

Observing the sun in the light of a single line has the advantage of permitting observation of certain features that ordinarily cannot be seen because they are overwhelmed by the light of the photosphere. But these features are often particularly bright at certain wavelengths. The spectroheliograph can be used to admit only these wavelengths, and blocks out the rest of the photospheric spectrum.

This technique also permits one to observe the various atmospheric levels of the sun. This is so because each layer produces light of characteristic wavelengths, determined largely by its temperature. Thus the lower chromosphere produces the blue line of ionized calcium, whereas the higher temperatures of the upper chromosphere produce lines of ionized oxygen. Therefore, if we wish to examine the structure of the chromosphere, from top to bottom, we would view it in the light of ionized calcium and ionized oxygen lines. Even higher solar regions may be studied in the light of ultraviolet and X-ray lines.

The spectroheliograph is a device for recording an image of the sun in light of a single wavelength.

A more recent development is the **monochromatic filter.** Like the spectroheliograph, this device can be incorporated into the optical system of an ordinary telescope. The filter is of the narrow bandpass type. This means that, like the spectroheliograph, it can isolate a single line from the solar spectrum, allowing only light of that wavelength to reach the observer. The chief advantage of the monochromatic filter is that it enables an observer to view the entire solar disk and its atmosphere in the light of a single line all at once, rather than slice by slice as is necessary with the spectroheliograph. A photograph of the image transmitted by the filter is known as a filtergram. The chief disadvantages are that such filters are relatively expensive, and each filter can be used only for its specific wavelength.

The Corona

At a height of about 5000 km, the chromosphere phases into the outer and hottest layer of the sun, the corona, or crown. This is an envelope of extraordinarily hot gases, so hot that all atoms are ionized. The corona, like the chromosphere, can ordinarily be observed only during a total solar eclipse. The corona is illuminated by reflected light from the upper reaches of the photosphere, and so is only about as bright as moonlight. Moreover, the relatively little light that does reach earth is scattered and diffused by the atmosphere, masking the corona from observers.

The Corona at a Glance	
Depth	20,000,000 km +
Temperature	2,000,000 K +
Density	10^9 atoms/cm³ (bottom) to near vacuum
Chief permanent features	solar wind

At the moment of total eclipse, the moon occludes both the bright disk of the photosphere and the scarlet crescent of the chromosphere. Only a halo of pearly light, extending from the sun's limb millions of kilometers into space, can be seen. During periods of intense solar activity, this light takes the form of graceful streamers and filaments. Recent measurements by spacecraft suggest that the outer boundaries of the corona may coincide with those of the solar system itself.

THE CORONAL TEMPERATURE Why should the corona, at such a great distance from the source of solar heat, be so hot? Since the temperature at the top of the chromosphere is about 100,000 K, the much higher coronal temperature—1 million K or more—would seem to violate a basic physical principle: heat cannot flow continuously from a cold body to a hot one. One popular hypothesis that attempts to account for this seeming contradiction could be called the "sonic

Figure 11-7
The corona, at times of minimum (left) and maximum (right) sunspot activity. At minima, the corona extends farther at the equator then at the poles; at maxima, it is roughly circular, and often exhibits prominent streamers (see chapter opener photo).

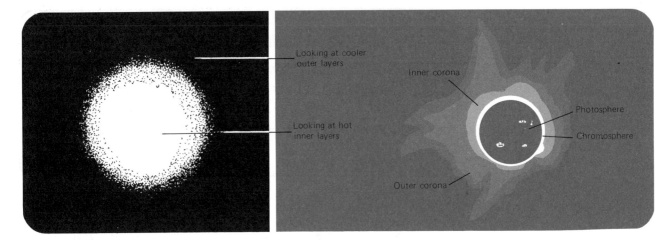

Figure 11-8
At left, the phenomenon of limb darkening. When we look at the edge of the sun's disc, we are looking through the cooler, outer layers. Nearer the center of the disc, we are looking more deeply into the hotter interior layers. Thus the center of the solar disc appears brighter than the edge, or limb. (Compare the white-light photograph of the sun on page 335.) At right, the structure of the solar corona.

boom" theory. It states that the hot gas in the photospheric granules breaks like ocean waves as it rises from the hot solar interior to the cooler surface. This breaking releases convective energy in the form of shock waves. The shock waves travel upward into the increasingly rarefied corona, picking up speed. Eventually, they accelerate to the speed of sound to produce sonic booms similar to those of a jet plane breaking the sound barrier. The supersonic shock waves provide the energy that heats up the gases of the corona.

Even with a kinetic temperature of 2,000,000 K, however, the corona does not actually hold much heat. The transparent gases there contain so few atoms that an object placed in their midst would scarcely be touched by them. The total energy of these fast moving atoms, therefore, is quite low. Indeed, if a human being could somehow be placed in the corona and shielded from the radiation of the photosphere, he would quickly freeze to death because there would not be enough heat energy to warm his body.

THE CORONAGRAPH As we have seen, the bright light of the photosphere obscures the corona so that it can be seen only briefly, for seven or eight minutes at most, during an eclipse. For this reason, very few extended studies of the sun's outer atmosphere could be made until techniques were developed to overcome this observational limitation. One of these techniques is the spectroheliograph. Another is the **coronagraph,** inverted in 1931.

The coronagraph is a modification of the ordinary telescope. It produces an artificial total solar eclipse, facilitating study of the chromosphere as well as the corona. The device is simply a solid disk with a diameter equal to the apparent size of the sun. The disk is positioned in such a way that it blocks out the sun's image inside the telescope so that the corona can be studied. The coronagraph has to be used in high

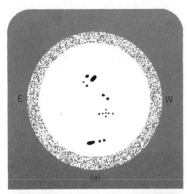

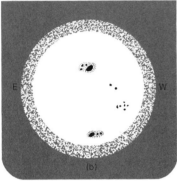

Figure 11-9
The sun is a fluid body, and its rate of rotation varies with latitude. At the equator, the period is about 25 days, near the poles, about 31 days. The solar rotation can be determined from the Doppler shifts of the sun's east and west limbs, or, more simply though less precisely, from observation of surface features such as sunspots.

mountains so that the atmosphere does not scatter the sunlight and thereby obscure the corona.

Solar Rotation and Magnetic Field

Although the rotation of the sun had been casually observed during the time of the Greeks, Galileo was the first to offer proof that the sun was a sphere and that it turned on its axis. Galileo observed that sunspots—irregular dark splotches on the photosphere—first appeared on one limb of the sun. Within a week, they moved toward the center, then appeared at the opposite limb about two weeks later. More precise proof of the sun's rotation has been obtained through spectroscopic analysis of the Doppler shift. When photographed with a spectrograph, for example, the spectral lines of the limb moving away from the observer shift toward the red end of the spectrum, whereas the lines of the limb approaching the observer shift toward the violet. From the difference between the two wavelengths, the overall rate of rotation can be determined.

This technique has yielded the interesting fact that the different latitudes of the sun rotate at different speeds. At the equator, the solar period of rotation is 25 days; it takes almost 28 days, however, for the sun's surface to make a full rotation at latitude 45 degrees north and south. The sun is able to rotate in this manner because of its gaseous nature. By contrast, all regions of the earth, moon, and other solid bodies must rotate in unison.

The rotation of the sun's interior is complicated, and its causes are not fully understood. However, this phenomenon may be the mechanism responsible for the sun's magnetic properties. To see why this is so, it is necessary to learn something more about the phenomenon of magnetism.

Since the time of the Greeks, men have known that the earth possesses magnetic properties. The use of the magnetic compass as a direction-finding device is a major technological achievement that arose from this knowledge. The earth's magnetism is easily demonstrated in a simple experiment. If a small piece of magnetized metal is suspended by a string so it is free to turn in any direction, the magnet will automatically orient itself so that its long axis parallels an imaginary line connecting the earth's two poles. Our suspended magnet aligns itself in this manner because the earth itself can be imagined to have a giant bar magnet lying in its interior and possessing two opposite poles where magnetic forces are concentrated.

Since one end of the suspended magnet points in a northerly direction, this end is called its north-seeking pole, or north pole. Similarly, the opposite end, which points toward the south, is called the south-seeking, or south, pole. A basic principle of magnetism is that unlike poles attract one another, whereas like poles repel one another. Thus, given two magnets, the north pole of one will be attracted to the south pole of the other, and vice versa. But the north poles of the magnets will repel one another, as will the south poles.

Surrounding every magnet is a magnetic field, the region of space within which the attractive and repulsive forces operate. A magnetic field is usually depicted as an array of lines of force, which indicate the strength and direction of the magnetic forces that would be experienced by a magnetically susceptible substance entering the area. It is easy to map the lines of force around a bar magnet, for example, by sprinkling iron filings on a sheet of paper held above the magnet and watching how the filings distribute themselves (Figure 11–10).

Astronomers have sought to understand the solar magnetic field by comparing certain properties of the sun with those of the earth. For example, the earth is believed to have a fluid outer core consisting of charged particles. The fluid is set in motion by the convective processes which arise from temperature differences in the core. The motion of the electrically charged fluid establishes electrical currents. These currents, together with the rotation of the earth, apparently produce our planet's magnetic field.

Something similar is thought to produce the sun's magnetism. In the solar interior, it is thought that the moving ionized gases—which by definition consist of charged particles—create electrical currents. The electric currents, which may be set in motion by a number of factors, including the sun's differential rotation, establish a magnetic field. The sun's overall magnetic field is believed to take roughly the same configuration as that of the earth. That is, the sun has a north pole and a south pole where the lines of force are concentrated. The lines that emanate from the north pole loop around both sides of the sun in a wide arc and reenter at the south pole. However, the sun also appears to possess an east-west magnetic field that changes in strength and position over certain well-established periods.

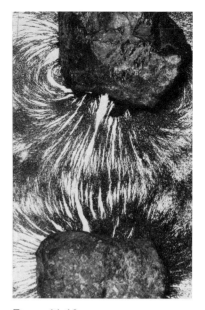

Figure 11-10
Iron filings align themselves along the lines of force between these two pieces of magnetic ore. The lines of force connect opposite magnetic poles.

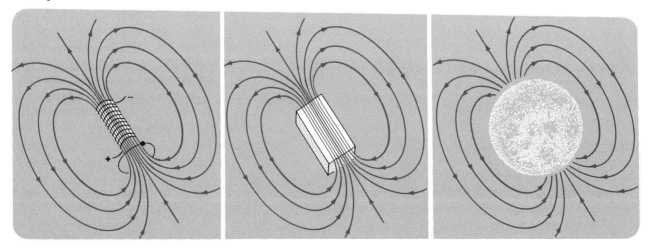

The existence of a solar magnetic field can be predicted from our theoretical model of the sun as a star. And, as we have seen, so can many other solar properties. But there are some solar features that the model does not predict. These features which take place in the sun's atmosphere, are often referred to as the solar weather, because they are disturbances of the solar atmosphere. They are aspects of the **active sun,** and they include **sunspots, plages, prominences,** and **flares.**

Why has the model failed to predict these phenomena? The physical principles involved are not new or unknown. We understand, for example, the physics and mathematics of convection, radiation, and gas flow in magnetic fields. What we don't understand, though, is the complex ways in which these phenomena interact to produce the individual elements of the solar weather. Nor do we understand how they affect the sun and the rest of the solar system. Another factor that hampers our comprehension is that many solar phenomena occur over vast stretches of space, up to 100,000 km and more. Because of their huge dimensions, these phenomena cannot be duplicated in any laboratory on earth.

A similar situation is seen in the science of meteorology. Like those of the solar weather, the elements of terrestrial weather cannot be isolated and studied in the laboratory because they occur over large areas. Meteorologists understand the physics of the various elements that make up the weather—high and low pressure zones, cold and warm fronts, convection currents, jetstreams, and the like. But they have difficulty predicting exactly how these huge air masses will interact over great distances to produce the weather in a given region for any extended period.

Sunspots

Despite these limitations, there are certain things we can say about each of the elements that make up the solar weather. Consider, for example, the most obvious features of the solar surface, sunspots, which can frequently be seen by the unaided eye. Indeed, as far back as 600 B.C., well before the invention of the telescope, Chinese observers reported having seen these dark areas on the surface of the sun. Sunspots were first reported in the Western world by the Greeks, some of whom thought that the spots might be small planets passing between the earth and the sun. By 1611, two years after the invention of the telescope, Galileo and other scientists studied and reported on these spots, confirming that they indeed occurred on the face of the sun. But the true nature of the spots remained a mystery. As late as the first decade of the eighteenth century, for example, some thought the spots were mountaintops which were periodically covered by tides of hot lava.

Today, we know that sunspots are irregular regions of the photosphere that appear dark only because they are cooler than their surroundings. Their temperature is about 4500 K, compared to 6000 K for

Figure 11-11
The sun's magnetic field has a north and a south pole, like that of the earth, or an ordinary bar magnet. An electric current can also create a magnetic field, as in an electromagnet. The motion of electrically charged plasma in the solar interior is thought to create the sun's magnetic field.

the rest of the photosphere. But even at this temperature, they emit many times the light of the full moon.

The average sunspot lasts about a day, but some live for only two or three hours, others for as much as two or three months. Spots also appear in a wide variety of sizes. Large ones may be as huge as 100,000 km across — several times the diameter of the earth. Almost invariably, however, sunspots occur in pairs. Indeed, clusters consisting of two large spots surrounded by numerous smaller ones are extremely common. Such groups may extend over more than a fifth of the sun's face at times.

Sunspots appear darker than their surroundings only because they are about 1500 K cooler.

Doppler analysis has shown that the gases comprising a sunspot move in definite patterns. The gases at the higher (chromospheric) levels, for example, flow inward toward the center of the spot at about 1 km/sec. Those at the lower (photospheric) levels, by contrast, flow outward at roughly the same speed.

An outstanding feature of the sunspot is its strong magnetic field. The strength of this field ranges from 500 gauss in small sunspots to 4000 gauss in larger ones. (A **gauss,** named after the nineteenth century German astronomer, Karl Gauss, is a unit that measures the overall strength of the magnetic field.) By comparison, the magnetic field of the sun is somewhat less than 1 gauss, and that of the earth is believed to be about the same. A field of 4000 gauss is extremely intense. For example, if a man were to enter such a field holding a pair of pliers, its force would literally tear the tool from his hand.

Scientists have been able to determine the existence and strength of the field because of its effect on the spectral lines emitted by the sunspot. When photographed with a spectrograph the lines are split into several separate components having different wavelengths. This is the Zeeman effect, discussed in Chapter 4. By examining the degree to which the lines are separated, observers can measure the strength of the magnetic field as well as its polarity.

THE SUNSPOT CYCLE Solar magnetic fields may also be linked to the cyclic appearance of sunspots. In 1843, the German astronomer Heinrich Schwabe discovered that the number of spots waxes and wanes over an 11-year period known as the sunspot cycle. The period during which large numbers of sunspots (as many as 100) appear is known as the sunspot maximum. The time of few or no sunspots, by contrast, is the sunspot minimum. Although the period between maxima averages about 11.1 years, it has varied from a low of 7.3 years to a high of 17.1.

The number of sunspots increases and decreases in a cycle of roughly 11 years.

From the Zeeman effect, astronomers have learned a number of important details about the magnetic fields of sunspots and their relation to the sunspot cycle. First, in groups dominated by two large sunspots, the two will be of opposite polarity, and the spots will be oriented along an east-west axis. The two are probably connected by magnetic lines of force. The Zeeman effect has also shown that during each cycle, the

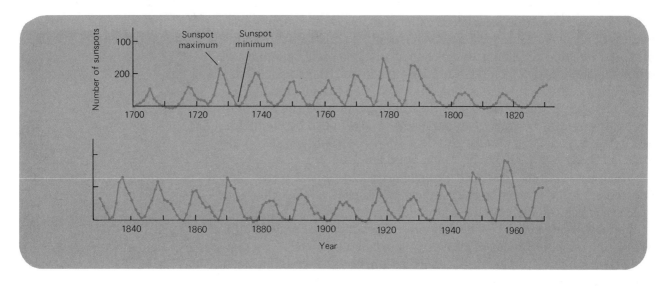

Figure 11-12
The sunspot cycle. The period, though it varies considerably, averages just over 11 years. Many other solar phenomena are linked to this cycle.

dominant sunspots in one hemisphere are of opposite polarity from their counterparts in the other hemisphere. For example, if the dominant spots in the southern hemisphere are of north-seeking magnetic polarity during a cycle, then the dominant spots in the northern hemisphere will be of south-seeking polarity. In the following 11-year cycle, the magnetic polarities of the dominant spots are reversed. Thus, in terms of magnetic polarity, the true length of the sunspot cycle is twice the 11-year cycle, or 22 years.

Other Phenomena

The sunspot cycle also appears to affect the corona—it assumes a more dramatic shape during periods of high sunspot activity. Streamers of luminous gas extend from it in all directions like the petals of a flower. During sunspot minima, by contrast, these bright streamers are replaced by a dim halo which projects further at the sun's equator than at the poles. Most of the other elements of the solar weather are similarly tied into the sunspot cycle. These elements are most active during sunspot maxima and are rarely observed during minima.

PLAGES   Associated with sunspots are bright granulated clouds of gas which appear near the chromosphere. These are plages (from the French word for "beach"). Plages are typically found near sunspots, but they can appear anywhere in the chromosphere where a strong magnetic field prevails. The initial appearance of a plage frequently presages the birth of a sunspot about a day or so later. The plage remains after the sunspot appears, growing in intensity and size during the life of the spot. Even after the sunspot dies, the plage may persist, lasting for an average of about two weeks.

PROMINENCES   Also associated with sunspots are prominences, streams of relatively dense hydrogen that jut out into the corona from the photosphere. Most prominences are connected to the photosphere by an umbilical-like thread of gas, but a few lack this thread and so float freely in the corona. A prominence is more luminous than the surrounding corona, although its gases are cooler. In fact, it is the lower temperature which increases the number of visible spectral lines emitted by the hydrogen in this region. Prominences can be immense, sometimes measuring 200,000 km long, 40,000 km high, and 6000 km deep.

Astronomers have classified prominences into several types. The first, the quiescent prominence, is, as the name implies, a relatively stable structure which can hover in the corona for months at a time. Apparently, magnetic fields in the corona support the prominence, preventing it from being pulled to the solar surface by gravitational forces. The relatively rare eruptive prominence is a burst of luminous gas that has been ejected high into the corona at speeds approaching 700 km/sec. Some astronomers believe that this type of prominence is triggered by a sudden change in the sun's magnetic field, possibly as a result of the activity of a nearby flare. A close relative of the eruptive prominence is the surge prominence, which jets into the corona at speeds of 1000 to 1300 km/sec.

Finally, there is the loop prominence. This structure gets its name from the shape that it forms as it arches out toward the corona, then loops back to the solar surface. Apparently, the shape is due to the dense clouds of luminous gas being trapped by magnetic lines of force that stream out into the corona. The loop prominence is important because, like the iron filings in our discussion of magnetism, it provides evidence for the existence of invisible magnetic fields in the solar atmosphere.

Prominences are masses of relatively dense, cool gas in the solar corona; as the atoms of the gas recapture electrons, they emit photons in the visible range.

SOLAR FLARES   Like plages and prominences, the solar flare also usually occurs near sunspots. The flare, which is a sudden brightening in the region above a sunspot, is the most spectacular element of the solar weather. The brightening is due to the excitation of a mass of gas by a huge outpouring of radiation, especially at short wavelengths — ultraviolet, X-rays, and gamma rays — but often at radio wavelengths as well. The flare releases a tremendous amount of energy in a very short time — as much as 10^{32} ergs in the space of a few minutes, or the equivalent of the energy liberated by the detonation of a billion hydrogen bombs. If the energy from a large flare could be captured, it would generate enough electricity to meet the world's needs for 100 million years. Not surprisingly, the temperature of a flare is extremely high, often approaching 20 million K.

Like other elements of the solar weather, flares are most numerous during a sunspot maximum, when they erupt three or four times a day. A large flare may have a diameter of some 100,000 km. Flares often eject quantities of luminous gas at speeds approaching 1000 km/sec.

PORTFOLIO

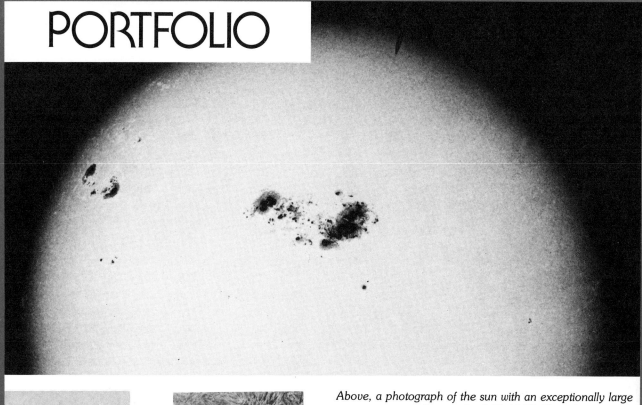

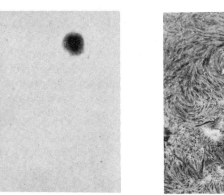

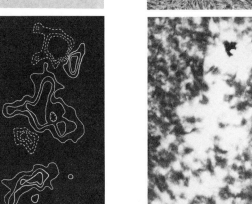

Above, a photograph of the sun with an exceptionally large sunspot group, taken on April 17, 1947. The other sunspot group is at the same latitude. Notice how the sunspots occur in pairs. The inner, dark region of the spot is called the umbra; it is surrounded by a lighter region called the penumbra. Limb darkening makes the edge of the sun darker and facilitates observation of bright active regions around the sunspot group at the left.

The four pictures at left are of the same active region. Clockwise from the top left, we have a photograph in white light, a photograph in the red line of hydrogen, a spectroheliogram in the light of calcium, and the contours of the magnetic field made with the Mt. Wilson magnetometer. Notice how the plage in the calcium spectroheliogram has the same shape as the magnetic contours in the adjacent picture.

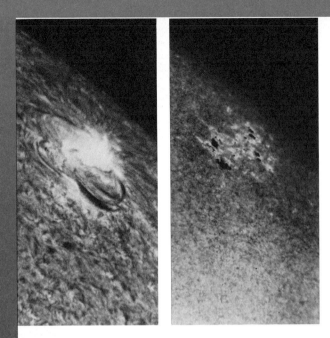

The two photographs at the top of this page are of the same active region of the sun. The white light picture on the right shows sun spots and faculae, features of the photosphere. The photograph on the left, in the red light of hydrogen shows the fine structures of the chromosphere above the sunspot region. Such structures are seen only in filtergrams made with filters of very narrow pass bands, in this case only .05 nm. The photographs at the bottom of the page show how material flows into a sunspot umbra from two points near it. The motion of the gas is strictly controlled by the magnetic fields, which explains its curious curved trajectory. The sequence of pictures covers about 20 minutes.

At the top of the facing page, a gigantic prominence at the edge of the sun. The motion of the gas is controlled by gravity and the local magnetic field. The latter is the agent behind the intricate, veiled structure that we see here. On the same page, below left, is a sequence of photographs, taken on August 11, 1963, showing the development of an eruptive prominence. The interval between the first and last photographs is just 34 minutes. In this time the prominence has risen to more than 200,000 km. The solar disc and the prominence were photographed on the same film, but through different filters. The photograph on the lower right of the facing page shows a solar prominence that is 160,000 km high. This photograph, in the light of hydrogen, was taken on June 12, 1972.

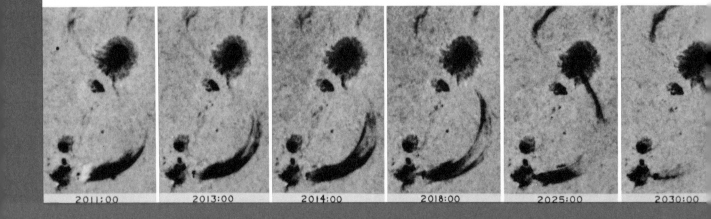

2011:00   2013:00   2014:00   2018:00   2025:00   2030:00

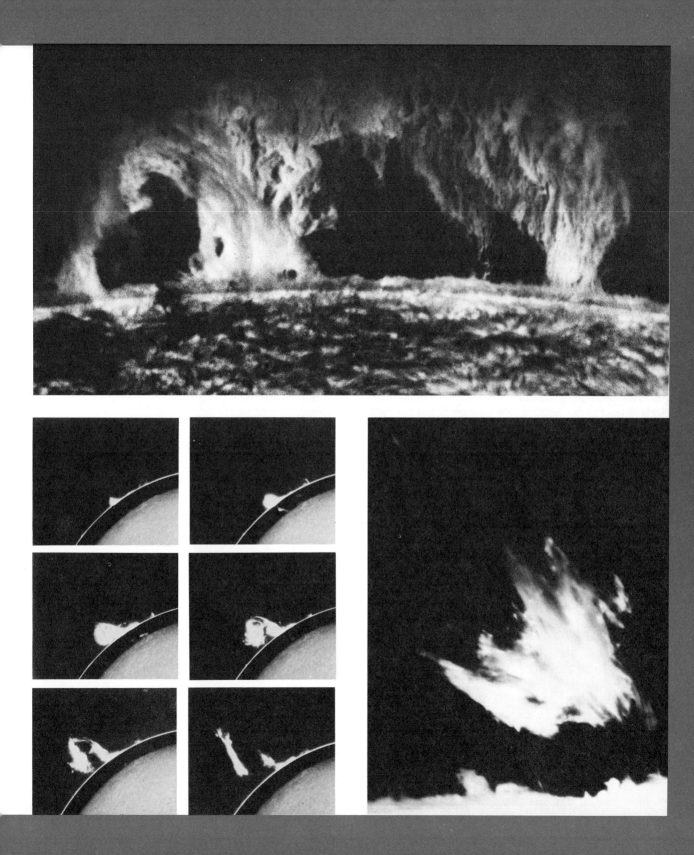

A Theory of Solar Activity

As we have seen, sunspots, plages, prominences, and flares are associated with strong solar magnetic fields. Moreover, all seem to appear with greater frequency during sunspot maxima. Is there any connection, then, between the elements of solar weather, the sun's magnetic field, and the sunspot cycle?

One American astronomer, H. Babcock, believes that there is. According to his theory, the magnetic field of the sun has two constituents which vary with the 11-year sunspot cycle. The sun, like the earth, has a polar magnetic field—a north pole and a south pole, from both of which emanate lines of magnetic force. In addition, the sun has a strong east-west field where lines of force become wrapped around the lower latitudes at shallow depths below the photosphere.

At the heart of Babcock's theory is the observation that the sun rotates faster toward the equator than at higher latitudes. This phenomenon is true for both hemispheres. As the sun rotates, the charged gases in the lower latitudes trap the weaker north-south magnetic lines of the sun's polar field. When gas particles attempt to cross the force lines of the field, they are deflected into spiral orbits around the lines. In this way, the magnetic fields become trapped in the moving gas. The gases drag the lines along with them, stretching them out. At the regions closest to the equator, the magnetic lines are pulled into loops. The loops become wrapped around the sun like string around a ball.

After several years and many rotations, the lines of force are wound closer and closer, at lower and lower latitudes. In these regions, the tightly wrapped lines create intensely strong magnetic fields measuring as much as 4000 gauss. This aspect of the theory explains three observations about sunspots: their east-west magnetic orientation, the strength of their magnetic fields, and their gradual migration from higher latitudes to lower ones over the course of the 11-year sunspot cycle.

Babcock's theory also offers a plausible explanation of the mechanism that gives rise to sunspots, plages, and prominences. According to his model, the gas in the sun's interior is normally in a state of thermal equilibrium: heat is flowing out toward the surface at the same rate as it is being produced at the core. The appearance of the strong magnetic fields, created by the winding lines of force at the lower latitudes, exerts pressure on the gas in these regions. This increased pressure upsets the condition of thermal equilibrium. As a result, the gas expands and its density drops. The buoyant gas rises and breaks through the surface of the photosphere in the form of a giant loop.

As it does so, the charged particles heat the gases in the chromosphere, causing a small part of it to become incandescent. The luminous cloud that forms as a result of this action is a plage. The magnetic loop continues moving out till it reaches the corona. Here, it ignites gases in the same way, forming a prominence. At the two points where the two sides of the loop leave the sun, two magnetic fields of opposite polarity are established. These become a pair of

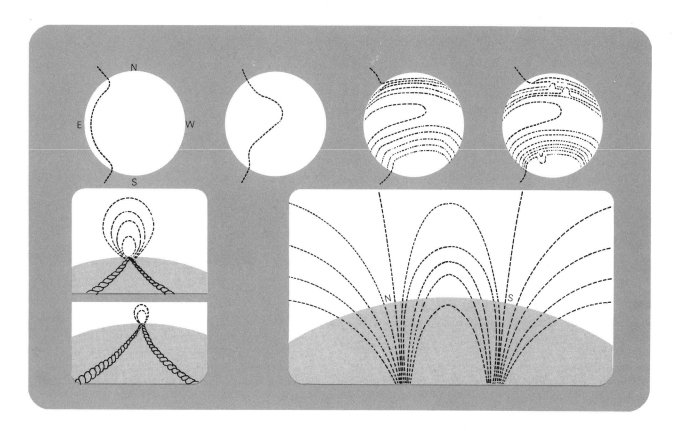

dominant sunspots. In this manner, the Babcock theory states, sunspots, plages, and prominences are born.

Roughly eleven years after the appearance of the first sunspot, the magnetic lines of force, winding tighter and tighter with each solar revolution, finally fuse. This causes neutralization of the strong magnetic fields that have been created at the lower solar latitudes. The neutralized lines then unwind, move out toward the corona, then loop back toward the surface. They enter the surface at about latitude 30 degrees, north and south, to start the sunspot cycle over again. When the lines reenter the surface, the general magnetic field of the sun reverses its polarity. As a result, the sunspots which appear in the following 11-year cycle will have the opposite polarity of their predecessors.

It should be mentioned that some astronomers don't agree with Babcock's theory because it fails to explain the mechanism responsible for the sun's differential rotation. Moreover, the magnetic fields measured by space probes have differed from those the model predicts will exist in interplanetary space. However, the Babcock model does provide a reasonable interpretation of many solar phenomena. No other model proposed thus far has managed to explain so much.

It is impossible to imagine what the earth would be like without the sun. For one thing, if there had been no sun there would be no earth, since the planets of the solar system are probably byproducts of the sun's own birth. Moreover, much of our planet's physical appearance has been shaped by the sun. It is the sun, for example, which energizes the atmosphere, producing such elements of our weather as wind, water evaporation, and rainfall. These in turn are responsible for erosion, which has given the earth its blanket of soil and molded its features.

The living things which populate our planet also owe their existence to the sun. In the early eons of the earth's history, before it acquired its protective atmosphere, high energy radiation from the sun bathed the earth's surface. Though these ultraviolet rays would be lethal to most

Do Sunspots Affect the Earth's Climate?

Since their discovery, sunspots have been linked, by scientists and laymen alike, to all manner of terrestrial activity. Some individuals, for example, have claimed that there is a positive correlation between sunspot maxima and the number of people entering mental institutions. Others have tried to relate sunspot maxima and minima to the ups and downs of the stock market. Still others have stated that sunspot activity influences the quantity of salmon caught in the Atlantic Ocean. Unfortunately, none of these alleged correspondences has ever been satisfactorily proven.

More recently, some astronomers have suggested that sunspot activity influences the earth's climate. This theory is really a resurrection and refinement of one that was born in the final quarter of the last century. At that time, two astronomers independently published findings pointing to a virtual lack of sunspot observations between the years 1640 and 1715 AD. The two scientists were Gustave Spörer of Germany and E. W. Maunder of Great Britain. Significantly, this seventy-year gap, called the Maunder minimum, happens to fall within the coldest years of the period known to historians as the "Little Ice Age." During this time, annual temperatures across Europe were unusually low, pointing to a possible link between the sunspot activity and the earth's climate.

The findings of the two scientists were not accepted at the time. They were thought to have relied too heavily on the lack of recorded observations in historical records and journals. Those who opposed the theory pointed out that the paucity of documented observations was probably the result of sloppy record keeping, rather than an actual lack of sunspot activity. Now, however, new evidence in support of the Maunder-Spörer hypothesis has been produced. This data reveals that there definitely were few or no observations of sunspots during this 70-year period, and during two others as well. It also indicates that there may be definite correlations between sunspot cycles and the climate of the earth.

The new evidence includes a re-check of documented solar observations in both Western and Oriental sources. Modern investigators have combed the ancient archives looking for reports of solar phenomena that are now known to be associated with sunspot activity, but which were not thought significant in the days of Maunder and Spörer. Thus scientists searched for sightings of auroras, which are usually produced when solar flares erupt during sunspot maxima. They also looked for reports of eclipses which described the streamers of luminous gas that extend outward from the corona. Like auroras, these streamers which follow the solar magnetic lines, can only be observed during periods of intense sunspot activity. The modern investigators

organisms today, they may have played a key role in the formation of complex chemical substances that eventually made possible the emergence of life.

Today, the sun's great gravitational force continues to hold the earth in its orbit, within range of the life-giving energy which it pours forth in such abundance. This energy is necessary for the biological process of photosynthesis, which enables plants to make their own food. The plants, in turn, provide nourishment for other living species, including humans. The sun has also provided us with vast stores of chemical energy in the form of coal and petroleum; these fossil fuels, which are the remains of ancient life, hold locked within them the energy of sunlight of past ages.

found that there were no recorded observations of auroras or coronal streamers for the period 1645 to 1715. In this way the first link in the chain of supportive evidence was forged.

The third piece of evidence is even more remarkable. It is based on an analysis of the carbon-14 content of tree rings. It is well known that cosmic rays interact with the nitrogen in the earth's atmosphere to produce a small amount of the radioactive isotope, carbon-14. The carbon-14 is absorbed by plants in the form of carbon dioxide, enters the food chain, and accumulates in the tissues of both plant and animal life. When organisms die, they no longer assimilate new carbon-14. The carbon-14 in their bodies gradually undergoes radioactive decay into normal carbon-12. Thus the ratio of carbon-14 to carbon-12 drops steadily from the moment of death, providing a way of dating any organic material, such as wood or leather.

It has been found, however, that intense solar activity reduces the number of cosmic rays reaching the earth's atmosphere. Thus the amount of carbon-14 deposited in tree rings ought to be relatively high during periods of low solar activity. The opposite, of course, is also true: greater solar activity means lower levels of carbon-14. In support of the Maunder-Spörer hypothesis, studies have shown that the amount of carbon-14 in tree rings formed during the

years 1640 to 1720 was indeed high. To clinch the argument, researchers compared several pieces of independent evidence on graphs. When plotted against the number of observed sunspots, the curves representing variations in carbon-14 levels and fluctuations in the severity of European winters agreed remarkably well.

In addition to the Maunder minimum, the sunspot and carbon-14 data show two other periods of abnormal solar activity. Between 1460 and 1550 AD, there was yet another period of low solar activity, known as the Spörer minimum. And during the entire twelfth century, and parts of the thirteenth, there was unusually high solar activity. Like the Maunder minimum, the Spörer minimum dovetails nicely with a cold spell of the Little Ice Age. Similarly, the period of high solar activity is coincident with the warm period known to have prevailed in Europe during the thirteenth century. While they do not show a causal relationship, these correspondences indicate that the phenomena involved do deserve more study. A number of scientists believe that further investigation will reveal, not merely fluctuations in solar surface activity, but in the total output of solar radiation as well. Such a finding, if confirmed, might well have far-reaching implications for our theories of solar energy production.

The Solar Wind

Electromagnetic radiation is the sun's most important "export," but not its only one. The sun also sends part of its own substance out into the solar system. Because of its enormously high temperatures, the corona is continuously expanding into space. The gases in its outer borders expand most rapidly, perhaps at the rate of 300 m/sec, because there is no pressure to halt them. Eventually, the gas may reach speeds as high as 400 to 800 km/sec. This rapidly expanding coronal gas, consisting chiefly of ionized hydrogen (that is, protons) and free electrons, is called the **solar wind.**

The particles of the solar wind race outward from the corona towards the earth and other planets, possibly reaching as far as Pluto. About two days after leaving the sun, they reach the earth, where they are deflected by our planet's atmosphere and magnetic field. Thus the earth can be thought of as a relatively cool island swimming through the hot coronal layer of the sun. The temperature of the particles of the solar wind is indeed high—about 160,000 K. By terrestrial standards, however, the wind is virtually a vacuum. Its density is about 10 to 100 particles/cm^3—only 10^{-17} that of the air at sea level. Thus the solar wind, like the corona proper, holds very little heat. The amount of mass lost to the escaping solar wind is also very small, compared to the sun's total mass—it is not enough to affect the structure or evolution of the sun. Thin as it is, however, the solar wind is still some 10 times denser than the matter of interstellar space, and it has some very significant effects on our planet.

The solar wind is a flow of charged particles from the corona out into the solar system.

Effects on Earth

One day in early August of 1972, a tremendous explosion occurred on a small area of the solar surface. Within five hours, a huge cloud of gas, measuring some 16 km wide by 75,000,000 km long, crashed into the earth's atmosphere at speeds up to 6500 km/sec.

The collision could neither be seen nor heard on earth, yet it touched off an enormous electrical and magnetic storm that violently affected the earth. Aboard airplanes and ships, for example, compass needles danced erratically, and long-distance radio communication was disrupted. In northern regions, the night sky glowed with brilliant red lights, and the electricity in rural areas flickered on and off as in the midst of a thunderstorm. The effects lasted for more than a week as material from the sun continued to bombard the earth, accompanied by bursts of solar X-rays and radio waves.

The terrific solar explosion was, of course, a flare. The flare was an unusually large one. But the typical flare has the same effects, only on a much smaller scale. As it explodes, a flare sends gust after gust of charged particles into the solar wind. The wind, with increased density and veloc-

DID YOU EVER STOP TO THINK WHAT THE WORLD WOULD BE LIKE IF THERE WERE NO SUN?

YES, AND IT'S AN INTRIGUING THOUGHT...ONE'S MIND IS SET TO REELING AT THE PROSPECT...THIS IS THE SORT OF PROPOSITION THAT CAN PRODUCE ENDLESS DEBATE

WHAT ARE **YOUR** VIEWS ON THE SUBJECT?

3-2

IT WOULD BE DARK!

ity, slams into the upper regions of the earth's atmosphere with great force. As a result, numerous and wide-ranging electrical and magnetic disturbances are produced.

Fortunately, these particles are blocked by the earth's protective atmosphere and magnetic field, and so most never reach the surface. However, both the magnetic field and certain parts of the upper atmosphere are affected. When they meet the magnetic field, some of the particles are deflected in spiral paths toward the north pole, whereas others are shunted in similar paths toward the south pole. This produces a change in the direction and magnitude of the magnetic field at the earth's surface. The **geomagnetic storm,** which causes compass needles to fluctuate wildly, lasts from one to three hours.

Some particles, however, do manage to break through the field. These become part of the Van Allen radiation belts, streams of fast-moving protons and electrons trapped into permanent orbits around the earth's lines of magnetic force. Occasionally, streams of particles accelerate out of the radiation belts near the earth's north and south polar regions, where the field is weakest. The escaping particles collide with atoms of oxygen and nitrogen in the atmosphere, exciting them to higher energy levels or even ionizing them (Chapter 4). When these atoms drop back down to lower energy levels again, they emit light of various wavelengths, corresponding to the particular atomic transitions taking place. The radiation from oxygen atoms lies in the red and green regions of the spectrum, that of nitrogen in the red and blue. The beautiful atmospheric glow that results is known as the **aurora:** the northern and southern lights.

Accompanying the disturbance of the geomagnetic field is a disturbance of the earth's upper atmosphere during daylight hours. This zone, known as the **ionosphere,** extends from 100 to 600 km above the surface of the earth. Its composition—chiefly ionized oxygen and nitrogen—enables it to reflect back to earth any short-wave radio signals which strike it. Thus the ionosphere facilitates the transmission of short wave radio signals between distant points on the earth. When the X-rays from a solar flare penetrate the ionosphere, they modify its electrical properties. This, in turn, interrupts radio signals, causing fade-outs and blackouts. The interruptions last anywhere from ten minutes to an hour.

SUMMARY

We cannot see the sun's interior, and the only direct messengers that reach us from that region are the elusive neutrinos. Nevertheless, with the aid of a computer, we can construct models of the sun, just as we do for other stars, giving the density, temperature, pressure, and energy production at every point of the solar interior. In general, the sun is a very typical star. It differs from all other stars, however, in that it is close enough for us to study its surface in detail. Much activity can be observed on the solar surface, both stable, long-term phenomena of the quiet sun, and short-lived, often violent events of the active sun (the solar weather). All belong to the 3 outer layers of the sun that together constitute its atmosphere: the photosphere, the chromosphere, and the corona.

The photosphere is the thin shell that constitutes the visible disc of the sun that we see with the naked eye. All sunlight originates here. We cannot see beneath the photosphere; its gases are rendered opaque by the presence of negative hydrogen ions, which absorb a wide range of wavelengths. Photons from the deeper layers of the solar interior are absorbed in this region, and reradiated —chiefly in the visible range, for the temperature of the photosphere is about 6000 K, so that its radiation peak lies in the yellow-green portion of the spectrum. The surface of the photosphere is mottled by dark granules, which represent convective regions where hot gas from the interior rises, cools, and descends again. The chemical composition of the sun is determined by studying the dark spectral absorption lines (Fraunhofer lines) produced by the gases of the photosphere. The chromosphere, which lies above the photosphere, is hotter (6000–100,000 K) and more rarified. Above the chromosphere is the sun's extended outer atmospheric layer of extremely hot (up to 2 million K), ionized gas, the corona.

The sun rotates, but not uniformly as a solid body does—the speed of rotation varies with latitude. Electrically charged particles in motion create magnetic fields, and the sun's gases, being highly ionized, are charged. Their motions, deriving from convection and the sun's rotation, are thought to create the sun's magnetic field. One component of this field is parallel to the sun's axis of rotation, like the north-south field of the earth. In addition, however, the differential rotation of the sun is thought to create an east-west field by wrapping the north-south lines of magnetic force tightly about the sun, especially at the equator. This process is thought to account for many phenomena of the active sun.

Sunspots are magnetically disturbed regions of the solar surface. They appear dark because they are somewhat cooler than the surrounding areas. Sunspots vary greatly in size and lifespan. Usually they occur in pairs, of opposite magnetic polarity, or in small groups. The number of sunspots varies in a fairly regular cycle which averages 11 years between successive periods of maximum sunspot activity. Many other aspects of the active sun are closely tied to the sunspot cycle, and are often associated with prominent sunspot groups. Plages, for example—bright clouds of gas in the chromosphere—often precede the appearance of a sunspot. Prominences are vast streamers of relatively cooler, denser gas that jut out into the corona from the photosphere. Flares are explosive outbursts of electromagnetic radiation, much of it at short wavelengths. This outpouring of energy excites large masses of gas to great brilliance, and may also result in the ejection of material at high velocity. Both prominences and flares tend to occur near sunspots.

Gas from the outer reaches of the solar corona is constantly expanding outward into the solar system. This material, chiefly protons and electrons, constitutes the solar wind. Violent solar events such as flares inject additional fast-moving particles into the solar wind. When these gusts reach the earth, they sometimes cause temporary disturbances in the earth's magnetic field. They may also disrupt the ionosphere (the upper layer of the atmosphere), interfering with long-distance radio transmission. Charged particles entering the ionosphere at high latitudes often collide with atoms of oxygen and nitrogen, exciting or ionizing them. As they return to lower energy levels, they emit the visible light that we call auroras.

EXERCISES

1. Why do we think that the sun is a star? How does it compare with other stars?
2. How do we know about conditions in the interior of the sun, if we cannot see into the sun or send instruments there?
3. What is the temperature of the solar core? What would conditions on earth be like if the surface of the sun had the same temperature as its interior?
4. How is it possible for the density at the sun's center to be so much greater than that of any element on earth?
5. Why is there less hydrogen at the center of the sun than in other regions?
6. What is the basis for our estimate that the sun is about 5 billion years old? What is its projected future life expectancy? What do you think will happen to the sun and the earth as the sun nears its old age?
7. How long does it take radiation emitted at the sun's center to reach the surface? How long would radiation take to travel the same distance in free space? How long do neutrinos require for this journey? Explain.
8. Name the various layers of the sun that we can study, and describe the special characteristics of each.
9. How deep is the photosphere? Why is it so opaque? What sort of radiation do we receive from it?
10. Describe the spectrum of the sun. What are Fraunhofer lines? How can we tell the chemical composition of the sun?
11. Why is the temperature of the chromosphere higher than that of the photosphere?
12. What is the corona? Why is it difficult to observe? If the temperature of the corona is above 1 million K, why isn't the earth burned to a crisp by the intense radiation that might be expected from such hot gases?
13. What are the advantages of photographing the sun in light of one particular wavelength? How can this be done?
14. Describe the sun's magnetic field. What role do magnetic forces play in such solar phenomena as sunspots and prominences?
15. Why are sunspots dark?
16. What is the sunspot cycle? What other phenomena can be linked to it? How does it affect the weather on earth?
17. What is the solar wind? How far does it extend? How does it affect the evolution of the sun? How does it interact with the earth's atmosphere and magnetic field?

CHAPTER TWELVE

MOTION IN THE SOLAR SYSTEM

Strangers who discover that they come from the same home town or have attended the same university are apt to exclaim, "small world!" Accustomed as we are to hearing news from China the day it happens, to crossing the ocean in a few hours, to seeing moonshot photos of our earth — a bright, fragile globe hung in the immensity of space — the phrase comes easily and naturally to our lips. In order to appreciate our changing conception of the solar system, however, we must free ourselves — at least momentarily — from these modern ideas. In ancient times, nobody said, "small world." The words would have made no sense, for by all available evidence the earth was vast. Merchants, soldiers, and sailors who visited strange and distant lands heard from the inhabitants of still more distant lands beyond. Travel was slow and hazardous. Most people knew only the small area surrounding their homes. Further than that, the world faded into myth and legend. Not only were people physically restricted, they were conceptually restricted as well. No one had ever looked through a telescope. No one had ever seen an aerial view. Whoever wished to attain a wider vision of the world was obliged to seek it on the wings of imagination. There was no other way.

Nowhere do these limitations show up as clearly as in the conceptions which ancient peoples had of the heavens. The earth was obviously huge and solid. The celestial bodies, on the other hand, appeared to be of an entirely different nature. By day the sun dominated the heavens—a fiercely glowing disc that traveled from east to west, changing altitude as the seasons progressed. By night, the sky grew dark, and the moon and stars appeared, moving in the same east to west direction. The moon changed shape according to a 29½ day cycle, going from a full circle, to a crescent, then down to nothing at all. The stars remained fixed in their places relative to one another—all except for five of the brightest ones which followed strange paths through a band of the sky. There was no way of telling how distant any of these objects were, or how big, or what they were made of. One thing, however, seemed self-evident: the heavens were in motion. Obviously, it was not the massive earth which was moving, but the sun, moon and stars.

Having so little real knowledge of these objects of the heavens, ancient peoples filled in the gaps with myth. The sun, moon, and stars became gods and goddesses. Stories evolved to explain their presence, their movements, their relation to the earth. To the ancient Egyptians, for example, the heavens were represented by the goddess Nut, whose star-bedizened body arched above the earth. Over her a river flowed, and each day the sun god, Ra, traveled the river from east to west, descending at sunset into the realm of the dead. To the ancient Hindus, the earth rested on four pillars which rested on the backs of four elephants which stood on the back of a giant tortoise. Encircling all was the great serpent to whose skin the sun, moon, and stars adhered.

These stories, fanciful as they might seem to be, represented an attempt to give the heavens meaning and purpose. Ancient astronomers, like astronomers of today, were interested in making sense of the things they saw in the sky. And the observation of the heavens had certain practical benefits. Many ancient civilizations constructed calendars based on the cycles of the sun or moon. These calendars told farmers when to plant their crops and specified the dates of religious festivals.

In order to use the heavens to accurately mark the passage of time, celestial observation of a rather high order is necessary. Ancient Chinese astronomers attained such a sophisticated understanding of the movements of the sun and moon that they were able to predict solar and lunar eclipses. They also kept records of unusual celestial events such as the visit of Halley's comet in 467 BC and the appearance in 1054 AD of a supernova which went unrecorded by Western astronomers. The ancient Babylonians too brought celestial observation to a fine art. Babylonian astronomers were able to predict both lunar and solar eclipses with startling accuracy. They also made stellar observations easier by separating the sky into constellations. The names which they gave to many of these star pictures, such as Gemini, Scorpius, and Taurus, are still used today.

Greek Astronomy

Impressive though such achievements are, it remained for another civilization—the ancient Greeks—to deal systematically with the problem of heavenly motion. Starting around 600 BC, there was a great flowering of Greek science, and much speculation about the physical universe. Of all the sciences, the Greeks held astronomy in the highest esteem.

Greek thinkers sought to understand reality in a way that was radically different from that of other civilizations. They brought new meaning to the concept of "explanation."

The Greeks expressed their ideas in terms of a *model*—a mathematical or physical representation of reality which was based on familiar principles and was logically consistent. But though the Greeks were rigorous thinkers and greatly advanced the method of logical analysis, as observers of reality they often left something to be desired. Most Greek thinkers believed that the mind was supreme, that ideas were the highest reality. The details of the physical world, on the other hand, the information that one received through the senses, was of a lower order and might, under some circumstances, be safely ignored. This idealistic, non-empirical attitude led Greek scientists at times to commit grave errors.

Greek astronomers were the first to seek an explanation of heavenly motions in terms of a consistent mathematical model.

The first Greek thinker who attempted to construct a coherent model of heavenly motion was Pythagoras (about 560–497 BC). Pythagoras' beliefs were a strange mixture of mathematics and mysticism. He theorized correctly that the earth, moon, and other heavenly bodies were spherical, but this belief seems to have been based on the conviction that the sphere is the most perfect of all geometrical shapes rather than on any objective proof. He proposed further that the heavenly bodies were attached to larger concentric spheres surrounding a stationary earth. The motions of the heavens were the result of these spheres rotating independently of one another. The motions of these spheres gave rise to heavenly harmonies, known as the "music of the spheres," which could be detected only by the most enlightened individuals. The Pythagorean belief that only the "perfect" sphere was worthy of describing heavenly motion was destined to become a characteristic theme in Western astronomical theory. As we shall see, it was not to be abandoned until the seventeenth century when Kepler devised a way of accurately plotting the orbit of Mars, and showed that it was not circular.

The early Greek thinkers failed in their attempts to explain heavenly motion because they were too concerned with making their models conform to certain mystical or idealistic notions, and not concerned enough with making them conform to reality. Yet we will better appreciate the difficulty of their task if we take the trouble to examine the sky ourselves. To do this, we must rid our minds of the preconceptions which centuries of scientific investigation have given us and try to see the sky the way the ancient astronomers saw it.

THE MOTIONS OF THE HEAVENS   If you stayed up all night and watched the sky, it would seem to you that the stars were points of light attached to a great sphere rotating nightly around a point marked by the star Polaris. After a number of hours, the rotation of the sphere would appear to bring the sun up over the horizon, and, overpowered by its intense glare, the stars would disappear. Thus, on the basis of one night's observation, you might come up with a simple, straightforward "one sphere" model of heavenly motion. However, if you bothered to watch the sky for a second night, the model would show certain defects.

You would notice that the sun came up about 4 minutes later than it did the morning before, causing it to appear against a slightly different background of stars. Since the sun and the stars move independently of one another, they cannot be attached to the same sphere. But even before you noticed the independent movement of the sun, you would notice that the moon too had changed positions with respect to the stars, and by an even greater angular distance than the sun. Therefore, after only two nights of observation, you would be forced to conclude the heavenly motion must be the result of at least three independently rotating spheres.

If you continued to spend your nights gazing at the heavens, you would notice other motions among the stars which required explanations of their own. You would see that five of the brightest stars change their positions in the sky just as the moon and sun do. The ancient astronomers observed these moving stars and called them planets, which means "wanderers." They also gave each one a name of its own: Mercury, Venus, Mars, Jupiter, Saturn. To account for the movements of the planets, 5 more spheres are needed, giving us an 8 sphere model.

But after you had logged several months of star gazing, you would begin to notice something which made the job of explaining heavenly motion even more difficult. You would notice that the five planets do not simply move eastward as the sun and moon do. Instead, they reverse their direction every few months, traveling westward for some time before resuming their eastward direction again. This apparent westward motion of the planets with respect to the stars, known as **retrograde motion,** causes the planets to describe loops of varying size and shape in the sky. The planets also change their brightness as they travel, becoming most brilliant when in retrograde motion. Obviously, an 8 sphere model in which the spheres simply rotate around a single axis would not account for such phenomena. More elaborate "machinery" is needed, and this is just what later astronomers attempted to furnish.

At regular intervals, the planets seem to reverse their motion against the background of the fixed stars, and move in a retrograde direction, from east to west.

THE HEAVENLY SPHERES   One of the first astronomers to make a serious attempt to explain planetary motion was a student of Plato named Eudoxus (about 408 – 355 BC). Like Plato, Eudoxus placed the earth at the center of the universe with the other bodies revolving around it. Unlike Plato, however, he assigned each planet not one sphere but four. The planet was attached to the innermost of the four spheres. The four spheres revolved within one another, but their axes were not in line. The axis of each sphere was pivoted at a point on the surface of the next outermost sphere, some distance away from *its* axis. By adjusting the placement of each axis and the speed with which each sphere revolved, Eudoxus was able to achieve an approximation of the actual motions of the heavens. The total number of spheres in Eudoxus' model was 27. The moon had three spheres to account for its phases, and the sun had three to account for its apparent movement along the ecliptic.

It should be pointed out that in all likelihood Eudoxus did not believe in the literal existence of his spheres. Since mortals could never hope to ascend to heaven like the gods, the real nature of heavenly movement would forever be a mystery to them. But because they had active, inquiring minds, humans could achieve a *possible* explanation, based on the data available. The Greeks approached the problem of heavenly motion somewhat in the manner of an intellectual puzzle. Considering the fact that all they had to work with were their eyes and their brains, the achievement of the Greek astronomers is impressive.

The spheres of Eudoxus were adopted by the philosopher Aristotle (384–322 BC) who was, like Eudoxus, a student of Plato. Aristotle accepted Eudoxus' model of the universe only after modifying it in accordance with his own investigations. Convinced that 27 spheres were an insufficient number to account for the movements of the heavenly bodies, he devised an alternative arrangement involving 55 spheres.

Aristotle's writings on astronomy are also notable for presenting logical proofs of the spherical shape of the earth. Aristotle argued that the earth must be a sphere because during a lunar eclipse the shadow of the earth which is thrown onto the surface of the moon is always circular in shape. Since only a sphere has a circular shadow no matter which direction the light is coming from, the earth must therefore be a sphere. Aristotle's second proof has to do with the fact that when people travel either to the north or south, new stars become visible at the horizon which could not be seen before. The only explanation for this phenomenon is that the earth is a sphere.

Aristotle's approach to scientific problems was very thorough, and he generally did not reject a theory without first presenting all the logical arguments against it. This procedure is evident in his treatment of the theory that it is the earth and not the heavens which is in motion. Aristotle's reasons for rejecting the theory of the moving earth are threefold: (1) he believed that the earth was too heavy to move; (2) there are no signs that it is in motion; and (3) if the earth revolved around the sun, we would be able to observe stellar parallax. Stellar parallax, as we learned in Chapter 2, is the apparent shift of the stars caused by the movement of the earth around the sun. Stellar parallax does exist, but because the stars are so distant compared to the size of the earth's orbit, the shift cannot be seen with the naked eye.

Early Greek astronomers thought that the heavenly bodies were fixed to enormous transparent spheres centered on the earth.

A Dissenting Voice

Not all the ancient astronomers were convinced by Aristotle's arguments against a moving earth. A few advanced the equally cogent idea that the sphere of fixed stars was far too huge to spin on its axis every 24 hours. It made far more sense to suppose that the movement was caused by the rotation of the earth. The first person to suggest a truly sun-centered or **heliocentric** model for the universe was Aristarchos, around 275 BC. Aristarchos believed that the earth, along with all the other planets, traveled around the sun. The main advantage of such a scheme is that it explains the phenomenon of retrograde motion without resorting to Eudoxus' complicated system of concentric spheres. If the earth and the other planets are orbiting the sun at different distances, then it is logical to assume that they take different amounts of time to complete one revolution — the greater their distances from the sun, the greater the time.

Let us take one of the outer planets as an example — say, Mars (Figure 12–1). In position 1 observers on earth would see Mars moving in an easterly direction relative to the background of fixed stars. In position 2 the earth is catching up with and overtaking Mars. At this point Mars would appear to stand still. In the third position, earth is leaving Mars behind. Mars now seems to be moving backwards. You can see the same effect when you pass a car on the highway if you observe it carefully with respect to objects in the background. The retrograde motion of Mars con-

Figure 12-1
The retrograde motion of Mars at a recent opposition. The earth, being closer to the sun than Mars, moves faster and has a shorter orbital period; the Martian year is 697 days long. As the earth catches up with and passes Mars, every 26 months, Mars appears to reverse its motion against the stars. For about 2 months, it moves from east to west, before resuming its normal course from west to east.

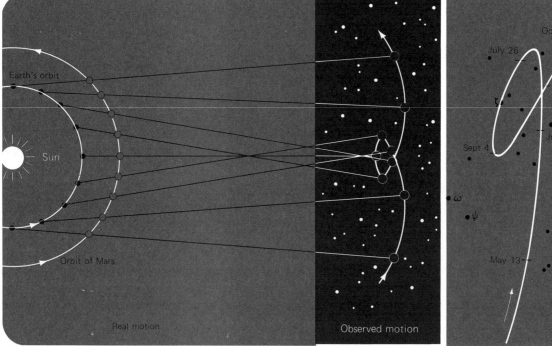

Aristarchos was the first
thinker to develop a sun-
centered model of the solar
system.

tinues until earth arrives at position 4, at which point the outer planet resumes its easterly motion once again. Aristarchos' model also explains
the periodic brightening of the planets. At the point where retrograde
motion begins the planet is closest to the observer and therefore appears
at its brightest.

But despite the advantages of Aristarchos' model, it failed to replace
the earth-centered or **geocentric** model of the universe. This was partially because, as far as we know, Aristarchos never published a treatise
formulating his ideas. More importantly, however, the heliocentric theory
fell by the wayside because it implied a view of the universe which at the
time most people were unwilling to accept. It followed from the stellar
parallax argument that if the earth was in motion, the stars were an almost inconceivable distance away. Remember that the ancient Greeks
had absolutely no idea that the stars were actually suns of various luminosities dispersed at various distances through space. To them the stars
were bright points of light fixed to a huge sphere. If Aristarchos was right,
then this sphere was of staggeringly vast dimensions, with the earth, sun,
moon, and planets stuck at the center of all that emptiness. What would
be the purpose of such isolation? No, it made far more sense to suppose
that the sphere of fixed stars was of a reasonable size and that it, along
with the other heavenly bodies, revolved around the earth—the motionless center of the universe.

While Aristarchos' geocentric system did not prove popular, much
of his other work had a considerable influence. Aristarchos lived at a time
when empirical approach to astronomy with emphasis on accurate observation and measurement was beginning to take precedence over de-
votion to idealistic concepts. One of Aristarchos' greatest contributions
was his calculation of the relative distances and sizes of the moon and
sun, calculations which were later refined by his successor Hipparchos,
who lived during the second century BC. Because of various erroneous
assumptions, both men miscalculated the size and distance of the sun.
The figures they obtained indicated that the sun is only about 20 times
further away than the moon. Since the two bodies appear to be about
the same size in the sky, this meant that the sun must really be about 20
times larger.

Actually the sun is almost 400 times more distant than the moon.
Nevertheless, Aristarchos' work represented a great step forward, in that
it established that the sun was in fact considerably farther away than the
moon. The method Aristarchos devised for determining the relative sizes
of the earth and moon was even more successful. When carried out, with
improvements, by Hipparchos, it yielded values remarkably close to the
truth. Another Greek astronomer, Eratosthenes, developed a way of
finding the diameter of the earth that also gave results very close to those
accepted today (see Box).

Two Greek Calculations

I. Using a method first developed by Aristarchos, Hipparchos found the size and distance of the moon in the following manner. He knew from observation that the angular diameters of the sun and moon in the sky are about $1/2°$. He also observed lunar eclipses, and from the time it took the moon to cross the earth's shadow, he estimated that the shadow cone is about $8/3$ the diameter of the moon. His third bit of information was the ratio of the sun's distance to that of the moon: about 20 to 1, from the calculations that Aristarchos had performed before him. As we have seen, this figure is seriously in error; fortunately, though, the error does not make a very large difference in finding the size of the moon.

Hipparchos constructed a diagram incorporating these relationships. First he drew two lines which intersect at an angle of $1/2°$. The point of intersection E represents the center of the earth. He then drew the sun and moon, making the sun 20 times further from the point of intersection E than the moon. Now, all that remained to be done was determine the relative diameter of the earth. Knowing that the size of the earth's shadow was $8/3$ that of the moon, he drew it

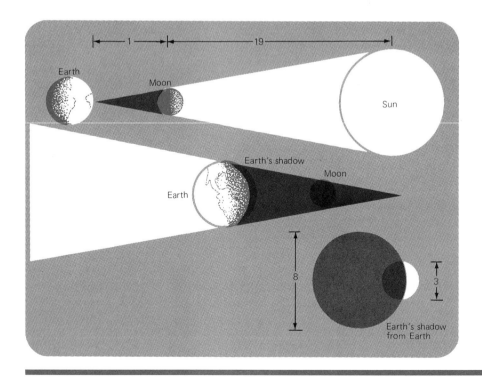

to scale at M, along with two lines tangent to the sun. The earth could then be represented as a circle tangent to these two lines with its center at E. In this way the relative diameters of the earth and moon could be found simply by measuring the diagram. Using this method, Hipparchos concluded that the moon was $\frac{1}{3}$ the size of the earth, (remarkably close to the true figure of .27), and that its distance was about 30 times the earth's diameter, which is almost exactly correct. As an exercise, try duplicating these steps.

II. To convert these relationships into actual units of distance, it was necessary to know the diameter of the earth. This had been determined with rather good accuracy using a method first introduced by Eratosthenes. Eratosthenes knew that at the city of Syene, Egypt, on the first day of summer, the rays of the sun struck the bottom of a deep well. This meant that the sun must be directly overhead. He then found that on the same day, in the city of Alexandria, 5000 stadia to the north, objects cast shadows, indicating that there the sun was **not** directly overhead. From the lengths of these shadows, he could easily calculate that the sun's rays made an angle of 7° with the vertical. Seven degrees is about $\frac{1}{50}$ of a full circle. Since the sun's rays were approximately parallel, due to the great distance of the sun, the distance between Syene and Alexandria must be $\frac{1}{50}$th of the earth's circumference (Figure 12–B). The earth's circumference must be 50×5000 stadia, or 250,000 stadia. The stadium was not a precise measurement, and authorities differ as to the value that Eratosthenes gave it. According to one theory, however, he used a stadium which was equivalent to $\frac{1}{6}$ km. This would have given a figure very close to the correct one of 40,000 km.

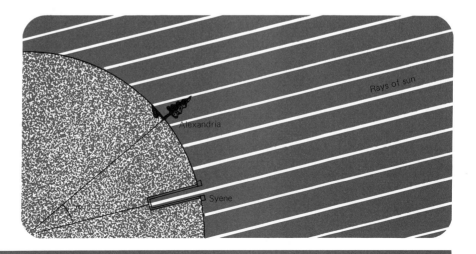

The System Perfected

Since the heliocentric model was unacceptable, virtually all notable astronomers after Aristarchos devoted themselves to perfecting the geocentric model. Apollonius and Hipparchos were two of the chief figures responsible for developing a convincing "machinery" to explain the paths of the sun, moon, and planets around the earth. The principal devices of this machinery were the deferent, and the epicycle. The **deferent** is a large circle whose center is located at the center of the earth. The **epicycle** is a smaller circle whose center is a point on the rim of the deferent. The planet is attached to the rim of the epicycle. The deferent revolves around the earth, causing the planet's gradual motion across the sky. Meanwhile, the epicycle also revolves, causing the planet's looping path (Figure 12–3a). The solution was a brilliant one since it successfully explained complex phenomena using a relatively simple mechanism.

Although the deferent and epicycle gave a fairly good approximation of heavenly motion, there was still much room for improvement. The task of perfecting the machinery of the geocentric system fell to a man whose name has become synonymous with the theory of a motionless earth: Claudius Ptolemy. Working with data Hipparchos and others had collected over the years, he succeeded in constructing a model of an

To make the earth-centered system work, it was necessary to introduce epicycles: circles whose centers moved on other circles.

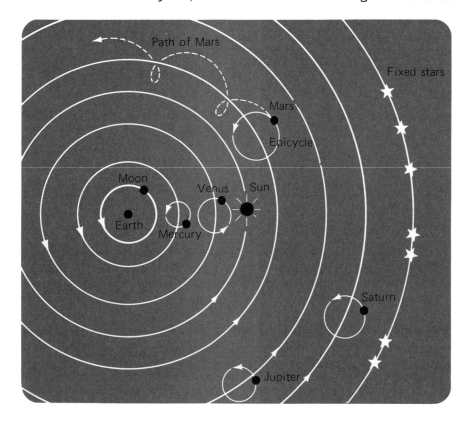

Figure 12-2
The Ptolemaic system (greatly simplified). All the heavenly bodies move about the earth. Epicycles and other geometrical complications (Figure 12–3) are introduced to account for the apparent paths of the sun, moon, and planets, while preserving the principle of uniform circular motion. The stars occupy a spherical shell outside the orbit of Saturn, so that they are all equally distant from the earth. The entire apparatus revolves about the earth every 24 hours, producing night and day. In addition, each body has its own individual motion.

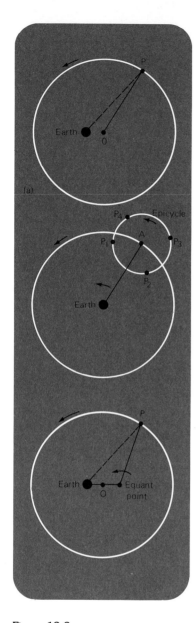

Figure 12-3
a) The eccentric. Seen from earth, the planets motion does not appear uniform. b) The epicycle and deferent. Because of the planet's movement on the epicycle, it exhibits retrograde motion during part of its orbit. c) The equant. The movement of the planet is not uniform with respect to the center of the circle (C) or the earth (E).

earth-centered universe which accounted with a high degree of accuracy for all of the observed motions of the heavenly bodies. Moreover, his model could predict the position which the moon, sun, or one of the planets would occupy at some date in the future.

Ptolemy paid a certain price for this success, however, in the elegance and consistency of his model. He had to accept some motions in the heavens that were not uniform — something that earlier Greek thinkers had been at pains to avoid. Hipparchos had introduced the device of **eccentrics:** circular orbits with the earth slightly off center. To explain the irregularities of planetary motion, Ptolemy was forced to go further. He made use of an innovation of his own, the equant. The **equant** was a point from which the motion of a planet would appear uniform, but located neither at the earth nor at the center of the planetary orbit (Figure 12–3c). The use of equants was a covert way of backing off from the principle of uniform circular motion in the heavens.

There is no doubt about the fact that Ptolemy's achievement was a great one. Looking back on his work, one is tempted to treat it condescendingly, to feel that he missed the main point and that, hence, all his effort was in vain. But this is not really so. In a purely geometric sense, the geocentric theory is just as "right" as the heliocentric theory. Scientists ask of a theory that it be true to the observed facts and that it have predictive power. Ptolemy's model fulfilled both of these requirements. It is only in the light of later findings that the geocentric model breaks down. The observations of Kepler and Galileo, as well as the gravitational theory of Newton, proved irreconcilable with Ptolemy's model, and the geocentric model had to be abandoned. But given the information available in 150 AD, the time that Ptolemy did his work in Alexandria, the geocentric model seemed the most logical solution to the problem of heavenly motion.

Ptolemy was the last great geocentric theorist, and his model was to enjoy a life of many centuries, during which time it was to grow from a theory into a dogma. Not only did the Ptolemaic model win the approval of common sense and scientific opinion, it also harmonized with a view of humanity's place in the universe which was rapidly gaining acceptance. With the growth of Christianity, spiritual concerns came to the fore. The earth came to be seen as a stage on which God and the devil strove for the possession of the human soul. It seemed appropriate for the earth to be the center of things physically as it was spiritually. To have the earth flung as if at random into the midst of an inconceivably vast cosmos would not at all have suited the mood of the times. In the thirteenth century, St. Thomas Aquinas incorporated the Ptolemaic model into his system of thought. St. Thomas' philosophy was soon accepted as official Roman Catholic doctrine. Hence, it became heresy to suggest that the earth went around the sun.

Yet, a mere 300 years later, the Ptolemaic system was to be supplanted by a new version of Aristarchos' heliocentric model. The overthrow of Ptolemy's venerable machinery was one result of a drastic change that had affected the thinking of the Western world. Many long-established ideas, in philosophy and the arts as well as in science, were profoundly shaken by this change. Today we call this great transformation, which swept over Europe from the 15th to the 17th century, the Renaissance.

One of the most important features of the Renaissance was a rebirth of the spirit of originality. Throughout the middle ages there was an essentially conservative character to most intellectual endeavors. Most learning took place under the auspices of the Church. There was a tendency to revere ancient authority and to discourage anything that would change it. As the Renaissance progressed, however, scholars grew more confident of their own intellectual ability. They began to ask what was preventing them from formulating their own answers to the questions of science and philosophy instead of meekly accepting opinions that were handed down to them.

One man who dared to hold an opinion which conflicted with official dogma was a fifteenth century cardinal, Nicholas of Cusa, who believed that the earth was in motion. Nicholas of Cusa never published his ideas, however. The first model of a heliocentric system to obtain widespread recognition was the work of another churchman—a physician and amateur astronomer who was born in 1473 in Torun, Poland. His name was Nicholas Copernicus.

During the renaissance, when many classical ideas were being rediscovered, Copernicus revived the earth-centered model of the solar system.

Copernicus

Copernicus' break with the past was by no means a complete one. Like Pythagoras and Plato before him, Copernicus was convinced that the sphere and circle were the most perfect of geometric shapes. Since the heavens were perfect, their motion must be the result of some combination of circular orbits. The feature of Ptolemy's system which most displeased Copernicus was the former's use of the equant to account for irregularities in the motions of the planets. The trouble with the equant, from Copernicus' point of view, was that it implied planetary motion which was not uniform, and uniformity of motion, along with perfect circularity in orbital shape, was still accepted as one of the "givens" of heavenly movement. As Copernicus remarked, in reference to the Ptolemaic models,

> For these theories were not adequate unless certain equants were also conceived; it then appeared that a planet moved with uniform velocity neither on its deferent nor about the center of its epicycle. Hence a system of this sort seemed neither sufficiently absolute nor sufficiently pleasing to the mind.

Figures 12-4, 12-5
The Ptolemaic and Copernican systems, as depicted in an early atlas of astronomy.

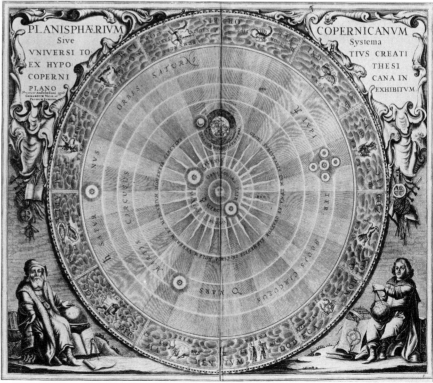

It is worth noting that Copernicus did not reject the Ptolemaic model because he felt that it was not consistent with the observed facts. As we have seen, the "facts," as they were understood in Copernicus' time, seemed, if anything, to support the geocentric view. Rather Copernicus began his reevaluation of Ptolemy's system because it was not "pleasing to the mind." Today scientists still consider this quality to be an important indication of a theory's validity.

Copernicus spent 40 years perfecting his system, which he described (in Latin) in his book *De Revolutionibus Coelestium,* or *On the Revolutions of the Heavenly Spheres.* There is reason to think that Copernicus feared the criticism that his theory would arouse, for he postponed publication of the book until the very end of his life. (In fact, the first copy was brought to him as he lay on his death bed in 1543.) As a further precaution against controversy, the book contained a preface (not written by Copernicus) which stated that the geocentric model was no more than a convenient method of calculating the future positions of the planets and not to be taken as literal fact. Nevertheless, Copernicus' model soon proved an intellectual bombshell, especially among mathematicians who were quick to recognize its elegant simplicity and explanatory power.

Copernicus assumed that the earth, rather than being the center of the universe, is merely one of a number of planets which are circling the sun. Making this supposition allowed him to explain the looping path of the planets in a much simpler way than Ptolemy had done. According to Ptolemy, the loops in the planets' orbits were the result of a system of epicycles. Copernicus believed, however, that the looping was only apparent, not real. Like Aristarchos before him, he explained that it was the product of the unequal rates of motion of the planet and the observer.

Copernicus' system allowed him to make certain deductions about the arrangement of the planets which were impossible for Ptolemy. He was able, for example, to place the planets in their correct order: Mercury nearest the sun, then Venus, earth, Mars, Jupiter, and Saturn. He was able to calculate with a high degree of accuracy each planet's orbital period—the time it takes to complete one revolution around the sun. The values he obtained correspond closely with those computed by present-day astronomers. Using this data, he was then able to calculate the distance of each planet from the sun. His unit of measurement was the astronomical unit—the distance from the sun to the earth. Since the astronomical unit had not yet been determined in Copernicus' time, his values for the distances of the planets were relative, not absolute. But they provided an idea of the proportions of the solar system.

Although the Copernican system was mathematically satisfying, it was, to many, philosophically repugnant. It assumed that the earth was one planet among several. What then of the other planets? Were they similar to earth? Were they made of rock, soil, water, rather than of some

Copernicus retained the ancient idea of uniform, circular motion for all the heavenly bodies.

Figure 12-6
The title page of Copernicus' epoch-making book outlining his model of the solar system. It appeared in 1543, just before his death.

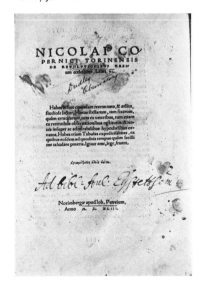

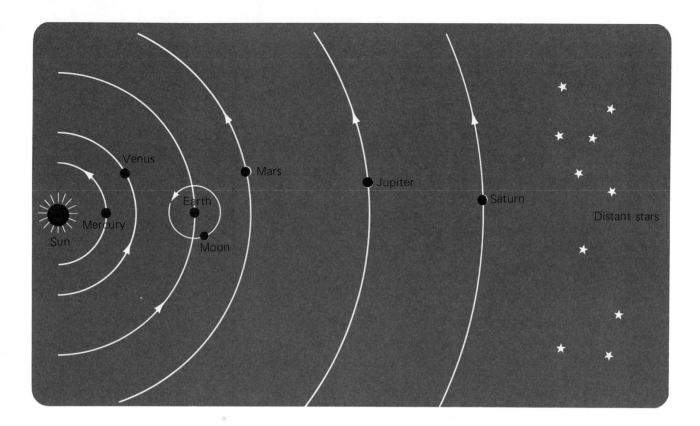

Figure 12-7
The Copernican system. The sun is at the center of the solar system. The earth rotates on its axis once every 24 hours, causing the heavenly bodies to rise and set, and also revolves about the sun, as do the other planets. The moon moves in an orbit around the earth. Copernicus believed that all the orbits were perfect circles—an idea that Kepler later disproved. As a result, Copernicus' model was also encumbered with geometrical makeshifts.

heavenly substance as had always been assumed? Did they have inhabitants? If so, were they Christians? These were serious questions to people of Copernicus' time, and the frightening implications of the heliocentric system were enough to cause many intelligent people to reject it. Martin Luther, for example, a man who had begun as great a revolution in the Church as Copernicus had in science, unsympathetically called Copernicus "the fool who would overturn the whole science of astronomy."

For more than half a century, the Copernican system was only the rival of the Ptolemaic, appealing to some, rejected by others. The final victory of the heliocentric model was delayed because there was still no proof that it conformed more closely to observed reality than the geocentric model. And due to Copernicus' erroneous assumption that the orbits of the planets are perfect circles, his system was no more accurate than Ptolemy's at predicting the planets' future positions. Copernicus began the revolution in astronomy, but the triumph of his system would not have occurred if it were not for the work of several men whose combined contributions placed the validity of the heliocentric model beyond doubt.

The first of these was the Danish astronomer Tycho Brahe (see Feature). Tycho is remembered not for his direct contributions to the debate over the Copernican system, but for his observations. Without question he was the greatest observer of the pretelescopic era. When he

Tycho Brahe was born in 1546, three years after the publication of Copernicus' book on the solar system. His family belonged to the Danish nobility. While still in school, Tycho developed a consuming interest in astronomy. One of the causes was a total eclipse of the sun that took place on August 21, 1560, when Tycho was 13. Even more miraculous than the event itself, to Tycho, was the fact that it had been predicted by the astronomical tables of the day. That the heavenly realm could be so precisely understood by the human mind made a lasting impression on him.

But while observing the stars might be acceptable as a hobby, it was not something which a nobleman did as his lifework. Consequently, the young Tycho had to pursue his interest in secret, risking the displeasure of his guardian-uncle. After his uncle's death, however, Tycho was able to spend his time and his money as he pleased. He set up an observatory which he equipped with quadrants, sextants, and other instruments which were the most precise and finely constructed of their day.

On November 11, 1572, Tycho saw something that surprised him. An object had suddenly appeared in the constellation Cassiopeia which was brighter than the planet Venus and was clearly visible in the daytime. In Tycho's words,

> . . . since almost from boyhood, I had known perfectly every star in the sky (there is no great difficulty in gaining this knowledge), it was obvious to me that there never had been a star at that place in the sky, not even the smallest, to say nothing of one so conspicuously bright as this: it was so astonishing that there was no shame in doubting the trustworthiness of my own eyes.

According to traditional beliefs, the heavens were eternal and unchanging. Since new stars were an impossibility, objects such as the one Tycho observed must not be part of the heavens, but rather phenomena within the earth's atmosphere. Tycho, however, decided to find out for himself; he tried to observe the parallax of the object. If it was truly close to the earth, it should shift position when viewed from

two widely spaced locations. The object exhibited no parallax. Therefore, Tycho concluded that it was a star. Changes could take place in the heavens after all. Tycho observed the new star for 16 months until it faded from sight. It is now believed to have been a supernova and is referred to as "Tycho's star."

Tycho wrote a book about the new star which, along with his other accomplishments, favorably impressed King Frederick II of Denmark. He rewarded Tycho by presenting him with a small island called Hveen and enough money to build an elaborate observatory complex on it. Indeed, Tycho's establishment was in many ways the forerunner of today's great scientific research institutes. Tycho later estimated that over a ton of gold had been spent on the facilities on Hveen.

The complex included two large buildings: Uraniborg, or castle of the heavens, and Sterneborg, castle of the stars. There were four large observatories, laboratories, machine shops, a library, a printing press complete with its own paper mill, living quarters for servants, assistants, and visiting scholars — even a jail. The room of Uraniborg were filled with works of art, including statues that could be made to talk (they were connected to hidden speaking tubes), and there was running water from a basement well, a great luxury in those days. Small wonder that the peasants of Hveen regarded Tycho as a wizard—an attitude still common today, as we can see from any number of films about "mad scientists."

At Hveen, Tycho designed and constructed instruments whose accuracy was unparalleled. They were larger than the instruments used by previous astronomers, more carefully calibrated, and more firmly mounted. Tycho even calculated their margin of error, something that makers of modern scientific apparatus now do as a matter of course. He was also careful to introduce into his observations carefully worked out corrections for atmospheric refraction: the bending of light rays as they pass through the layers of air above us.

During more than 20 years of work at Uraniborg, Tycho single-mindedly amassed the most complete and exact body of astronomical data which the world had known up to that time. In 1577, he witnessed the appearance of a bright new comet. He checked it for parallax as he had the new star, and found that it too was a denizen of the "unchanging" heavens. Once again, doubt was cast on the ancient beliefs; change **had** taken place in the astronomical realm.

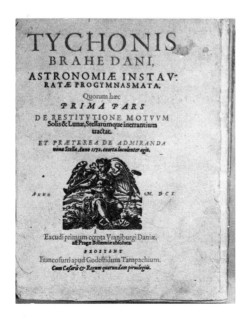

In addition to being a superb observer, however, Tycho was an arrogant and quarrelsome man. As a young man, he had the misfortune to lose his nose in a duel and was subsequently obliged to wear an artificial one made of silver and gold. His irascibility may have been the deciding factor which made the new king, Christian IV, discontinue Tycho's royal stipend after the death of Frederick II in 1597. Another contributing factor may have been the discontent of the peasant tenants of Hveen, from whose rents Tycho's observatory was supported. Evidently Tycho, like many landlords, was not very conscientious about providing needed maintenance work on the island. The peasants seem also to have resented his taking a local peasant girl as his common law wife. In any event, Tycho was forced to abandon Uraniborg and seek another haven where he could continue his studies. Finally, Emperor Rudolph II of Bohemia made him court astronomer, and Tycho was able to resume his work, this time in Prague. Tycho died 3 years later in 1601—of a burst bladder during a night of extravagant feasting and drinking.

died in 1601, Tycho left behind him a great volume of remarkably accurate data, especially on planetary motions. Fortunately for the history of astronomy, in the year before his death Tycho had employed a gifted young mathematician named Johannes Kepler. By a stroke of good luck, many of Tycho's records fell directly into Kepler's hands. Building on these observations, Kepler was able to free the heliocentric theory from Copernicus' erroneous assumption that all planetary motion had to be uniform and circular. This proved to be the decisive breakthrough.

Kepler

Kepler had been hired by Tycho on the strength of a book he had written, **Mysterium Cosmographicum,** or **Mystery of the Universe.** In the book, Kepler set himself the task of discovering why the orbits of the planets were arranged as they were. Kepler was struck by the fact that there were 6 (known) planets and 5 regular geometric solids. He believed that there ought to be some way of fitting the cube, the dodecahedron, the tetrahedron, the octahedron, and the icosahedron inside one another in such a way that the distances between their outermost limits would correspond to the distances between the planets. After much juggling with these shapes, Kepler succeeded in arranging them so that they fit Corpernicus' calculations of the planetary orbits to within about 5 percent. Kepler was enormously pleased with his solution to this problem, although, as we realize today, the correspondence of orbits to geometric shapes is entirely accidental.

Although Kepler's work with geometric solids is not what he is remembered for, it does tell us something about the man and about the age he lived in. First, it tells us of Kepler's admiration for the Copernican system. Copernicus' heliocentric solution to the problem of heavenly motion filled Kepler with "incredible and ravishing delight." Second, the fact that such a project could be admired by Tycho, the foremost astronomer of the day, shows us how little scientific thought had changed from ancient times. Like Pythagoras and Plato, Kepler was convinced that the heavens must be based on perfect geometric shapes. Later he was to alter this view.

The first project to which Kepler was assigned when he came to Prague was to use the observations made by Tycho in order to accurately plot the orbit of Mars. It took Kepler a year and a half to convert Tycho's figures into a coherent orbital shape. The shape which he obtained, however, was a puzzling one. It was definitely not a circle. Another astronomer might have attributed the discrepancy to observational error, but Kepler knew that he could not take this way out. Tycho's observations were the most accurate ever made. They simply did not contain mistakes of this magnitude. Kepler was forced to conclude that, in spite of the assumptions of 2000 years of astronomy, the planets did not travel in perfect circles.

THE LAWS OF PLANETARY MOTION   If the orbit of Mars was not a circle, what was it? Kepler found that the orbit of Mars conformed to a shape which had been known and studied since the time of the ancient Greeks. The shape, as we saw in Chapter 5, was an ellipse. Kepler published the results of his study of Mars in 1609. The book contained his first law of planetary motion. It states that *the orbits of the planets are ellipses with the sun at one focus.* It was indeed fortunate that Kepler began his study with the planet Mars. Of all the planets for which complete data was available, Mars was the one whose elliptical orbit is most pronounced. The orbits of the other planets are so close to circles that Kepler might never have noticed their elliptical nature.

In addition to plotting the shape of Mars' orbit, Kepler also calculated the planet's speed at various points along its orbital path. He noticed that as Mars drew closer to the sun, its speed increased. When it travelled away from the sun, its speed decreased. Kepler expressed the relationship between distance from the sun and orbital speed in the second of his laws of planetary motion, called the law of areas. It states that *a line drawn from the planet to the sun sweeps out areas which are proportional to the time intervals* (Figure 5–4).

With these two simple laws, Kepler could do something which no astronomer had ever been able to do before. Given accurate observational data, he could plot the future position of any planet with absolute precision. But even with this great achievement behind him, Kepler was not satisfied. He still had not found a law which described the relationship between the motions of the different planets. It was the same problem he

Figure 12-8
a) Kepler's method of plotting the orbit of Mars. After laborious calculations, he determined that the orbit of Mars was an ellipse rather than a circle. b) The oppositions of Mars for the period 1971–1986. Notice how much more eccentric is Mars' orbit compared to that of the earth. As a result, the distance of closest approach varies between 89 million km and 51 million km.

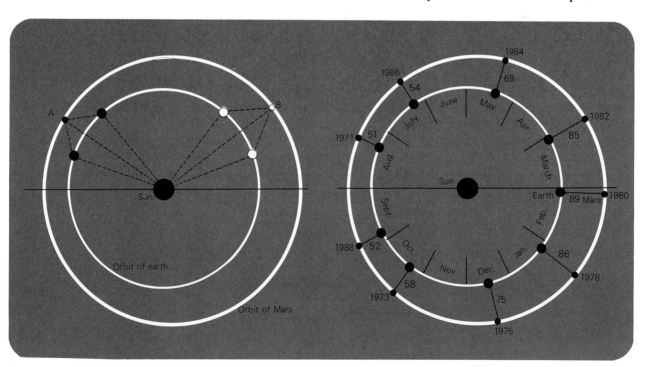

had once tried to solve by trying to relate the orbits of the planets to the five regular solids. Now, however, he wanted to discover a relationship that would be consistent with the laws he had formulated as the result of his study of the planet Mars.

Kepler had no computer to help him in his search, and so he had to learn the needed mathematics and work each lengthy computation out on paper—a very time consuming process. At last, after many years of work, his trial and error method yielded a simple algebraic formula relating the period of a planet to its distance from the sun. The formula appeared in his book, *Harmony of the World,* published in 1619. It is generally referred to as the ''harmonic law.'' It states that *the squares of the periods of the planets are proportional to the cubes of the semimajor axes of their orbits.* The law is usually expressed in algebraic terms as

$$P^2 \propto a^3, \text{ or } P \propto a\sqrt{a}$$

In this equation, P is the planet's period, or the time it takes to revolve once around the sun, and a stands for the semimajor axis of the planet's orbit. The semimajor axis is simply half the major axis, and is equal to the average distance of the planet from the sun. Astronomers have found that Kepler's third law is also valid for comets, asteroids, and even, as we saw in Chapter 5, for double stars.

Kepler modified Copernicus' scheme by showing that the planets moved in elliptical orbits which could be described by mathematical laws.

Kepler's work gave the science of astronomy a concise formulation of the laws governing heavenly motion. In his insistence that these laws correspond, without deviation, to observed reality, Kepler is the personification of the modern scientist. Yet there was an aspect of Kepler which remained deeply rooted in the mysticism of previous times. A large portion of his book *Harmony of the Worlds,* for example, was concerned with a notion which hearkens back all the way to the doctrines of Pythagoras, namely that of the ''music of the spheres.'' Kepler tried to find parallels between the numerical relations of the solar system and the notes of the musical scale. He believed that each of the planets ''plays'' a characteristic melody. These celestial tunes are written out in his book in standard musical notation.

It may be that beliefs such as this survived intact into Kepler's day because, essentially, the sky was still the same unreachable dome of wonders it had been in ancient times. The planets were still bright points of light tracing out elaborate patterns among the stars. The moon was still a bright disc of crystal; the sun, a fiery deity dispersing heat and light. Kepler, by dint of great intellectual ability and greater persistence, had at last sketched out the plan of the solar system as it actually was. His effort was extremely convincing to those who were able to follow his reasoning and computations, but for the rest the universe had not greatly changed. In order to alter the average person's mental picture of the universe, something more was needed besides Kepler's abstract theorizing. That ''something more'' was provided in large measure by a contemporary of Kepler, an Italian mathematician and physicist name Galileo Galilei.

Galileo

Galileo was born in Pisa in 1564. Beginning his studies as a medical student, he soon changed to mathematics, for which he proved to have an unusual aptitude. As a young man, Galileo gained a reputation as a difficult, quarrelsome person. His contemporaries gave him the nickname "wrangler" because he demanded proof of scientific assumptions and theories which, at the time, were considered self-evident. Nevertheless, his talents earned him great renown. He was given important posts at some of the best universities in Italy and became famous throughout Europe for his scientific knowledge.

Galileo was particularly interested in the nature of motion and the effect of force on bodies—topics which fall under the category of **mechanics.** This field was still dominated by the theories of Aristotle. Galileo, however, went about discovering mechanical principles in a new way, a way that was similar to Kepler's approach to planetary motion. Instead of accepting the age-old assumptions about force and motion, Galileo performed experiments himself, then drew general conclusions from the results. Aristotle, for example, had said that heavy objects fall faster than light objects. People had accepted this rule for centuries simply because it had been stated by Aristotle. Galileo was the first person to subject it to an empirical test. He dropped heavy and light objects and observed their descent. If one made allowances for the drag of the air on the lighter object, both hit the ground at the same time. Experimenting further, Galileo discovered the rate of increase in the speed of a falling body. He observed pendulums and found that the period of their swing depends only on their length. He was the first person to observe that a force is needed to make an object stop moving just as one is needed to make it start. And like Kepler, he formulated his discovery in precise, algebraic terms.

Galileo was an experimental scientist, and also a skillful popularizer of his ideas.

Galileo accepted the heliocentric model of the solar system—an unpopular opinion to have in sixteenth century Italy. The Ptolemaic system was still the official doctrine of the Roman Catholic Church. For over 1500 years Catholicism had identified itself with the view that the earth is the center of the universe, and it was not about to change without a struggle. Supremely convinced of the truth of his own opinions, Galileo was quite prepared to give it one. He only needed some extra proof of the heliocentric model to swing the argument over to his side. Then in 1609 he found the proof he was looking for.

GALILEO'S TELESCOPE   A Dutch lens maker had constructed a device through which distant objects "could be seen as if nearby." Hearing of the device, Galileo quickly reproduced it, perfecting the system of lenses until he had achieved a magnification power of 30 times. Then he turned the device skywards at the moon, the planets, and the stars. The era of the telescope had begun.

With this small instrument Galileo probably made more basic discoveries about the heavens than any astronomer since. He discovered that there were many more stars in the sky than could be seen with the naked eye. The Milky Way, which had always appeared to be a faint white glow, proved to be made of multitudes of stars, too dim to be seen with the naked eye. Turning his telescope on the moon, Galileo discovered that that body was not perfectly smooth as had always been thought, but covered with rough craters, mountains, and dark areas he believed were seas.

Two discoveries were of prime importance in the debate between heliocentrism and geocentrism. First, he found that the planet Venus goes through phases like the moon, changing from crescent to full. This

Figure 12-9
The phases of Venus, first observed by Galileo with his primitive telescope in 1610. The fact that Venus exhibits phases was strong evidence for the Copernican theory.

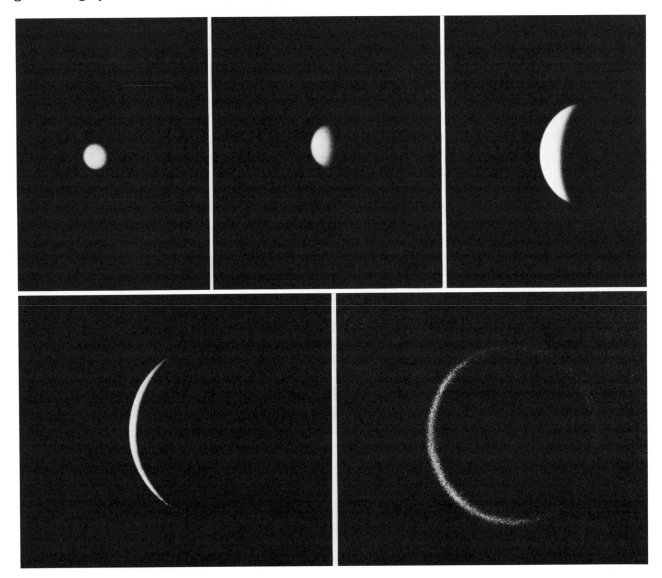

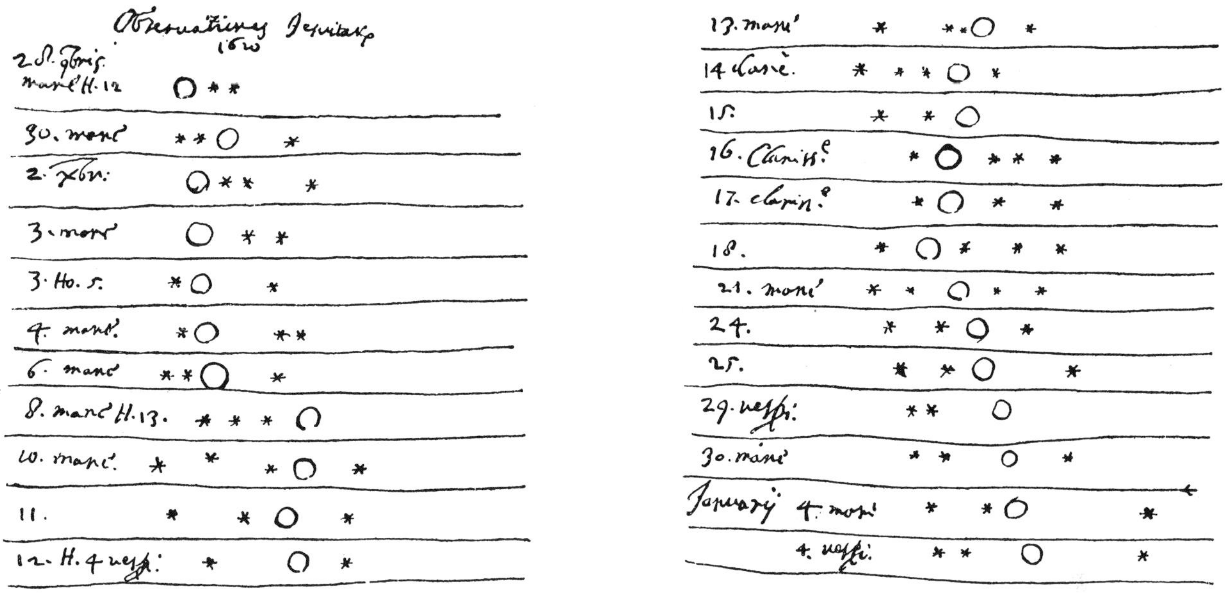

Figure 12-10
The motions of the 4 largest moons of Jupiter, discovered by Galileo, as recorded in his notes of 1610. The fact that a moving body could itself be a center of motion was another support for the Copernican model.

Galileo's observations of the phases of Venus and the motions of Jupiter's moons did much to win support for the Copernican model.

could only happen if the planets went around the sun. If Venus traveled around the earth, it would always appear as a crescent. Second, Galileo discovered the four largest moons of Jupiter. One of the chief arguments against the idea that the earth was in motion was that if the earth went around the sun the moon would be left behind. The discovery of Jupiter's moons proved that an object which moved could itself be a center of movement.

Galileo published his findings in a little book called **Sidereal Messenger.** His revelations about the heavens were startling, and the book sold many copies. It also aroused numerous objections. Adherents of the Ptolemaic system attacked Galileo bitterly. At last the Church stepped in. A decree of 1616 forbade anyone to ''hold or defend'' the Copernican hypothesis.

Despite this setback, Galileo was by no means vanquished. He obtained permission from Pope Urban VIII to publish a book which, he claimed, would explain in detail both the Ptolemaic and Copernican systems, leaving it up to the reader to decide which was correct. The book, **Dialogue on the Two Great World Systems,** appeared in 1632. Galileo had written it in Italian rather than Latin so as to make it more accessible to the average reader. The two systems are set out during the course of a long discussion between three philosophers. It is obvious almost from the first that Galileo's sympathies lie with the defender of the heliocentric model. He is the most quick-witted and completely demolishes the absurd arguments of the Aristotelian philosopher who defends the geocentric system. The third character, who starts out being uncommitted, is soon won over to the Copernican view.

The Church was angry with Galileo, not only for holding the Ptolemaic model up to ridicule, but for using fraud to procure the opportunity to defend his views. Galileo was tried by the Inquisition and forced, under threat of torture, to plead guilty to the charge of holding doctrines which were false and in conflict with Divine Scripture. It is said that as Galileo publicly swore his allegiance to the doctrine that the earth is stationary, he was heard to mutter under his breath, "But still it moves." His book, along with those of Copernicus and Kepler, was placed on the Church's *Index of Prohibited Books.* Galileo himself, already an old man, was placed under house arrest until his death in 1642. *Dialogue on the Two Great World Systems* was not removed from the *Index* until 1835, but by then the new concepts of the heavens which Copernicus and Kepler had formulated and Galileo had popularized had long since spread throughout the world.

SUMMARY

A number of early civilizations made careful astronomical observations, kept records of heavenly motions, and even learned to predict celestial events such as eclipses. But the Greeks were the first to bring to the study of astronomy a high degree of mathematical ingenuity. With their geometrical skill, Greek astronomers successfully measured the sizes of the earth and moon and the moon's distance with considerable accuracy. More importantly, the Greeks were not content merely to observe and to measure the motions of the heavenly bodies. They also tried to explain these motions. They devised models of the solar system that might account for the appearence of the skies seen from earth.

Though Aristarchos (about 275 BC) proposed a model of the solar system in which the earth moved about the sun, most Greek astronomers preferred an earth-centered model. One reason for this was the widely-held belief that if the earth were in motion, objects on its surface would be hurled off into space. Another was the failure to detect any stellar parallax. To account for the observed motions of the heavens, using an earth-centered system, an elaborate geometrical machinery of circles revolving on other circles was devised. The system was given its definitive form by Ptolemy about 150 AD, and remained unchallenged thereafter until the Renaissance in Europe.

In 1543 Nicolaus Copernicus revived the sun-centered theory of Aristarchos. Copernicus persisted, for philosophical reasons, in clinging to the ancient idea that all planetary motion must be uniform and circular. In the second half of the sixteenth century the greatest observer of the age, Tycho Brahe, amassed a remarkably accurate body of data on planetary motions. Working from this data, Johannes Kepler was able to show that the planetary orbits were not circles, but ellipses, with the sun at one focus. He also discovered two other laws describing the motion of the planets. His second law, the law of areas, states that a line drawn from the sun to a planet as it moves in its orbit sweeps out areas proportional to the time interval. Kepler's third law, the harmonic law, states that the square of a planet's period of revolution is equal to the cube of its average distance from the sun.

Though Kepler's mathematics was convincing to many thinkers, dramatic proof of the correctness of his scheme was lacking until Galileo built his first telescope in 1609. Of Galileo's many astronomical discoveries, two had particular impact on the debate over the solar system. One was his observation of the phases of Venus—easily explained by a sun-centered model, but inconsistent with an earth-centered one. The other was the discovery of four satellites of the planet Jupiter, which showed that a moving body could itself be the center of motion. Galileo's writings resulted in his being disciplined by the Church, but they did much to help popularize the model of the solar system worked out by Copernicus and Kepler.

EXERCISES

1. Why did people think that the heavens moved rather than the earth? Try to trace your own changing conceptions of the earth and its situations in the universe from the time you were a child.
2. How did ancient astronomers try to understand the universe? What do you think was their motivation? Were there any practical benefits to be gained from an understanding of heavenly motion?
3. How did the Greek approach to astronomy differ from that of peoples who had come before? What were the merits and the shortcomings of the Greek approach?
4. How did Aristotle show that the earth is a sphere? Why did he believe that it was unmoving? What experimental evidence did he cite to support this belief?
5. How did Eratosthenes measure the circumference of the earth?
6. Why were the planets so named? How would you recognize one through careful naked eye observations of the skies over a period of time? How many planets do you think you could find by this method?
7. What are epicycles? Why were they necessary in the Ptolemaic system? What were the advantages of this system?
8. How did Aristarchos' heliocentric model of the solar system explain the retrograde motion of the planets? Why was this model rejected by later astronomers?
9. What were the achievements of Tycho? How did they pave the way for Kepler's work?
10. What are the three laws of planetary motion discovered by Kepler? Why were they of such crucial importance in the history of astronomy? In what way did they represent a major modification of Copernicus' ideas? Why is it lucky that Kepler began his work with the study of Mars' orbit?
11. What were the contributions of Galileo to astronomy? How did his investigations greatly strengthen the heliocentric theory? What role did the telescope play in this work?
12. Which of the planets do you think has the fastest orbital velocity? The slowest?

CHAPTER THIRTEEN

PLANETS AND OTHERS

It is a curious paradox of astronomy that far more is known about the remote stars than about our planetary neighbors right here in the solar system. This is true despite the fact that human beings have walked on the moon, and space probes have photographed several of the planets at close range. To see why this should be so, suppose you were asked which of two people, A or B, you knew more intimately. A has been a friend and a pen-pal for years. You have read many of his letters, in which he has revealed his innermost thoughts and feelings, and told you much of the story of his life. Yet you have never met him in the flesh — in fact, you have seen only a rather blurry snapshot of him. B, on the other hand, lives in your neighborhood. You have seen him on the street many times, you have noted his comings and goings, observed his appearance and clothing — yet you have never exchanged more than a few words of polite conversation. Which person could you say you knew more about?

The stars are like the person we have called A. Because they are hot, they radiate electromagnetic energy, from which we can learn a great deal about them. Moreover, the stars are gaseous, and therefore obey relatively simple laws. The planets, by contrast, are cool, opaque, and solid. It is much harder to learn anything about their make-up and internal structure merely from observing their surfaces. Even our own earth, which we sometimes think that we know so well, is still a relatively mysterious body as far as its innermost structure is concerned.

There is still another problem which thwarts our attempt to understand the origin of the solar system: its uniqueness. There are many stars, of all ages, visible in the sky. There are only nine planets, though, all apparently the same age. We can neither observe the formation of our own solar system, nor compare it with any other systems, though we now have some evidence that other systems probably do exist. In Chapter 14, we will see how we can use the limited evidence available to reconstruct at least the broad outlines of an event that happened more than 4½ billion years ago—the formation of the solar system. First, however, we must describe the individual bodies that make up the system, as well as their relationships and interactions.

The most important object in the solar system is the sun itself. It serves as a gravitational anchor for the other bodies; it is the source of nearly all the light and energy; and it contains nearly 99.9 percent of the system's total mass. The sun, however, is a star, and thus is radically unlike the other bodies of the system. The smallest speck of meteoric dust has more in common with the earth than either has with the sun. When we speak of the bodies of the solar system, therefore, we generally begin with the next most important class of objects—the planets.

There are nine planets in the solar system. A comparison of their basic characteristics (Table 13-1) shows that they fall into two distinct groups. Those closest to the sun—Mercury, Venus, the earth, and Mars—are small, but have high densities, and rotate relatively slowly. Jupiter,

THE PLANETS

FIGURE 13-1
The solar system, drawn approximately to scale. The orbits of the outermost planets extend far beyond the right hand edge of the figure. The sizes of the sizes of the sun and planets, however, are *not* to scale (see Figure 13-3).

Table 13-1A   The Planets: Orbital Data

Planet	Semimajor Axis AU	Semimajor Axis 10⁶ km	Orbital Period	Mean Orbital Speed (km/sec)	Eccentricity of Orbit	Orbital Tilt	Period of Rotation	Tilt of Equator to Orbit
Mercury	0.39	57.9	88 days	47.8	0.206	7°	58.6 days	0°
Venus	0.72	108.2	225 days	35.0	0.007	3.4°	242.9 days	0°
Earth	1.00	149.6	1 year	29.8	0.017	0°	23^{hr}56^{min}	23°27'
Mars	1.52	227.9	1.88	24.2	0.093	1.85°	24^{hr}37^{min}	24°
(Ceres)	2.77	414	4.60	17.9	0.077	10.62°	?	?
Jupiter	5.20	778	11.86	13.1	0.048	1.31°	9^{hr}50^{min}	3°
Saturn	9.54	1427	29.46	9.7	0.056	2.49°	10^{hr}15^{min}	27°
Uranus	19.18	2870	84.01	6.8	0.047	0.77°	10^{hr}45^{min}	98°
Neptune	30.06	4497	164^{4}8	5.4	0.009	1.77°	16^{hr}?	29°
Pluto	39.52	5912	248.4	4.7	0.249	17.17°	6.4 days	?

Table 13–1B   The Planets: Physical Data

Planet	Diameter km	Diameter earth = 1	Mass (earth = 1)	Mean density (g/cm²)	Surface gravity (earth = 1)	Escape velocity (km/sec)	Moons
Mercury	4,878	0.38	0.055	5.4	0.38	4.3	0
Venus	12,112	0.95	0.82	5.3	0.91	10.3	0
Earth	12,756	1.00	1.00	5.5	1.00	11.2	1
Mars	6,800	0.53	0.11	3.9	0.38	5.1	2
Jupiter	143,000	11.2	317.9	1.3	2.64	60	13
Saturn	121,000	9.5	95.2	0.7	1.13	35	10
Uranus	47,000	3.7	14.6	1.6	1.07	22	5
Neptune	45,000	3.5	17.2	2.3	1.41	25	2
Pluto	5,000?	0.5?	0.1?	~5?	?	?	0

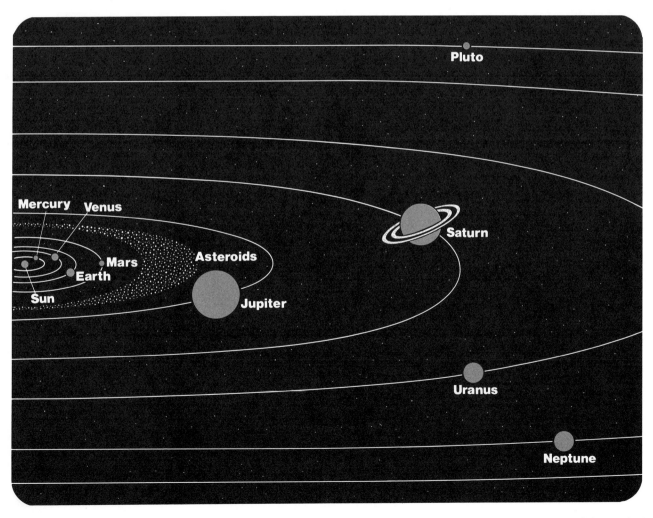

Saturn, Uranus, and Neptune are far larger, but less dense, and rotate faster. The inner planets are generally known as the terrestrial planets, the outer ones as the Jovian planets. Pluto, the ninth planet, seems to be in a class by itself, but much less is known about it than any of the others.

Earth and Moon

The earth is unique in many ways. It is the largest of the terrestrial planets—the biggest of the dwarfs. It is the only planet in the solar system whose surface is largely covered by water. It is the only planet, so far as we know, whose physical properties have been altered by the activities of living creatures. And it appears to be the most geologically active planet of all the terrestrial bodies.

Although it is not obvious to the casual glance, something is always going on, internally and externally, to change the features of our planet. The earth's surface is continually being torn down and built up. Vulcanism, which transports molten rock beneath the surface, throws up volcanos and produces lava fields. Diastrophism, the force by which huge masses of rock in the earth's crust are shifted vertically or horizontally, causes earthquakes and pushes up mountains. And erosion, the whittling away of mountains and other land forms by wind and water, gives our landscapes their familiar appearance.

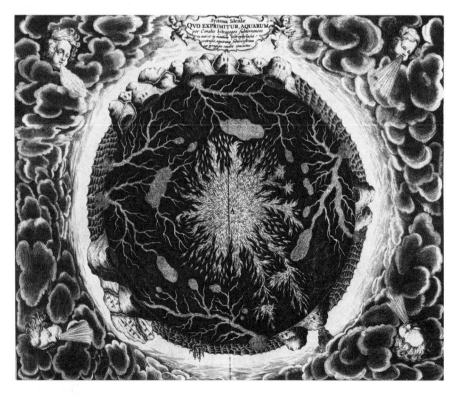

Essentially, our planet consists of two parts: a core composed chiefly of iron and nickel; and a mantle composed of lighter rocky materials. The interior of the earth is quite hot, both on account of simple pressure and the decay of radioactive materials. The part of the earth that we live on is called the crust. It consists of a layer of granite and basalt — rocks which are much lighter than those composing the mantle. The crust extends, on the average, to a depth of less than 32 km. In proportion to the size of the earth, it is no thicker than the skin of an apple. Yet all the continents, ocean beds, mountain ranges, islands, canyons, river basins, valleys, caves, and deserts of the earth are part of the crust. Human beings have yet to penetrate below it.

ATMOSPHERE AND HYDROSPHERE The atmosphere is essentially a gaseous envelope surrounding the earth. It extends to a height of about 300 km, although the great bulk of it is forced down by gravity into a layer about 8 km high at the poles and 15 km high at the equator. The atmosphere and hydrosphere (the oceans) are intimately connected.

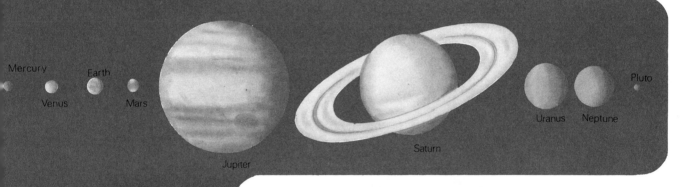

Mercury
Earth
Venus
Mars
Pluto
Uranus Neptune
Saturn
Jupiter

Figure 13-2 (left)
A seventeenth century conception of the earth's structure. Notice the central fire that reaches the surface in volcanos, the subterranean lakes and streams, and the hollow mountains with their underground springs.

Figure 13-3 (above)
The relative sizes of the sun and planets.

It is only because of the atmosphere that liquid water can exist on the earth's surface. Water can remain in a liquid state only when there is sufficient atmospheric pressure. When the pressure is absent or insufficient, it vaporizes. The pressure of the earth's atmosphere — in effect, the weight of the entire envelope above us — is about 14 pounds per square inch at sea level.

Both the earth's oceans and its atmosphere came into being early in its history, when heat produced by the decay of radioactive elements caused water, carbon dioxide, and nitrogen to be expelled from the crust. This process is called **outgasing**. Methane, ammonia, and hydrogen are also thought to have been present in the earth's primitive atmosphere. Oxygen, however — the most important component for animal life — was missing. It was produced later, after the evolution of photosynthetic plants, which use carbon dioxide and expel oxygen as a waste

Continental Drift

Geologists classify all large-scale movements of the earth's crust under the heading of tectonics. One of the most important agents of geological change is the tectonic process known as **continental drift.** Continental drift was suspected early in the twentieth century, when it was noticed that the east coast of South America seemed to conform remarkably closely with the west coast of Africa. In the last thirty years, however, more concrete evidence has come to light confirming the theory that the continents are in motion. About 150 million years ago, the continents of the world were crowded together into a single land mass to which we now give the name Pangaea. Since that time, the continents have slowly drawn apart. Their present positions, too, are only temporary. In another 150 million years, the earth will present a very different picture than it does today.

The process of continental drift is possible because the earth's crust, together with a small portion of the underlying mantle, is divided into a number of massive structures known as plates. The size of the plates varies, from the huge one that carries the entire Pacific Ocean, to a relatively tiny one about the size of Texas. The layer of the mantle immediately below these plates is plastic — that is,

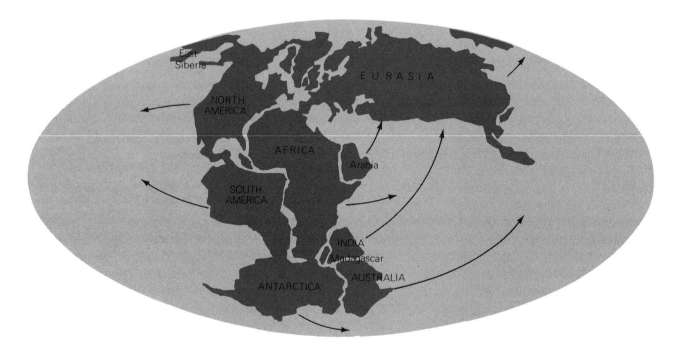

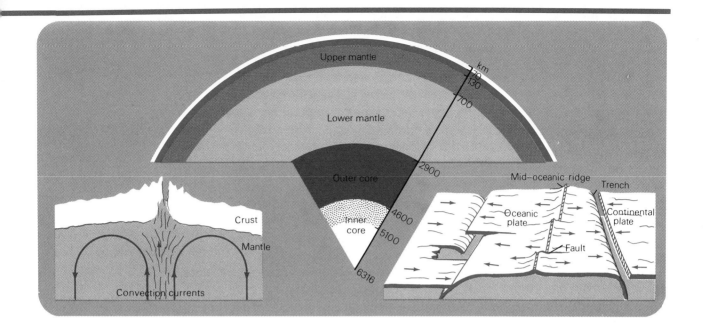

it can be stretched and deformed, like putty. This enables the plates to slide over it. The movements of the plates are thought to be controlled by heat-driven convection currents in the mantle, as magma (molten rock) surges up from lower regions.

The surface birthplaces of the plates are the mid-ocean ridges on the sea floors. These mark the seams in the earth's crust, through which new material is continually being extruded. Magma flows out, cools, and solidifies, pushing plates apart as it joins them. For example, running roughly north and south under the Atlantic Ocean is a 16,000-km-long mountain range called the mid-Atlantic ridge. Molten rock spews up from the top of the ridge, to cool into new ocean bed on either side. As a result, the Atlantic Ocean grows wider at a rate of about 1 inch per year.

In places where two plates are being pushed together, they may buckle into mountain ranges, or one may fold under the other. Such activity is often accompanied by earthquakes. California, for example, experiences so many quakes because of its location on the San Andreas Fault—the meeting place of the North American and Pacific plates. The coming together of two plates may have results even more spectacular than California's occasional catastrophes. The subcontinent of India, for example, lies on a plate which once occupied a position at the tip of Africa, alongside Madagascar. Over millions of years, it drifted northward to its present position. At the place were it made contact with the continent of Asia, the earth buckled, and the mighty Himalayas were born.

product. After millions of years, oxygen produced by plants built up in the atmosphere until it reached its present concentration of about 20 percent. It remains at this level because it is utilized by animals, as well as in the processes of decay and combustion, as fast as it is provided by plant life.

THE MOON   Compared to the earth, the moon is a dead world, not only because of the absence of biological organisms, but also because of its relative geological inactivity. When the Apollo astronauts walked on the moon, they left bootprints which will probably remain intact for at least a million years. Nothing changes on the moon's surface because there is no weather. The moon has no atmosphere, and hence no wind; no water, and hence no rain. Without weather there is no erosion — the slow, abrasive force which slowly wears down surface features on the earth. The reason the moon is so pitted with craters is not only that so many impacts have scarred it, but also that the scars have never healed.

Geologically, the moon is a dead body.

Indeed, the presence of ancient craters — or any other old land forms — on a planet is the first indication that it is geologically inactive. The earth, for example, was also heavily bombarded and cratered early in its history. But only the more recent impacts have left marks that are still visible, such as the Barringer meteor crater in Arizona. The great majority of craters have been erased by the earth's erosive and tectonic forces.

Like several of the terrestrial planets, the moon has two types of terrain — highlands and lowlands. The lowlands comprise the **maria** (singular **mare**), or seas. These are relatively smooth, dark areas that are actually basins of solidified lava. They are generally circular, and connected to one another. The highland regions are more rugged, with numerous craters and jagged mountains, some of which are joined together in chains. Most of the craters were formed during a period some 4 billion years ago, when all the terrestrial planets were heavily bombarded by debris (including bodies as large as a good-sized asteroid) left over from the formation of the solar system. As a result of these impacts, floods of lava flowed onto the moon's surface. The biggest impact basins became the maria. Since then, there has been little change in the moon's sterile surface, save for the creation of some additional craters by meteor bombardment.

The lunar highlands are rugged and mountainous; the darker, smoother "seas" are basins of solidified lava.

There is very little activity inside the moon as well. The average density of the moon is less than that of the earth, indicating that it lacks a heavy metallic core. Its crust, which is composed chiefly of loose, rubble-like material, is believed to be much thicker than the earth's crust. The moon does seem to possess high-mass areas, called **mascons,** scattered at random at shallow depths below the surface. Instruments were left on the moon during the Apollo missions to relay back information about possible moon quakes. The findings show that such activity is very rare and feeble on the moon. When it does take place, it is usually associated with tidal forces exerted by the earth.

PORTFOLIO

Photograph of the lunar far side taken by Lunar Orbiter V on August 14, 1967. This side of the moon is marked by a paucity of maria. The large crater on the left is perhaps the only one which could so qualify. It was discovered by the Soviet spacecraft Luna 3, in 1959, and is called Mare Moscoviense.

Also invisible from the earth is Mare Orientale, at right. This three-ringed crater preserves, frozen, the waves in a pool of once molten rock, melted by the impact with the incoming object.

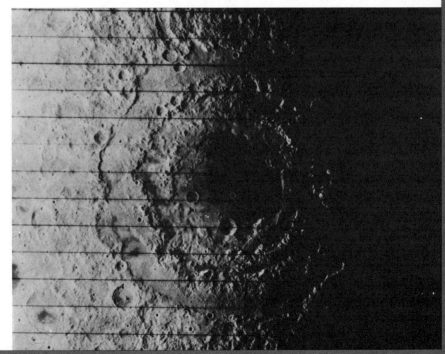

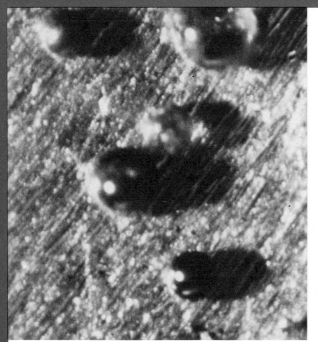

Tiny glass spheres, about 1/100-inch in diameter, in the sample of lunar material collected by Apollo 11.

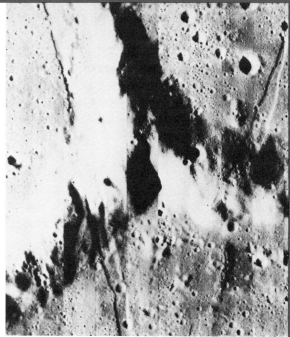

Lunar rilles, above right, are river-like structures. They may have been cut by lava flows. Below is the first close-up photograph of Copernicus. The walls of the crater are 1000 feet high. Lunar Orbiter II was 150 miles south of this point.

Mercury

Mercury has the smallest volume and the least mass of any planet in the solar system. It is also closest to the sun and, in accordance with Kepler's third law, it has the greatest orbital velocity. Mercury's proximity to the sun has long proven a serious obstacle to telescopic observation of the planet. Mercury never appears more than 28 degrees from the sun. This means that it can only be seen near the horizon, as a morning star just before sunrise, or an evening star just after sunset. Although it is a fairly bright object, its low position in the sky makes it difficult to study from earth, since it must be observed through a thicker layer of turbulent, distorting atmosphere. Consequently, astronomers have generally chosen the lesser of two evils and observed Mercury during the daytime, when it is higher in the sky. So little detail can be seen under these conditions, however, that until recently astronomers had not even been able to make a correct estimate of Mercury's period of rotation.

Our knowledge of Mercury increased many times over when Mariner 10 passed less than 10,000 km above its surface and sent back more than 2000 photographs to earth. The surface features revealed by these photographs very closely resemble those of the moon. Mercury's terrain is harsh and bleak, pitted with craters large and small. And like the moon, Mercury has two faces. One side is mountainous, dusty, and heavily cratered with huge basins, some measuring 12 miles across and filled with lava, like the lunar maria. The other half of the planet is flatter, like the lunar highlands, with plains and regions that show evidence of large lava flows. Mercury also contains a number of unusual and unexpected features. For example, huge scarps or cliffs, one of which towers nearly two miles above the crater-pocked landscape, run sinuous courses for hundreds of miles. Geologists think these immense structures may be the result of compressive forces that crumpled the surface as it cooled and shrank some 4 billion years ago. Another unusual feature, appropriately called "weird terrain," is a 5800-square-mile-region consisting of hills and craters whose walls have collapsed. This phenomenon lies exactly opposite a basin known as Caloris, which measures some 1300 km in diameter. The positions of the two features suggest that they may be causally related. Geologists think that the tremendous impact of the meteor or asteroid that created Caloris may have sent huge shock waves through the planet. At their point of convergence, diametrically opposite the point of impact, the shock waves tumbled existing land forms, creating the "weird terrain".

Mercury's surface, like that of the moon, is barren and pockmarked with numerous craters.

The similarity between Mercury and the moon is only skin deep, however. The average density of the moon is 3.34 times that of water, suggesting that it is composed chiefly of rocky material. The average density of Mercury, by contrast, is 5.4 times that of water. This fact, together with the existence of a planetary magnetic field, indicates that Mercury probably has a metallic core, like that of the earth.

Figure 13-4
Mercury, photographed in September, 1974 by the Mariner 10 spacecraft from a distance of 65,000 km. The jagged, cratered surface of the planet resembles that of our moon. The thin, dark arc in the upper left portion of the picture is a scarp, or cliff, more than 300 km in length. This and similar features are thought to have been caused by compressive forces in the planet's crust.

Mercury has only traces of an atmosphere. This is not surprising, considering its small size and proximity to the sun. With just 6 percent as much mass as the earth, Mercury has a rather weak gravitational field, and a low escape velocity. Lightweight gas molecules, traveling rapidly at the planet's high surface temperature (nearly 1800° F during the Mercurian noon) could easily fly off into space and be lost. Without an atmosphere to hold the sun's heat, the planet's surface cools rapidly when turned away from the sun. As a result, Mercury's temperature varies dramatically, falling to −30° F during the planet's long nights.

Venus

Aside from the sun and moon, Venus is the most brilliant object in the heavens. At its brightest, Venus can cast a shadow on a very dark night, and can be seen in the daytime if the observer knows where to look for it. Venus is such a striking sight to the naked eye, and so close to earth (only about 40 million km at its closest approach) that one might expect telescopic observation to reveal fascinating details of its surface. Unfortunately, Venus turns out to be one of the most frustrating and least rewarding of telescopic objects. Seen through a telescope, Venus appears as nothing but a featureless, yellowish disc. Almost nothing about the planet can be learned from observations in the visible spectrum. The reason is that Venus is shrouded in thick clouds, which are impenetrable to visible wavelengths of light.

Since its average distance from the sun is only about .7 that of the earth, Venus receives more radiation than earth does, but for a long time astronomers believed that its thick, light-reflecting clouds might keep its surface relatively cool. When the actual surface temperature was first measured, therefore, many people found the results quite surprising. As we saw in Chapter 4, the temperature of any body can be determined from the intensity of its thermal radiation at various wavelengths. By analyzing Venus' radiation at radio wavelengths, which are relatively easy to measure, scientists succeeded in plotting its black-body curve. It suggested that the planet's surface temperature was about 800° F. This estimate was later confirmed when the two Soviet Venera space probes successfully deposited instrument packages on the Venusian surface. Temperatures approaching 900° F were recorded.

Why is Venus so hot? It appears that the cloud layer, once credited with keeping Venus cool, is actually responsible for making it an inferno. Only about 20 pecent of the sun's radiation penetrates Venus' clouds, making the Venusian day about as bright as a dull, rainy day on earth. The temperature of a planet, however, is determined not only by how much radiation it receives, but also on how much it reflects or reradiates back into space. Since planets are relatively cool, most of their thermal radiation lies in the infrared. The clouds of Venus are extremely opaque to infrared rays, and let almost no infrared radiation escape. The trapped radiation causes heat to accumulate in the lower regions of the Venusian atmosphere. The principle is actually identical to that utilized by greenhouses, the glass panels of which admit visible light but prevent heat from being radiated away in the infrared. For this reason the heat-trapping ability of Venus' atmosphere is known as the "greenhouse effect."

The chemical composition of Venus' yellowish clouds has long been a mystery. The most widely accepted theory at present is that they are

The carbon dioxide in Venus' atmosphere traps the sun's heat, keeping the planet's surface temperature extremely high.

Figure 13-5
The greenhouse effect that keeps Venus so hot. Radiant energy from the sun passes through the planet's atmosphere and warms its surface. The surface in turn reradiates some of this energy, chiefly in the infrared portion of the spectrum. If this radiation could escape into space, the temperature on Venus would not be much greater than that of earth. But carbon dioxide in the planet's atmosphere absorbs the infrared rays, converting them back to heat.

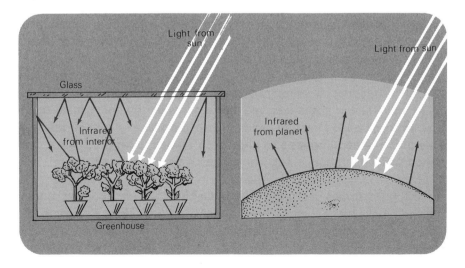

composed chiefly of sulfuric acid. One of the Soviet probes also made a partial chemical analysis of the 40-mile-thick Venusian atmosphere. It found that 90 to 95 percent of the atmosphere consists of carbon dioxide. Water vapor and oxygen were also detected, although there is some doubt about the validity of the measurements. The instruments found that the atmospheric pressure at the planet's surface is about 100 times as great as on earth, and that there seemed to be almost no wind.

With the exception of its cloud layer, Venus appears to be almost a duplicate of the earth. Indeed, Venus has often been called the earth's "sister planet." It is slightly smaller and slightly less massive than earth, but its average density is almost the same. This suggests that, like earth, Venus is probably composed of a rocky crust, a mantle, and a core of iron or nickel. Radar mapping shows that the surface of Venus is cratered like the moon and Mercury. The most recent radar images (Figure 13–00) show a huge impact basin, some 1000 by 1600 km, or roughly the size of Hudson Bay in Canada. This feature may be the scar of a past collision with a massive body—perhaps a good-sized asteroid. Similar events are thought to have created the **maria,** or seas, on the moon, the Hellas plain on Mars, the Caloris basin on Mercury, and possibly certain features of our own planet, such as Quebec's Manicouagan Reservoir (Figure 13–9 and Color Portfolio II).

Such radar studies have also turned up evidence that Venus may have experienced extensive tectonic activity. For example, investigators have found a huge gash in the Venusian surface that measures 1400 km long and 150 km wide. This crack, geologists believe, may be the result of tectonic forces pulling the surface apart. Another radar image shows a mountain range with a fault displacement—one part of the range has shifted relative to the other. On earth, this phenomenon is considered a distinctive sign of tectonic activity.

Radar images of Venus suggest that the planet may be geologically active.

Other signs of such activity include radar images showing structures that appear to be huge volcanos. A large, bright area about the size of the state of Oklahoma, probably a lava flow, is also visible in the latest radar pictures. It seems to be a relatively new feature, overlaying older terrain —a sign, apparently, that internal volcanic activity has been a part of Venus' recent past. Long, parallel ridges crossing the flow suggest that mountain-building forces may also be at work.

Until recently it was believed that erosion by the hot, dense, corrosive atmosphere of Venus would smooth the planet's surface features. One of the Venera probes sent photographs back to earth that were consistent with this theory. They showed an ancient mountain chain, with indications of past lava flows. The rocks in the picture were rounded, evidently by the forces of erosion. The other Venera craft, however, photographed rocks that appeared to have sharp, jagged edges, indicating that they have undergone very little erosion. One possible explanation for this surprising discovery is that the rocks are of relatively recent origin—

A highly reflective region of Venus, named Maxwell, is seen in the picture above. It is about the size of Oklahoma. It is a detail of the picture below, obtained by bouncing radio waves off the planet using the 1000-foot radio telescope at Arecibo. The picture reveals impact basins, lava flows and signs of tectonic activity. The opaque atmosphere of the planet prevents us from visually studying these features.

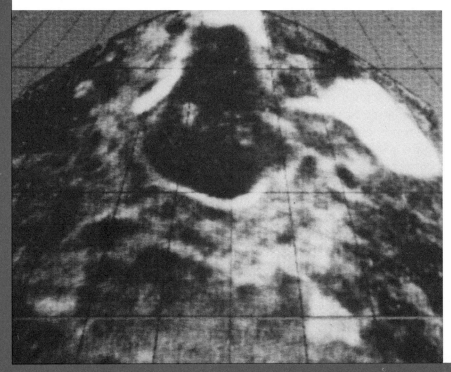

Sharp edged rocks strewn around the Soviet spacecraft, Venera 9, are seen in the picture above taken from the surface of Venus.

created, perhaps, by quakes or volcanic activity. This is still another hint that Venus may be far more geologically active than had previously been thought.

Radar has revealed still another surprise about this surprising planet. Venus rotates in a retrograde direction—that is, clockwise when viewed from the north. Its period of rotation is 243 days, 18.3 days longer than its year. A curious feature of Venus' rotation is that each time the planet is at inferior conjunction (lined up with the earth on the same side of the sun) the face it presents to the earth is the same. This fact seems to suggest that Venus is locked into a pattern of synchronous rotation by the tidal forces of earth's gravity, just as the moon is. At Venus' distance from the earth, however, the gravitational forces at work do not seem strong enough to have such an effect, and the phenomenon is still not adequately explained.

Mars

Mars is the only one of the terrestrial planets whose surface can be seen clearly through a telescope. Except for Venus, it comes closer to us than any other planet; during opposition, it may approach within 56 million km of the earth. It is much smaller than the earth, having less than 1/9 the mass and only about 1/2 the radius. However, it is remarkably like the earth in other ways. Its period of rotation is almost the same as earth's, making a Martian day just 41 minutes and 19 seconds longer than an earth day. The inclination of its axis is also almost identical to earth's. Mars, therefore, has seasons just as the earth does. But because Mars' period of revolution is nearly twice as long as earth's, each of its seasons lasts for six months instead of three.

Through the telescope, Mars appears as an orange globe. At the north and south poles are the gleaming white polar caps. There are areas of green or grey which lighten and darken with the seasons and occasionally change shape. Until recently, astronomers speculated on the possibility of these color changes being the result of the seasonal growth of vegetation, but it is now thought that their cause is shifting masses of dust carried by the high Martian winds.

Much of what we now know about Mars, we owe to the unmanned space probes of the past decade, especially Mariner 9, which orbited the planet in 1971, and Viking 1 and 2, both of which landed instrument packages on Mars in an attempt to determine whether it might harbor life (see Interlude IV). The Martian atmospheric pressure has been measured and found to be less than 1 percent that of earth. The atmosphere is composed chiefly of carbon dioxide with some water vapor and very small amounts of oxygen. Because of the thin atmosphere, Mars does not retain heat very effectively. Although the temperature at the equator rises to about 80° F at noon, at night it drops to less than −100° F.

The thin, dry Martian atmosphere holds little of the sun's warmth.

Even before the two Viking spacecraft reached Mars in the summer of 1976, the astronomer Carl Sagan remarked, "Every time we have looked at Mars in finer detail, we have discovered entire new sets of astonishments and delights, things no one ever guessed would be there." This was certainly true of the Mariner 9 photographs, and proved even more true of the pictures taken from orbit by the Viking 1 and 2 mother ships. Together, these three expeditions have produced thousands of photographs, giving us a remarkably complete and detailed view of the planet's surface. Details as small as a few hundred feet — about the size of a football field — can be distinguished in the pictures.

Much of the Martian surface is pitted with craters similar to those found on Mercury and the moon. Unlike these, however, the Martian craters have been weathered by the action of strong winds. There is a huge chasm, similar to the Grand Canyon, but longer than the United States from coast to coast, 70 km wide, and some 6 km deep. If Mt. McKinley were placed at the bottom, its summit would barely project above the walls of this enormous gash in the planet's surface. Other Martian surface features are on a similarly gargantuan scale. There are several large volcanos. The biggest, Olympus Mons, covers nearly as much area as the state of Texas, and towers 24 km — nearly 3 times the height of Mt. Everest.

Most of the surface geological features of Mars are much larger than those of earth — perhaps a sign of the absence of tectonic activity.

Like Mercury and the moon, Mars is two-faced. One hemisphere is marked with a few old volcanos, and with numerous craters. The other consists largely of lower-lying plains dotted with only a few craters, and younger volcanos. It is thought that the deposition of sediment in the lowlands during past eras may be responsible for this difference. Another factor that shapes the Martian landscape is the fierce wind, which originates at the north pole and often sweeps before it a 50-km high wall of dust that rushes across the planet at speeds approaching 500 km/hr. The dust whittles away at volcanos, crater walls, and other elevated landforms, producing unusual features such as plumes, pyramids, and dune fields.

An unresolved, but fascinating, question is whether Mars may be a geologically active planet, like the earth. Numerous fault lines have been detected — places where one area of the planet's surface has been displaced relative to an adjacent region. There are also indications of past earthquakes (or, more properly, Marsquakes) and lava flows. However, none of the many Martian volcanos appears to be active at present. There is also no evidence that Mars possesses moving crustal plates like those of our planet. The enormous size of the younger volcanos, such as Olympus Mons, may be attributable to the lack of plate motion, which would carry away newly-formed surface features before they could reach such proportions. Additional evidence can be found in the fact that Mars has very few compressed land forms. Chains of mountains, the tell-tale signs of compression on earth, are absent on Mars.

The Canals of Mars

One of the most heated controversies in the history of astronomy appears to have originated in part from a linguistic misunderstanding. In 1863, Father Pietro Angelo Secci of Rome published color sketches of Mars showing streaklike features. Similar markings were recorded by a few other observers. About 1869, Father Secci referred to some of these features as *canali,* which simply means narrow bodies of water and is best translated as "channels." The choice was consistent with the common practice of naming dark areas of the moon and planets after oceans and other natural bodies of water. The very first telescopic observer, Galileo, had initiated the usage by calling the dark plains of the moon *maria,* or seas. But the similarity of *canali* to the English word "canal" was later to cause trouble.

The term was popularized by Giovanni Schiaparelli, director of the Brera Observatory in Milan, after the close approach of Mars to Earth in 1877. Schiaparelli charted and named many more canali than had previously been observed, and he defended their existence when other astronomers did

not confirm his observations. Moreover, his drawings represented these features with striking geometrical regularity. He showed the familiar dark areas of the planet connected by lines and circular arcs crossing the lighter "desert" areas. In some cases two *canali* appeared to run parallel for hundreds of miles. During the close approach of Mars in 1886, astronomers in Italy, France, England and the United States claimed to see the canals, as they had come to be called. But many others remained skeptical.

As the years passed, the believers slowly grew in number, and their confidence increased. The maps that they drew were bolder; the canals appeared ever straighter and narrower. The obvious possibility that they might be the creations of a Martian civilization occurred to many, and was hotly debated. Schiaparelli himself remained neutral. "Their being drawn with absolute geometrical precision," he wrote, "as if they were the work of rule or compass, has led some to see in them the work of intelligent beings, inhabitants of the planet. I am very careful not to combat this supposition, which includes nothing impossible." But he was also quick to point out that such symmetry often occurs in nature.

It was the American astronomer Percival Lowell who emerged to champion the idea of artificial waterways, and captured the popular imagination. Lowell was a member of a prominent and wealthy Boston family, the brother of the president of Harvard University and of the poet Amy Lowell. After ten years as an American diplomat in Korea and Japan, Lowell retired to devote himself to his passionate interest in astronomy. Using his own money, he constructed an observatory at Flagstaff, Arizona, a mile and a half above sea level. The leading ophthalmologist of the time told Lowell his eyesight was the sharpest he had ever tested. In the thin, steady air of Flagstaff, Lowell believed, his observations would establish the existence of the canals beyond all doubt. He was convinced that he was destined to be Schiaparelli's successor. And indeed, from his early observations, he charted four times as many canals as Schiaparelli.

Lowell was also an accomplished author, having previously written several books about the Orient. In 1895 he published *Mars,* an account of his observations and theories, which was sensationally

persisted about the existence of the canals as a natural phenomenon—perhaps cracks in the planet's surface. Photography could not settle the issue, since the turbulence of the earth's atmosphere prevented a sufficiently steady image for the required exposure time. Nor could the human eye be entirely trusted, since it has a tendency to fill in detail and to create patterns out of disconnected, irregular features which are on the borderline of resolution. It remained for the Mariner space probes, the first of which reached Mars in 1965, to settle the question.

When the first photographs returned from the spacecraft, it became clear that there are no features corresponding to the long straight lines drawn by Schiaparelli and Lowell. After analyzing the maps prepared at Lowell Observatory in the early years of this century, two recent investigators conclude:

> The vast majority of the canals appear to be largely self-generated by the visual observers of the canal school, and stand as monuments to the imprecision of the human eye–brain–hand system under difficult observing conditions.

They are also (at least to some extent) monuments to the unconscious power of expectation and wishful thinking over observation—a constant problem in any scientific research.

In a few cases, however, there *are* irregular ridges or grooves on the Martian surface that roughly correspond to the classical drawings. The most notable of these occurs in the Coprates region where one of the largest canals was thought to be. It consists of a rift valley three thousand miles long, sometimes sixty miles wide, and a mile deep.

It is often the case in science that a hypothesis which is proved wrong also turns out to have had some element of truth, or that a correct theory turns out to have been right for the wrong reasons. Running into the Coprates valley are hundreds of sinuous channels, with second- and third-order tributary systems, which could only have been cut by running water. This water has long since disappeared. "One day in the future," Dr. Carl Sagan has remarked, "perhaps, the channels will again be filled with water, and for all we know, with visiting gondoliers from the planet Earth."

received by the public and which intensified the debate among astronomers. Lowell's canals were too long, straight and elaborate to be accounted for by any known natural phenomenon. He proposed that they were irrigation canals lined with vegetation, constructed by a highly advanced civilization to distribute precious water form the melting polar caps to the cities of a drying, dying planet.

Three years later H. G. Wells published his novel ***The War of the Worlds,*** in which horribly ugly Martians invade the earth to destroy its population. The red planet had entered the popular consciousness, spawning generations of extraterrestrial creatures, intrepid rocket pilots and exotic weapons. Mass fascination with Mars culminated in the 1938 radio dramatization of Wells' novel by Orson Welles, which thousands of frightened listeners mistook for a news broadcast of an actual invasion. In the ensuing panic, several terrified people committed suicide. Roads near the supposed site were jammed with heavy traffic in both directions—those who were fleeing the Martians, and those who were flocking to get a look at them.

But even as Martian intelligence thrived in the novels of Edgar Rice Burroughs and other writers of the 1930's, professional astronomers paid less and less attention to Lowell's hypothesis because so few were able to see the canals. Yet some debate

PORTFOLIO

The 15-mile-high Martian volcano, Olympus Mons, photographed by the Viking 1 Orbiter from a distance of 8000 km. The picture, above, shows it wreathed in clouds. The volcanic crater is 80 km. across and far surpasses, as the mountain does, any corresponding structure on earth.

The Mariner 9 picture, below left, show a meandering "river". Such features are not uncommon on Mars. They lead us to suspect that there was liquid water on the Martian surface in the past. The Viking 1 picture, below right, shows braided channels cut in the past by flowing water. The shore is at the lower right of the picture.

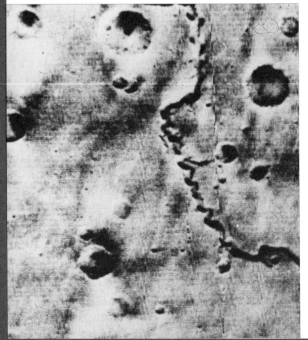

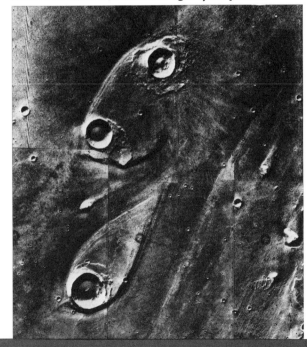

At right, an oblique view of Mars across the Argyre Planitia, obtained by a TV camera on Viking Orbiter 1. The picture below shows the northern ice cap receding with the approach of summer. The ice cap consists of frozen water and CO_2. The complex structure may be due to the prevailing winds. The picture at the bottom shows a huge equatorial canyon. This complex structure was probably caused by the wind and by the alternate freezing and thawing of ground ice.

Probably the most surprising, and exciting, feature of the Martian surface are the many sinuous, braided channels, resembling dry river beds, that meander across this nearly waterless world. There are thousands of such channels, with deltas, islands, and tributary branches, the largest of them measuring some 1000 km. Every indication points to their having been made by running water. For example, small spits of land located in some of them have their narrow ends pointing downstream, and the braided silt patterns characteristic of streams that have overflowed their banks are also clearly visible.

There seems to be no liquid water on the Martian surface today, however. What has happened to it? Traces of it have been detected in the planet's atmosphere. Some may be imprisoned in the rocks, or lying beneath the permafrost—the thin layer of permanently frozen ground that blankets the planet. The most intriguing theory suggests that Mars may now be in the grip of an ice age, and that most of its water—along with the rest of the planet's primeval atmosphere—is locked in the polar ice caps. These were long thought to consist entirely of frozen carbon dioxide (dry ice) but recent evidence shows that water is abundantly present as well. In the past, the axis of Mars may have been tilted at a different angle, and temperatures on the planet's surface may have been warm enough to melt the polar ice. There has been speculation that thousands of years from now, the process may repeat itself, and Mars will be a warm, moist planet once again.

Mars appears to have had running water on its surface in past ages.

Mars has two moons, named Phobos ("fear") and Deimos ("panic")—appropriate companions for Mars, the god of war in classical myth. These satellites are so small that they were not discovered until 1877. Both are highly irregular in shape, and heavily chipped and pitted by meteor bombardment. Phobos, the larger of the two (about 22 km in diameter) orbits at a distance of only about 9000 km from Mars. In accordance with Kepler's second law, this small orbit results in a very rapid period of revolution. In fact, Phobos is unique in the solar system in that it revolves around its primary faster than the planet itself rotates on its axis. It makes a complete trip around the planet in about 7½ hours. To an observer on Mars, therefore, Phobos would appear to rise in the west and set in the east three times in the course of a Martian day.

Jupiter

With Jupiter we come to a group of planets which are of an entirely different kind from those we have studied thus far. The Jovian planets are sometimes called the major or giant planets, and these names serve to draw attention to their most obvious characteristic: their size. Compared with them, the terrestrial planets look like insignificant debris. Jupiter, the largest, has 318 times the mass of the earth and over 1300 times its volume. Its mass is almost 2½ times the mass of all the other planets put together. Yet, in spite of its size, Jupiter spins faster than the earth. A day on Jupiter is less than 10 hours long. Jupiter spins so fast that there is a visible bulge at the planet's equator, and a noticeable flattening at the poles — the result of centrifugal force. In fact, the rotational velocity of a point on Jupiter's equator is in excess of 12 km/sec, whereas the corresponding velocity of earth is only .46 km/sec.

It is possible to time Jupiter's rotation quite easily by observing the distinct bands and spots of the planet's visible disc. In fact, the combination of rapid rotation and varied surface markings makes Jupiter one of the most fascinating planets to look at with even a medium-sized telescope, for its appearance changes markedly in the course of just a few hours of observation. Careful study shows that Jupiter's equator spins faster than regions at higher latitudes. This indicates that when we observe Jupiter we are not seeing a solid surface, but rather a thick layer of opaque, turbulent gases.

Spectroscopic analysis, along with the data gathered by the Pioneer 10 and Pioneer 11 spacecraft in 1973 and 1974, show that Jupiter's atmosphere is composed chiefly of hydrogen and helium — the two most common substances in the universe. Methane, ammonia, and water are also present. Unlike the gaseous envelopes that surround the terrestrial planets, however, Jupiter's atmosphere does not stop at any definite point. In fact, except for a small, rocky core (the existence of which is not entirely certain), most of the planet is either liquid or gaseous, and the

Except for a possible small, solid core, Jupiter appears to be entirely liquid and gaseous.

Figure 13-6
At left, the structure of Jupiter, as it has been deduced from theoretical studies based on the data obtained during the Pioneer 10 and 11 flybys. The temperature at the planet's center is thought to be in the neighborhood of 30,000 K. At right, a cross-section of Jupiter's atmosphere, showing the many layers of clouds.

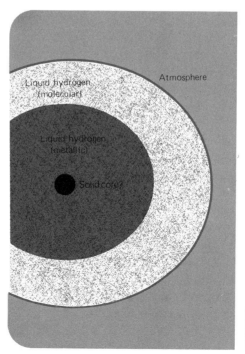
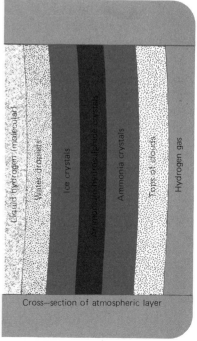

transition between the two states is probably gradual rather than clearly marked. At a depth of several hundred km, the gaseous hydrogen of the atmosphere gives way to liquid hydrogen. At even greater depths, the liquid hydrogen is thought to take on metallic properties—the ability to conduct electricity, for example.

Since Jupiter has such an extensive atmosphere, it is not surprising to find that its weather patterns are quite complex, and of great interest to scientists. Like the earth's atmosphere, Jupiter's is full of clouds. The Jovian clouds, however, seem to contain layers of ammonia droplets, ammonium hydrosulfide droplets, and ice crystals as well as the water droplets of earthly clouds. Many of the cloud patterns in Jupiter's atmosphere are extraordinarily stable, persisting relatively unchanged for years, or even centuries. These spots and bands seem to be the result of convective processes. Unlike the weather of earth's atmosphere, which is energized by radiant heat from the sun, Jupiter's weather patterns seem to be controlled by heat from below—from the interior of the planet itself. In fact, it has been found that Jupiter radiates considerably more heat than would be expected if it were merely reflecting solar radiation. It seems likely that this heat is left over from its formation, more than $4\frac{1}{2}$ billion years ago. Another possibility is that Jupiter is still slowly contracting, converting gravitational potential energy into thermal energy, as stars do in the early stages of their lives. In fact, the planet's internal temperature is thought to be as high as 30,000 K. This is far short of the temperature needed to ignite nuclear reactions, however. To truly become a star, Jupiter would have to be some 80 times more massive than it now is.

A great deal of Jupiter's heat comes not from the sun, but from the planet's center.

An extreme example of the stability of Jupiter's atmospheric features is the Great Red Spot. First described in 1660, the Red Spot is large enough to encompass three earths. During the 300 years it has been observed, it has changed shape and color, but never entirely disappeared. Curiously, however, it does not rotate at the same speed as the surrounding regions, but lags slightly behind—enough to lose one full rotation in about 10 years. The Red Spot is now generally believed to be a storm of long duration raging in the upper atmosphere of the planet.

At least one scientist has suggested that the red color is caused by a concentration of organic molecules, created in the atmosphere by electrical discharges similar to the lightning of terrestrial thunderstorms. This theory points up the fact that conditions in Jupiter's atmosphere today closely resemble those thought to have existed on earth some $4\frac{1}{2}$ billion years ago—shortly after its creation, and before the emergence of life. There has been much speculation about whether the same processes that led to the evolution of the first living things on earth might be going on in the Jovian atmosphere. Further investigations of Jupiter may lead to a clearer understanding of this greatest of scientific mysteries.

The convective features of Jupiter's weather, similar to storms in earth's atmosphere, are remarkably stable; one of them, the Great Red Spot, has been observed for centuries.

Jupiter has an intense magnetic field, similar to the earth's, but about 30 times stronger. On their approach to Jupiter, the two Pioneer

PORTFOLIO

Jupiter and its satellite Ganymede, photographed by Pioneer 11 about 9½ hours before it swung past the planet and raced of to a 1977 rendezvous with Saturn. At left, a Pioneer 11 photograph of the Red Spot on Jupiter. Notice the many smaller spots, each a storm system large enough to swallow the earth. The picture was received on earth at 7:23 a.m. December 2, 1974, more than half an hour after it was transmitted.

spacecraft passed through zones of intense radiation—thousands of times the lethal level for humans, in some regions. This radiation consisted of charged particles—chiefly protons and electrons—captured from the solar wind, and trapped by the powerful Jovian magnetic fields. Jupiter's radiation belts are not only more intense, but also much larger than the similar Van Allen belts that girdle the earth.

THE GALILEAN SATELLITES   Jupiter has 13 satellites, more than any other planet. The four largest, ranging in diameter from about 3000 to 5300 km, were discovered by Galileo in 1610. They can be seen easily with a small telescope, or even with a good pair of binoculars. The satellites change their positions with respect to Jupiter from night to night, and it is particularly interesting to observe their transits (movements across the face of the planet) and occultations (disappearances into the planet's shadow) with a telescope. As we saw in Chapter 3, it was by such observations that Roemer first measured the velocity of light.

The four Galilean satellites have been found to differ considerably among themselves in their characteristics. Io and Europa, the innermost pair, are probably composed chiefly of rock, and have densities about 3½ times that of water—comparable to that of our moon. Ganymede and Callisto, by contrast, have densities between 1½ and 2 times that of water, suggesting that they contain a great deal of ice. The surfaces of the four also must differ considerably, for while Io and Europa reflect nearly ⅔ of the sunlight that falls on them, Callisto reflects less than ⅕.

Io, the innermost satellite, is particularly interesting in a number of ways. In size and mass it resembles our moon, and it is about as far from Jupiter as the moon is from earth. It revolves about Jupiter, however, in about 1¾ days, compared to the more than 27 days required for our moon to complete its orbital path. This higher speed in an orbit of approximately the same size is a result of Jupiter's being so much more massive than the earth. Scientists have recently discovered that Io is surrounded by a vast cloud of hydrogen, sodium, and potassium, the origin of which is not known. The position of Io has also been found to be closely related to the emission of bursts of radio waves from the vicinity of Jupiter. Presumably the passage of Io through the Jovian radiation belts somehow triggers these emissions, but here too the exact mechanism is not completely understood. There is much more to be learned about Jupiter and its satellites, but our knowledge will probably have to wait until a space probe is put into a long-lived orbit about the planet sometime in the future.

Saturn

Saturn is the next largest planet after Jupiter. Like Jupiter, it is composed largely of hydrogen and helium, compressed toward the center into liq-

Perhaps because of its great mass and rapid spin, Jupiter's magnetic field is far stronger than the earth's.

Jupiter's four largest satellites were discovered by Galileo in 1610; they are easily visible with good binoculars.

Figure 13-7
A photograph of Saturn, taken at a time when the rings were most, "open." Twice every 15 years, we see the rings edge-on. At these times they disappear, for although the ring system is some 275,000 km in diameter, it is probably no more than a few meters thick. (see page 414). In this picture, Cassini's division—the dark band separating two of the rings—is clearly visible.

Saturn is less dense than water.

uid form. Saturn, however, is considerably less dense—only .7 times the density of water, in fact. This means that, if a body of water large enough could be found, Saturn would float like a cork. Saturn spins almost as fast as Jupiter, and because of its lower density, it is even more markedly flattened by its rotation: the planet's diameter is fully $1/10$ less from pole to pole than at the equator. The extent of this flattening not only tells us that Jupiter and Saturn are fluid, but also enables us to determine the distribution of mass within them, thus making possible the construction of models of the planets' interior.

Saturn has a total of 10 known satellites. One of them, Titan, has the distinction (along with Ganymede) of being one of the two largest in the solar system—both are about $1\frac{1}{2}$ times the size of our moon. Its density is quite low, however, and it is probably composed largely of icy materials. Titan is also the only satellite in the solar system to possess an atmosphere (if we exclude Io's mysterious hydrogen cloud). Methane and hydrogen have been detected spectroscopically in Titan's atmosphere, and some investigators have speculated on the possible presence of organic compounds there as well. The density of the gases is considerable, and the atmospheric pressure may approach that at the earth's surface.

Uranus and Neptune

Like the earth and Venus, Uranus and Neptune are nearly twins. They are quite similar in size, mass, period of rotation, and probably in composition as well. Both also share many characteristics with the larger gaseous giants, Jupiter and Saturn. They differ slightly in their chemical make-up, however. In addition to hydrogen and helium, the atmospheres of Uranus and Neptune appear to contain large quantities of methane. It is probably this gas which gives Uranus its characteristic greenish color.

Uranus is unique in the solar system in that its axis of rotation is tilted 82° to the plane of the ecliptic. This means that the planet is, in effect, lying on its side. It also rotates from west to east, a peculiarity it shares only with Venus. Uranus has five known satellites. They revolve in the equatorial plane of the planet, so that their orbits are approximately perpendicular to the plane of the solar system. Neptune's rotation, by contrast, is in the normal east to west direction, and its axis is tilted only 29° to the ecliptic. At a distance of 30 AU from the sun, Neptune moves so slowly that it requires 165 years to circle the sun; it has not yet completed a revolution since its discovery in 1846 (Chapter 5, Feature). Neptune has two satellites. One of them, Triton, is larger than our moon, and revolves in a retrograde direction. The other is much smaller, and is unnamed. It revolves in the normal direction, but in an orbit of great eccentricity, and is probably a captured asteroid.

Pluto

Very little is known with certainty about Pluto, the most distant of the planets. It is so far away, and so small, that telescopic observation is not very rewarding. Determining the mass of Pluto has been particularly difficult. Since Pluto has no satellite, so far as we know, the only way to do this is through the study of the perturbations that it causes in the orbits of Uranus and Neptune. But since both of these planets are quite distant from Pluto, and move very slowly themselves, it is hard to achieve much accuracy through this method.

The size of the planet has been equally hard to pin down, since it shows no measurable disc. One method that has been used depends on timing the period of occultation when Pluto passes in front of a bright star. Another method involves estimating the planet's size from its brightness. To do this, however, it is necessary to take a guess at the reflectivity of the planet's surface, which we have no way of knowing with any degree of accuracy. Recently, spectroscopic evidence has pointed to the existence of methane ice on Pluto. Since this material is probably very bright, it is possible that Pluto may be considerably smaller than had previously been thought.

At present, both the mass and the size of Pluto are believed to be about $1/10$ those of earth. This means that its density must be about the same as earth's. Pluto, therefore, is a rocky rather than a gaseous planet. How it acquired its present location, out beyond the Jovian planets, is

Figure 13-8
Uranus and its five satellites.

unknown. The orbit of Pluto is tilted more than 17° to the plane of the ecliptic. It is also very eccentric—so much so that during part of its year it dips inside the orbit of Neptune. These facts, along with Pluto's incongrous physical characteristics, have persuaded some astronomers that the planet may once have been a moon of Neptune which broke away from its gravitational bonds to revolve around the sun on its own.

THE LESSER BODIES AND DYNAMICS OF THE SOLAR SYSTEM

So far we have discussed the major bodies of the solar system, the nine planets. We have also said a few words in passing about some of their more interesting satellites. Though the sun and planets account for most of the system's mass, they are vastly outnumbered by the lesser members: 33 satellites, thousands of asteroids, billions of comets, and trillions meteoric particles. In describing these objects, we will also have occasion to talk about certain aspects of the dynamics of the entire system—how the various bodies affect each other's movements through the operation of gravitational forces.

Bode's Law

Before Kepler formulated his three famous laws, he attempted to solve another problem of heavenly motion. As we saw in Chapter 12, it was the question of why the planetary orbits are spaced as they are. Kepler attempted to answer this question by proposing that the spacing of the orbits of the six planets corresponded to the relationship of the five regular geometric solids. Kepler's model, however, was not convincing, and the later discovery of three more planetary members of the solar system rendered it totally useless as an explanatory device.

The question which Kepler asked, however, was an interesting one. Why do the distances of the planets from the sun form one particular set of proportions rather than some other? One person who tried to solve his problem was Johannes Titius, a German mathematician. He worked out a formula which expressed the numerical relationships between the planetary distances with remarkable, accuracy. His formula was published in 1772 by Johann Bode, director of the Berlin Observatory, and has since become known as the Bode-Titius law, or simply **Bode's law.**

The relationship that Titius found is surprisingly simple. To derive his formula, start with the number sequence 0, 3, 6, 12, 24, and so on. Each number (except for 3) is obtained by doubling the previous one. Adding 4 to each number gives us another series: 4, 7, 10, 16, 28, and so on. Dividing these numbers by 10, we find that the resulting progression corresponds quite closely to the actual distances of the planets from the sun, expressed in AU (Table 13–2).

At the time the formula was published, there was no known planet at a distance of 2.8 AU, and Uranus, Neptune, and Pluto were still undis-

Bode's law is a mathematical progression that corresponds closely to the spacing of the planets in our solar system.

Table 13–2
Bode's Law

Planet	Bode's progression	Actual distance from sun (AU)
Mercury	$(0 + 4)/10 = 0.4$	0.387
Venus	$(3 + 4)/10 = 0.7$	0.723
Earth	$(6 + 4)/10 = 1.0$	1.000
Mars	$(12 + 4)/10 = 1.6$	1.524
(Asteroids)	$(24 + 4)/10 = 2.8$	~2.8
Jupiter	$(48 + 4)/10 = 5.2$	5.203
Saturn	$(96 + 4)/10 = 10.0$	9.539
Uranus	$(192 + 4)/10 = 19.6$	19.191
Neptune		30.071
Pluto	$(384 + 4)/10 = 38.8$	39.518

covered. The "law," therefore, applied only to six known planets. Even for these, it did not work perfectly, for it seemed to predict the presence of a planet about 2.8 AU from the sun, between Mars and Jupiter, where no planet was known to exist. In 1781, however, William Herschel discovered Uranus. The appearance of the new member of the solar system was startling, not only because it was the first planet discovered since ancient times, but because its distance from the sun corresponded closely to the next number in the Bode formula. Astronomers were inclined to take Bode's law more seriously. Many supposed that there might really be a planet between the orbits of Mars and Jupiter, hitherto overlooked, and the search for the missing planet soon got under way.

On the night of January 1, 1801 a Sicilian priest, Father Giuseppe Piazzi, discovered a small object in the constellation Taurus which appeared to change its position among the stars from night to night. At first he thought it was a comet, but after observing it for a while, he became convinced that it was traveling in an orbit between Mars and Jupiter. Before enough observations could be made to compute a definite orbit for the new object, it approached too near the sun to be seen, and was lost. Fortunately, the great mathematician Karl Friedrich Gauss had devised a new method of calculating orbits, which he applied to Piazzi's observations. On the last night of the year 1801, the missing object was found again, in the position predicted by Gauss' calculations.

The asteroids are found in the region where Bode's law predicts a planet.

The orbital radius of the new member of the solar system was very close to the 2.8 AU predicted by Bode's law. Convinced that he had found the missing planet, Piazzi named it Ceres, after the patron goddess of Sicily. But like a person who steps forward to declare himself the inheritor of a fortune, Ceres did not long remain the sole candidate for the honor of being the eighth planet. During the next few years, three other objects were discovered at approximately the same distance from the sun. They received the names Pallas, Juno, and Vesta.

The Asteroids

Astronomers continued to look for more of these **minor planets,** or **asteroids,** as they came to be called. In the early nineteenth century, however, observations were made by looking through the eyepiece of a telescope. Searching for asteroids in this way was extremely unrewarding, with the telescopes then in existence. The four asteroids discovered in the first decade of the century were the brightest ones; a fifth was not seen until 1845.

With the construction of larger telescopes, and the introduction of photography, the search became easier. One way of finding an asteroid is to take a fairly long-exposure photograph of a star field near the ecliptic — the plane of the planetary orbits. The telescope will of course move to follow the stars as they rise and set, so that their images will appear as dots. If there is an asteroid in the field while the exposure is made, it will move against the background of the stars, and its image will appear as a streak. To date, 1816 asteroids have been identified; about 1000 were discovered during the long search for Pluto (Feature, Chapter 5). Tens of thousands more can be seen with the largest telescopes; in most cases, there is no particular reason for anyone taking the time and trouble to plot their orbits.

To call the asteroids minor planets is giving them the benefit of the doubt. The total mass of all the asteroids can only be roughly estimated, but it is probably no more than about $1/1000$ that of the earth. The largest, Ceres, is only some 1000 km in diameter, and there are just ten or a dozen known to be larger than 160 km. Many of the asteroids that show up on photographs are probably about 1 km across. The lower limit, if there is one, is uncertain; there may be thousands of asteroids no bigger than boulders. The asteroids are thought to be made of rocky materials, similar to the moon, or the earth's crust. Three or four of the most massive asteroids have gravitational fields strong enough to make them spherical; the rest are probably jagged and irregular in shape.

The asteroids range in size from pebbles to minor planets.

Most of the asteroids revolve in a belt lying between 2.3 and 3.3 AU from the sun. There are others, however, which stray from this area. The minor planet with the smallest orbit is Icarus. At perihelion (its closest approach to the sun), it passes well within the orbit of Mercury. The largest orbit belongs to Hidalgo, which at aphelion (its furthest distance from the sun), travels beyond the orbit of Jupiter and approaches that of Saturn. One group, known as the Apollo asteroids, approach quite close to the earth. In 1937, Hermes passed within 650,000 km of our planet — less than twice the distance of the moon. The earth has doubtless been struck by these minor planets in the past, probably with a frequency of about once every million years or so. Such features as the Manicouagan Reservoir in Quebec (Color Portfolio) bear witness to these encounters. The effect of a collision with even a small asteroid would cause more destruction than the explosion of many thousands of hydrogen bombs.

Figure 13-9
The Manicouagan Reservoir in Quebec, the remains of a giant crater caused by the impact of an asteroid-sized body some 200 million years ago. (See also Color Portfolio II.)

JUPITER'S INFLUENCE   The asteroids are relatively close to Jupiter —the most massive member of the solar system, after the sun—and show the effects of its gravitational force in several ways. One family of asteroids, known as the Trojans, revolve in orbits which have the same radius as Jupiter's. They form two groups, one of which travels ahead of Jupiter, while the other trails it. If lines are drawn connecting Jupiter, the sun, and the Trojan asteroids, they form two equilateral triangles (Figure 13–10). The reason for this configuration was explained, before the discovery of the Trojans, by the French mathematician Lagrange in the course of his work on celestial mechanics. At the two Lagrangian points, the gravitational forces of the sun and Jupiter combine to produce unusually great orbital stability. Other asteroids in the vicinity of Jupiter are likely to have their orbits disrupted by the effects of its gravity, but the Trojans occupy positions of relative security, oscillating in a complex dance about the Lagrangian points.

Jupiter also tends to make certain other possible asteroid orbits particularly unstable. The period of revolution of an asteroid will depend on its distance from the sun, in accordance with Kepler's third law. Orbits of certain sizes will result in periods of revolution that are a simple fraction of

Jupiter's gravity influences the orbits of the asteroids, making some unusually stable, others unstable.

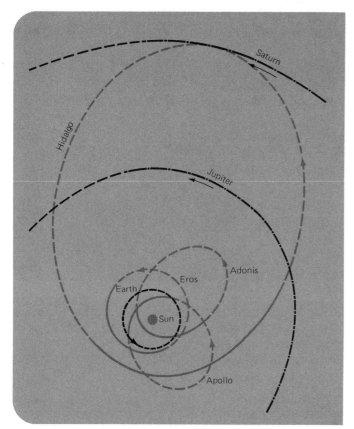

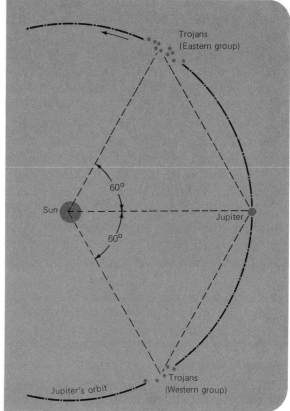

Figure 13-10
At left, the orbits of four atypical asteroids. Apollo and Adonis are members of the relatively small group that pass inside the earth's orbit. Eros is among the asteroids which approach closest to earth; Hidalgo's orbit takes it farther from the sun than any known asteroid. At right, the Trojan asteroids. The two points about which they cluster are known as Lagrangian points, where the gravitational forces of the sun and Jupiter combine to produce orbits of unusual stability.

Jupiter's period. An asteroid whose orbital radius is about 3¼ AU, for example, would have a period of 5.9 years—almost exactly half that of Jupiter. Every two trips around the sun, therefore, it would find itself near Jupiter—at the same point in its orbit. Thus whenever Jupiter's gravitational influence on the asteroid was at its strongest, it would always be operating in the same direction. The repeated tugs in the same direction would eventually be enough to pull the asteroid into a different orbit. In fact, such a gap in the asteroid belt exists, as do others where the orbital radius would result in periods ⅓ that of Jupiter, ⅖ that of Jupiter, and so on. These empty regions are known as **Kirkwood Gaps.**

Comets

Until recent times, comets were regarded as bad omens, signaling the coming of disastrous events. With its glowing head and long, luminous tail, a comet seems to be a spectacular object. Physically, however, most comets are quite unimpressive. The basic component of a comet is its nucleus. This consists of dust and fragments of rock, mixed with ice and other frozen gases. Indeed, a comet has often been compared to a dirty iceberg.

The outstanding physical feature of comets is their extremely low average density. They contain a very small amount of matter spread over a large volume. Estimates of the mass of comets range from 1 billionth to perhaps 1 trillionth of the mass of the earth. We know that comets have very little mass not by how they affect other bodies in the solar system, but by what they fail to do. When Brooks' comet, for instance, passed very close to Jupiter, its own orbit was greatly changed but it had no measurable effect on even the very smallest of Jupiter's satellites. Since comets are large, with tails that often reach 150 million km in length, the volume through which their low mass is spread is huge. This results in an extremely low average density. Figuratively speaking, one could take all the matter in the tail of a comet, pack it into a suitcase, and walk off with it. The density in the nucleus is about 1 millionth that of air at sea level, but even this is much greater than the density of the tail.

How can an object of such low density and such great size, containing such a small quantity of material, be so spectacular? As a comet approaches the sun, the rising temperature causes the frozen gases of the nucleus to evaporate and become fluorescent. Instead of merely reflecting sunlight, the comet absorbs some of the solar radiation and re-emits it, producing a bright-line spectrum of its own. Emission lines characteristic of many radicals (fragments of molecules) are found in cometary spectra. These provide a clue as to the origin of the comets, for while these radicals are found in interstellar space, they are generally absent within our solar system. As the gases of the nucleus expand, they form the comet's head, or coma, which may eventually reach 100,000 km in diameter. The tail — which always points away from the sun, even when the comet is receding from the sun — is formed chiefly by the pressure of radiation and the solar wind. Generally there are markedly different tails, one composed primarily of gas, the other of dust particles (see Portfolio, opposite).

Until its frozen gases are vaporized by the sun's radiation, a comet is a rather small object.

There is great variation in the orbits of comets. A few are relatively small and only moderately elongated. Others follow more elongated paths that take them out to the fringes of the solar system, and back to swing close to the sun at perihelion. There are about 200 known comets of this type, about 85 of which have periods of less than 200 years. The most famous of all comets belongs to this group: Halley's comet (Figure 13-11). In 1705, while studying the orbit of the comet of 1682, Edmond Halley noticed its surprising similarity to the orbits of two other bright comets, those of 1531 and 1607. Noting the common interval of approximately 75 years between these appearances, he predicted the return of the comet in 1758. Although he did not live to see his prediction come true, the comet was picked up on Christmas night in 1758 and passed perihelion in the spring of 1759.

With one exception, returns of Halley's comet at intervals of about

Comet West, above, was the fourteenth comet of the year; it is therefore designated 1975n. Notice the two distinct tails. In this picture, taken on March 12, 1976 the angular extent of the comet is about 25 times that of the moon. Below, the Comet Ikeya-Seki, a spectacular object seen in 1965. This photograph, taken with a 35-mm camera, shows it in the sky above the City of Los Angeles. The comet approached the sun very closely and as a result became very bright.

75 or 76 years have been recorded ever since 240 BC. It was Halley's comet that blazed in the sky when the Normans conquered England in 1066. It last appeared in 1910; the early actually passed through its tail, occassioning great anxiety on the part of some nervous people, but without producing any observable effects. It passed aphelion, beyond the orbit of Neptune, in 1948, and will be back again in 1986.

The great majority of comets, however, have periods so great that they have been seen only once in recorded history. Comet Kohoutek, for example, which passed through the inner regions of the solar system in 1973–1974, is believed to have a period of about 5 million years. At aphelion, Kohoutek was about 56,000 AU (8.4 billion km) from the sun —about 1400 times the mean distance of Pluto. Comets such as Kohoutek travel so far from the sun that there is some question as to whether they can be considered members of the solar system at all. Most astronomers now believe, however, that the majority of comets were formed from the outermost remnants of the cloud of gas and dust whose center coalesced to form the sun and planets. There may be as many as a hundred billion comets at vast distances from the sun, awaiting the moment when minute gravitational perturbations of distant stars will send them accelerating inward toward the sun.

Figure 13-11
At left, part of Jupiter's family of comets. There are several dozen known members in all, only a few of which are shown here. These comets were captured by the gravitational force of the massive planet and deflected into relatively small orbits. Membership in the family may not be permanent, however.

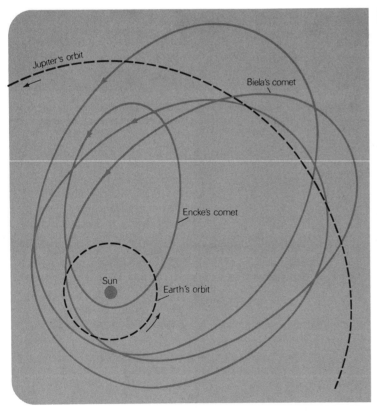

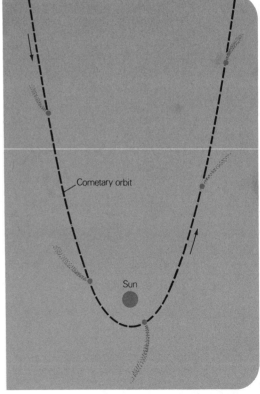

Kepler's second law implies that a planet will travel faster when it is near the sun than when it is far away from it. Since the orbits of most of the planets are close to perfect circles, the planets' changes in velocity are relatively slight. Because of their extreme eccentricity, cometary orbits provide the best illustration of Kepler's law in our solar system.

At perihelion, comet Kohoutek came within 21 million km of the sun. At this point, it was traveling at about 400,000 km/hr. At aphelion, a line connecting Kohoutek with the sun would be some 400,000 times longer than a corresponding line at perihelion. For this line to sweep out equal areas in equal intervals of time, the comet must move extremely slowly—about 1 km/hr, or about as fast as a toddler can walk. When Kohoutek is traveling at this tortoise pace, the filament of gravity connecting it with the sun will be very weak indeed. Slight gravitational disturbances caused by nearby stars disrupt the comet's orbit and prevent it from ever paying a return visit to the inner regions of the solar system.

The huge mass of Jupiter may also be responsible for changing the orbit of a comet during its trip sunward. Depending on Jupiter's position, it may either be accelerated so that it leaves the solar system altogether—perhaps never to return—or deflected into a new, smaller orbit, dominated by Jupiter's gravitational influence. A whole family of comets has been captured by Jupiter in this way.

Satellites and their Orbits

Comets are not the only bodies in the solar system whose orbits may be affected by the gravitational forces of the planets. We have already seen how the asteroids are influenced by Jupiter. Another example of such influence can be found in the planetary satellite systems. Of the 33 satellites belonging to the earth, Mars, Jupiter, Saturn, Uranus, and Neptune, 10 are believed to be captured asteroids which were pulled into orbit by the gravitational attraction of the parent planet.

In most cases, it is easy to tell which of the moons are "natural" satellites, formed at the same time as their parent planet, or **primary**, and which are captured asteroids. The natural satellites tend to revolve fairly close to their primaries. The generally revolve in a counter-clockwise direction when viewed from above the planet's north pole. Their orbits lie in or near the equatorial planes of the primaries, and tend to be approximately circular. Finally, the natural satellites of Jupiter, Saturn, and Uranus exhibit relationships similar to Bode's law, so that in their spacing they resemble miniature solar systems. Captured satellites, on the other hand, often are quite distant from their primaries. They may revolve in either a counter-clockwise or a clockwise (retrograde) direction. Their orbits may be tipped at an angle to the primary's equatorial plane, and may also be highly eccentric. They do not conform to Bode's law.

It is possible to guess from their orbits which satellites were probably formed with their primaries, and which are probably captured asteroids.

Jupiter, the planet which possesses the largest number of satellites (13 by the latest count) is an excellent example of a primary surrounded by both natural and captured satellites. The four Galilean satellites, described previously, all occupy relatively small, almost circular orbits close to Jupiter's equatorial plane. One more satellite, Amalthea, revolves very close to Jupiter—almost within the Roche limit (discussed below). This extremely small moon is probably also a natural satellite. The rest of Jupiter's moons, eight in number, are probably captured asteroids. They are all very small, and much farther from Jupiter than the Galilean satellites. The outermost one, known only as J-VIII, follows an eccentric orbit that takes it over 32 million km from the center of Jupiter. Some of the eight outer moons have orbits inclined steeply to Jupiter's equator, while the four most distant revolve in a retrograde direction.

What of our own moon? It may come as something of a surprise to learn that the moon, perhaps the most familiar object in the night sky, is still quite mysterious in this respect. Despite the fact that it is the only celestial body on which men have set foot, we still do not know very much about its origins. One of the most unique things about our moon is its mass. Though several of the other satellites in the solar system are more massive in absolute terms, the moon is some 10 times more massive **in relation to its primary** than any other satellite. The moon is also quite far from earth compared with other satellites, and its orbit is fairly eccentric. Moreover, the moon revolves, not around the earth's equator, but in an orbit very close to the plane of the ecliptic—the plane in which the planetary orbits lie.

These characteristics have led many astronomers to speculate that the moon is a captured rather than a true satellite. But this theory too has its weak point, for it would be nearly impossible for an object as large as the moon to be captured by a planet as small as the earth. Thus the earth-moon system is often characterized as a double planet rather than as a satellite and primary.

The origin of the moon is still mysterious; some astronomers regard the earth-moon system as a "double planet."

Tidal Forces

Everyone is familiar with the tides that affect the water level at the seaside. People had been aware for centuries that these tides were somehow related to the moon, but it was not until Newton formulated his theory of gravitation that the connection was made explicit. As Newton showed, the gravitational force exerted by one body on another is inversely proportional to the square of the distance between them. Since, at any given time, the moon is appreciably closer to some parts of the earth than to others, the gravitational force it exerts on the earth is not uniform (Figure 13–12). The moon's gravitational influence is greatest at point A and least at point B. These are the two points which experience high tide—point A because it is being pulled most strongly by the moon,

The gravitational pulls of the sun and moon cause earth's ocean tides.

point B because it is experiencing the least attraction. As the earth rotates, points A and B circle the globe, and as a result, each part of the earth experiences two high tides and two low tides per day.

The tidal forces exerted by the moon are not confined to the oceans. The earth's atmosphere experiences tides, as does the solid body of the earth itself. The point on the earth closest to the moon is pulled toward it most powerfully. This creates a stretching effect on the earth. Since the earth is not perfectly rigid, it responds to the the moon's influence by distorting slightly in shape. The distortion, though, is very small, amounting to no more than seven or eight inches. Every body in the solar system exerts a tidal force on every other body. Whereas simple gravitational force obeys an inverse square law, it can be shown that tidal forces increase inversely as the cube of the distance between two bodies. In the case of bodies which are relatively distant from one another, as for example the earth and Mars, the tidal forces are so small as to be negligible. In the case of bodies which are close, however, tidal forces can be considerable. The moon, whose gravitational force on the earth is only $1/150$ that of the sun, exerts a tidal force twice as great as the sun's.

Figure 13-12
The moon's gravitational attraction is greatest at point A, which is closest to the moon, and least at point B, on the opposite side of the earth. As a result, the earth's oceans bulge out on both sides of the planet. The earth turns under this bulge, which thus moves around the world, so that each location on earth experiences two high and two low tides every day. Because of friction with the ocean bottoms, however, the bulges align themselves slightly off the earth-moon axis; this effect is shown exaggeratedly in the first figure. When the gravitational forces of the sun and moon operate in conjunction, the result is an unusually high tide (spring tide). When they act at right angles they oppose each other, producing a lower tide (neap tide).

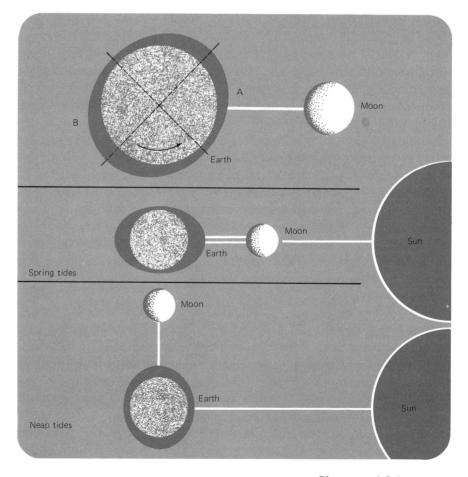

THE ROCHE LIMIT AND SATURN'S RINGS   The moon is far enough from the earth so that the tidal forces created by the earth's gravitational field are not strong enough to do it any damage. But what if the moon were closer? Theoretically, there should be a point at which the stretching effect of the earth's gravity should pull the moon apart. This problem was investigated by E. A. Roche. In 1850 he published a paper showing that the minimum distance at which a satellite could orbit its parent planet without being torn apart by tidal forces is 2.44 times the planet's radius. (This result assumes that the two bodies are of equal density, and that the satellite is held together by its own gravitational forces. It does not apply, therefore, to artificial earth satellites, which are held together by the strength of the metal of which they are made.) This distance has come to be known as the **Roche limit.** The radius of the earth is 6378 km, making the earth's Roche limit 15,562 km. The average distance of the moon from the earth is 384,000 km, placing it well outside the Roche limit.

Six of the 9 planets in the solar system have moons, making a total of 33 moons in all. None of these moons is inside the Roche Llimit. There is, however, one object, or collection of objects, circling a planet within the Roche limit: the rings of Saturn.

Saturn's rings were first noticed by Galileo, but his small telescope had insufficient resolving power to allow him to see them clearly. They were described accurately by Christian Huygens some years later. We now know that the rings consist of numerous particles of matter, each revolving independently around the planet. Radar and spectroscopic studies show that the particles are probably made of ice, and vary in size from roughly an inch to a foot. The are widely spaced enough so that stars can be clearly seen through them. The rings are also very thin, perhaps no more than a few feet. When the planet is positioned in orbit in such a way that the rings are edge-on to us, they become invisible even through the most powerful telescopes. Although there is some disagreement about how the rings came into being, the most frequently proposed explanation is that Saturn once possessed a small moon which wandered inside the Roche Limit. Tidal forces tore it apart, shattering it into tiny fragments which spread out to form the rings.

Saturn's rings may have been formed when a satellite wandered inside the Roche limit and was shattered by tidal forces.

SYNCHRONOUS ROTATION   As anyone who observes it over a period of time can see, the moon always presents the same face to the earth. The other side of the moon (erroneously called the dark side—though in fact it receives as much light as our side does) remained unseen until the U.S. Lunar Orbiter circled the moon in August, 1966, and sent back photographs. The moon keeps one side turned towards us because it rotates once for each revolution about the earth. This phenomenon is known as **synchronous rotation.**

Synchronous rotation is caused by the tidal forces exerted on the moon by the earth. It is believed that at some point in the past, the moon rotated more rapidly than it does today. The earth's tidal influence, however, caused slight bulges in the solid body of the moon, just as the moon does in the earth. As the moon revolved around the earth, the bulges would have had to move across the surface of the moon, in a direction opposite that of its spin, so as to remain always aligned with the earth. This process acted as a brake on the moon's rotation, causing it to slow down until it became locked in step with its revolution about the earth. Under these conditions, the bulge is always in the same place—facing earth—and need not move.

To a much lesser degree, the moon is acting as a brake on the earth's rotation. The earth is spinning slower and slower, an effect which causes the day to become about a thousandth of a second longer per century. This increase may seem negligible, but over many centuries, it can add up to a considerable amount. For example, during the Devonian period, about 400 million years ago, when the first fishes evolved on earth, the day was only about 22 hours long. We may expect the slowing of the earth's rotation to continue in the future.

The earth and moon are not the only bodies in the solar system whose rotation is affected by tidal forces. For a long time, astronomers believed that Mercury's rotation, like that of the moon, was synchronous, and that it kept one face forever turned toward the sun. Subsequent observations by radar and unmanned probes have shown that Mercury, instead, exhibits a closely related effect known as spin-orbit coupling. It rotates three times for every two revolutions that it makes about the sun. Like synchronous rotation, spin-orbit coupling is brought about through tidal forces, but in Mercury's case, they are exerted by the sun.

Tidal forces may cause the period of rotation of a planet or satellite to become synchronized with its period of revolution.

Meteors

Perhaps because of their long, luminous tails, which give the impression of rapid motion, many people think of comets as blazing, white-hot objects that streak through our skies and are gone in the blink of an eye. Comets do move fast—but they are so far from the earth that their motion is only apparent to the naked eye over a period of several nights. A bright comet may remain visible for a period of weeks. In short, comets, as Tycho was the first to show, belong to the heavens, not to the earth's atmosphere. When people think of objects shooting through our skies, what they really have in mind are **meteors.**

The terminology connected with meteors is a bit confusing. **Meteoroids** are fairly small fragments of rocky or metallic material that travel through interplanetary space. When they strike our atmosphere at high speeds, they are usually heated to incandescence by friction with the air, and burn up within a few seconds. The result is a meteor, also known as a "shooting star" or "falling star." The streak of light that the meteor creates during its passage is known as a **meteor trail.** If the meteoroid survives its passage through the atmosphere and strikes the earth's surface, it is then called a **meteorite.**

Apparently, the formation of the solar system was not a neat, orderly process, in which a cleanup followed the main event. Considerable debris appears to have been left in the spaces between the planets, and the meteoroids that strike our atmosphere probably originated in this way. Their number must be in the trillions, or more. It has been estimated that about 100 million meteor trails bright enough to be seen with the unaided eye against a dark sky occur all over the earth each day. Since half of them appear in daylight, and because most of the earth's surface is uninhabited, the majority are unobserved. The earth is thought to gain about a million kg of mass each day from meteorites and meteoric dust that reach the surface of our planet.

Meteoroids be divided into two categories. Some, like the asteroids, occupy orbits that are only moderately elliptical, like those of the earth and other planets. The velocity of such meteoroids with respect to the earth will never be very high, since they are moving in the same direction around the sun as the earth, at rather similar velocities. Because they strike the earth's atmosphere with such low speeds, these objects do not burn up as rapidly, or as high in our atmosphere, as faster-moving meteoroids. They thus give rise to meteors that are unusually brilliant and long-lived; their visible passage through the atmosphere may last several seconds, whereas most meteors have vanished almost before one has a change to look at them. Occasionally, they even produce audible sonic booms, like those of jet aircraft exceeding the speed of sound. These extraordinarily bright meteors are known as **fireballs.**

The vast majority of meteoroids, however, move in highly elongated orbits, resembling those of comets rather than those of asteroids. Bodies in such orbits are always very near the escape velocity needed to leave our solar system forever. In the vicinity of the earth, the escape velocity is about 42 km/sec. Since the earth's own orbital velocity is about 30 km/sec, it is possible to predict the range of velocities of meteoroids of this type quite precisely. The maximum velocity will occur when the earth and the meteoroid meet head-on; it will be about $30 + 42 = 72$ km/sec. The minimum will occur when a meteoroid overtakes the earth; the velocity then will be $42 - 30 = 12$ km/sec. Few if any meteoroids are found to strike the earth at velocities greater than 72 km/sec, indicating that these objects are truly members of the solar system, rather than intruders from interstellar space.

On an average clear night, an observer can count on seeing about 7 meteors in an hour. At times, however, the earth encounters a large group of meteoroids traveling together in almost identical orbits. The result is an unusually intense bombardment of meteors, known as a **meteor shower.** If the meteoroids are strung out over their orbit in a **stream,**

Figure 13-13
Meteor crater in Arizona. This scar, more than a kilometer in diameter, was probably made fairly recently— within the last 50,000 years. It has not had time to be effaced by the forces of erosion and tectonic activity (Compare Figure 13-9). The meteorite that made it was apparently shattered on impact, for tons of meteoric material have been found in the vicinity.

Most meteors travel in elongated orbits similar to those of comets.

rather than bunched together in a **swarm,** the meteor shower that results will be an annual event, occurring each time the earth crosses the orbit of the meteoroids (Figure 13–14). A good example is the Perseid shower, which is visible for two or three nights about August 11 or 12 of each year. During an ordinary shower some 30 to 60 meteors may be visible per hour, but in exceptional instances thousands, or even hundreds of thousands may be seen in the course of an hour. Such a shower occurred in the southern U.S. in 1835 — famous ever since as the year the "stars fell on Alabama."

Meteors that follow orbits resembling those of comets are in fact frequently associated with comets. It often happens that a comet breaks up into several fragments during its passage around the sun, probably from a combination of vaporization of its frozen material and the sun's tidal forces. Sometimes the comet simply does not show up at all at its expected return. Instead, when the earth intersects the comet's orbit, there is a meteor shower, suggesting that the comet has disintegrated into a swarm of meteoric particles.

The odds of a meteoroid surviving its trip through our atmosphere depend in part on its size. The meteors that we commonly see are created by particles ranging in size from a grain of sand to a pebble. These are generally vaporized in their entirety as they streak through the upper layers of the atmosphere. Smaller particles, however, are quickly slowed up by air resistance — often before they have become hot enough to vaporize. These tiny meteoroids then drift down through the air and settle, unnoticed, on the earth's surface like ordinary dust. At the opposite extreme, the largest meteoroids, which may be the size of huge boulders, may also survive their blazing fall to earth. Though they may lose much of their substance along the way, enough will often be left to reach the gound. The odds against anyone being hit by a falling meteoroid are very great, but, there have been a few recorded cases, and some near misses.

Meteor showers often occur when the earth crosses the orbit of a comet that has disintegrated.

Figure 13-14
The disintegration of a comet often produces a large number of meteorites. If the particles are dispersed around the comet's orbit, forming a stream, there will be a meteor shower each year when the earth crosses the orbit. If the group remains in a compact swarm, however, a shower will result only when the earth and the swarm happen to arrive simultaneously at the point where their orbits intersect.

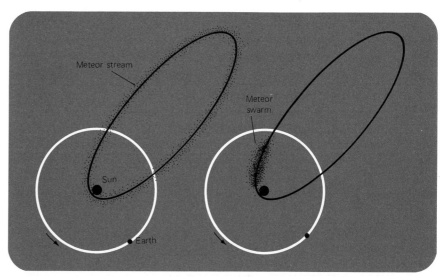

The sun contains nearly 99.9 percent of the solar system's total mass. The next most important bodies in the solar system are the nine planets. Mercury, Venus, the earth, and Mars are called the terrestrial planets. They are closer to the sun, rotate more slowly, and are smaller but denser than the Jovian planets—Jupiter, Saturn, Uranus, and Neptune. Little is known about Pluto.

The earth has a heavy core of iron and nickel, heated by pressure and the decay of radioactive materials. On top of the core is a lighter mantle composed of rock, and on the mantle is an even lighter crust, which is less than 32 km deep in most places. The hydrosphere (oceans) and the atmosphere surround the crust to a height of about 300 km.

Mercury is the smallest planet, the closest to the sun, and the fastest in revolving around the sun. It has only traces of an atmosphere and surface features resembling those of the moon. Mercury is denser than the moon, however, indicating the presence of iron or other heavy metals. The planet closest to the earth both in distance and size is Venus. Thick clouds covering the surface of Venus create a greenhouse effect that keeps the temperature there at 800° F or above. Tectonic forces seem to have shaped Venus's surface.

Mars is much smaller than the earth, having less than one-ninth the mass of our planet. Its atmosphere is very thin and does not retain much heat; although the temperature at the equator rises to 80° F at noon, at night it drops to $-100°$ F. There are huge craters, chasms, and volcanoes on Mars; and a fierce wind blows a 50-km high wall of dust across the planet at speeds approaching 500 km/hr. Photos taken from space probes show that many details of the dry riverbeds of Mars look just like such features on the earth—yet there seems to be no liquid water there. One theory suggests that Mars is currently going through an ice age, with most of its water and atmosphere frozen in the polar ice caps.

Jupiter is the largest of all nine planets, with 318 times the mass of the earth and over 1300 times the volume. Yet Jupiter spins faster than the earth, and as a result it bulges at the equator. Except perhaps for a small rocky core, Jupiter may be all liquid and gas. At a depth of several hundred kilometers below the apparent surface, Jupiter's gaseous hydrogen gives way to liquid hydrogen. Jupiter has a high internal temperature, and a very strong magnetic field. The planet's cloud-filled atmosphere contains many stable convective structures (such as the Great Red Spot, a giant storm) that are easily visible from earth.

Saturn, the next largest planet after Jupiter, is also composed largely of gases, compressed into liquid form toward the center of the planet. Uranus and Neptune, though smaller than Saturn, are larger than any of the terrestrial planets. They are also gaseous planets. Pluto is very far away and very small. Its orbit is tilted more than 17° to the plane of the ecliptic and is very eccentric. During part of its year, in fact, it comes inside the orbit of Neptune. Pluto is believed to be a rocky rather than a gaseous planet, about one-tenth the size of the earth.

Bode's law (actually a formula worked out by Johannes Titius) expresses the numerical relationships between the distances of the planets from the sun. According to this law, however, there should be a planet between Mars and Jupiter. In 1801, one was found—and then, in rapid succession, three more, all at about the same distance from the sun that Bode's law predicts. Since then, thousands more have been discovered and are now known as the minor planets or asteroids. The largest, Ceres, is only about 1000 km in diameter, but it and two or three others have gravitational fields strong enough to make them spherical. The asteroids are believed to consist of rock, like the moon.

Comets, with their glowing heads and luminous tails that may reach 150 million km in length, look spectacular in the night sky but actually have very little mass. They consist of dust and fragments of rock, mixed with ice and other frozen gases. As a comet approaches the sun, the rising temperature causes the frozen gases to evaporate and become fluorescent. The tail of the comet, driven by the solar wind, always points away from the sun, even when the comet is receding from the sun.

If a satellite comes close enough to its planet, tidal forces may cause it to disintegrate. The point at which this will happen is known as the Roche limit. No objects are found within the Roche limits of the planets except for the rings of Saturn, which are believed to have been formed when a small satellite wandered inside the Roche limit and was shattered by tidal forces.

EXERCISES

1. Describe the two planetary "families."
2. What is continental drift? Why does it make some persons reluctant to settle down in California?
3. Which of the terrestrial bodies do we believe to be geologically active?
4. What is the greenhouse effect? Why do we believe that it is responsible for the surface temperature of Venus?
5. Why do we think Mars may once have had running water? Where is the water now?
6. How does the composition of Jupiter differ from that of the earth? What accounts for the Great Red Spot and other markings on its "surface?"
7. Why are the planets harder to understand, in some ways, than stars?
8. Explain Bode's law. What role did it play in the discovery of the asteroids?
9. Describe two phenomena which show Jupiter's gravitational influence on the asteroids.
10. What are comets made of? Why do they change their size and appearance so dramatically as they approach the sun?
11. Most comets seem to travel in closed orbits. Why are only some of them called "periodic?" Why are 5 to 10 "new" comets discovered each year?
12. What is the period of a comet whose aphelion distance is 20,000 AU?
13. How would you distinguish the "natural" from the captured satellites? Illustrate your discussion with the satellites of Jupiter. Why do we think the satellites of Mars are probably captured asteroids?
14. Why does the moon always show the same face to observers on earth? How did this situation come about?
15. What is the Roche limit? How does it help explain the rings of Saturn?
16. What causes meteor showers? How are they associated with comets?

CHAPTER FOURTEEN

ORIGIN AND HISTORY OF THE SOLAR SYSTEM

In the last chapter we have examined the properties of various members of the solar system—planets, asteroids, comets, meteorites. Now we are faced with the task of tying together many isolated pieces of information into a unified picture of the solar system, in order to understand its structure, origin, and evolution. Our aim will be to put the solar system into perspective, with an emphasis on making generalizations and seeing what patterns and relationships exist among members of the planetary system. In this way we hope to set the stage for a discussion of how the solar system originated. As we will see there is a good deal of evidence to support several different theories of origin.

The solar system includes nine planets: Mercury, Venus, earth, Mars, Jupiter, Saturn, Uranus, Neptune and Pluto; their 33 satellites, or moons; and a large number of asteroids, comets, and meteoroids. The nine major planets revolve around the sun in elliptical orbits that are almost circular. Indeed, one of the most striking features of the system as a whole is the regularity of the planetary orbits. Not only do they deviate but little from circles, but all of the orbits lie in nearly the same plane. Except for Pluto, which has a tilt of 17 degrees, no other planet tilts from the plane of the solar system at an angle of greater than 7 degrees. This means that if a scale model of the entire planetary system out to Neptune were packaged in a box 100 cm across, it would need to be only 2 cm high. Moreover, all the planets move in one direction around the sun, and the sun also rotates on its own axis in the same direction.

Thus the planets appear to conform to certain rules. These rules are also followed, to some extent, by the other bodies of the solar system. But the massive planets seem to obey them better than the less massive objects such as comets and meteoroids, and the larger planets, such as Jupiter, observe them more completely than the smaller ones such as Mercury and Pluto. The larger planets, for example, move in orbits that are more nearly circular than the smaller ones, while the most elongated orbits belong to the comets and meteoroids.

STRUCTURE OF THE SOLAR SYSTEM

We do not yet have an explanation for all of the observed regularities of the solar system. It is by studying such patterns that we hope to arrive at a theory that will explain its origin. Even in the absence of a completely satisfactory theory, however, one thing seems certain: the order of the solar system represents an efficient means of self-regulation. The nearly circular orbits of the planets allow for the safest flow of traffic. Bodies which disobeyed these traffic regulations and did not conform to the regular orbital patter would sooner or later approach too close to other bodies. As a result of such encounters, they would either be torn apart by tidal forces, or accelerated out of the solar system entirely—fates which often befall comets. Even if the solar system had once been less orderly than it now is, some order such as the one we now see was bound to evolve.

Figure 14-1
The solar system is remarkably flat; all the planetary orbits lie in nearly the same plane. The chief exception is Pluto, whose orbit is inclined about 17° to the plane of the system as a whole. Pluto's orbit is also the most elongated; it actually passes within the orbit of Neptune at one point.

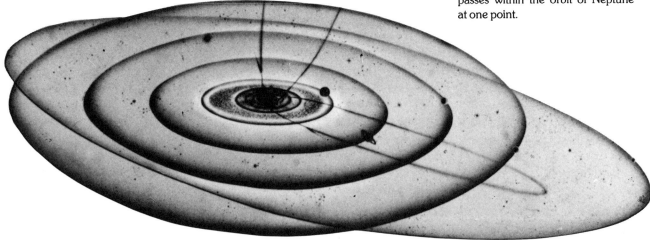

Orbital Periods and Rotation

In addition to the fact that the planets orbit around the sun in the same plane and direction, there are curious regularities in their orbital periods. When the periods of Jupiter and Saturn, the two largest planets, are compared, it is found that they are in the ratio of 3 to 5. The same is true when other planets are compared with respect to the time it takes them to revolve once around the sun. Uranus and Neptune have periods that approximate the ratio of 1 to 2; Jupiter and Uranus, 1 to 7; Uranus and Pluto, 1 to 3. The interesting point is that the ratios are simple — that is, they can all be expressed in terms of small, whole numbers. We do not find ratios such as 19 : 31 or 21 : 57. The periods of the planetary satellites also exhibit similar patterns. In the case of Uranus, for example, simple ratios can be found between the orbital periods of any two of its five satellites.

Another thing common to all the planets is that they rotate, or spin about their axes. Most of the planets rotate in the same direction in which they orbit around the sun. The exceptions are Uranus and Venus, which rotate in a retrograde, or reversed manner. The periods of axial rotation also exhibit intriguing numerical relationships. When the rotation of the innermost planets is compared to their revolution, around the sun, simple whole number ratios are found to exist. Venus, for example, rotates in such a manner that each time it passes the earth, it always presents the same face to us. Mercury rotates exactly three times on its axis for every two revolutions about the sun. These patterns have yet to be fully explained.

Seven of the nine planets rotate on their axes in the same direction in which they orbit the sun.

Angular Momentum

Rotation and revolution endow any body with angular momentum. As we saw in Chapter 5, the angular momentum of a body depends on its mass, its velocity, and the distribution of its mass in relation to the axis of rotation or revolution. The mass of the sun is far greater than that of the planets. But the planetary orbits are large, and the planets travel at high velocities. As a result, the planets actually have far more angular momentum than the sun, rotating slowly on its own axis. In fact, the sun, with over 99 percent of the mass of the solar system, accounts for only about 2 percent of its angular momentum. This is an important fact, which must be taken into account in any explanation of the origin of the solar system.

The sun has 99 percent of the solar system's mass, but less than 2 percent of its angular momentum.

Bode's Law

The relative distances of the planets from the sun have been known with considerable accuracy since the time of Kepler. As we saw in the previous chapter, Bode's law, discovered in the latter half of the eighteenth century, expresses the spacing of the planets extremely well. At the time it was proposed, only six planets were known. The discovery of Uranus (1781) and Ceres (1801) — both at distances from the sun predicted by the Bode formula — greatly enhanced the law's prestige. The position of Neptune, discovered in 1846, does not accord with the next number in the Bode series — it is about 30 AU from the sun, rather than 38.8. However, if we skip over Neptune, we find that Pluto, discovered in 1930, does have a distance closely approximating the Bode figure; its average orbital radius is 39.5 AU. Neptune, however, which does not obey the formula, is a large and important planet, while Pluto, which does, may be only an escaped satellite of Neptune.

We mentioned in the previous chapter that instead of the planet called for by Bode's law, 2.8 AU from the sun, we find the asteroids, many of which move in nearly circular orbits. It has been suggested that a planet once revolved between the orbits of Mars and Jupiter, and was shattered to fragments when its inhabitants precipitated a great nuclear war. A more sober theory proposes that the planet was broken up by Jupiter's gravitational force. Neither of these explanations is very likely, however. The mass of all the minor planets together would add up to no more than about $1/1000$ that of the earth — hardly enough to make a good-sized satellite, let alone a planet. Moreover, even massive Jupiter could not generate the enormous tidal forces necessary to tear apart a planet once it had formed.

It seems far more likely that the minor planets are chunks of matter which for some reason never coalesced to form a planet. As such, they may represent an early stage of planet formation. Other planets too may once have consisted of such swarms of "planetismals," which later came together through collision or gravitational attraction. In the case of the asteroids, there may not have been enough mass for planet formation, or perturbations caused by Jupiter may have been responsible for the failure. It has also been pointed out that the asteroid belt forms a demarcation between two different types of planets: the inner, rocky ones with thin atmospheres, and the outer, gaseous ones. This suggests that there may have been two sets of planets formed by two different processes, while in between neither process was strong enough to achieve the creation of a planet.

The asteroids may represent material that failed to condense to form a planet.

There is a very real question as to whether Bode's formula should be called a "law" at all. No physical reason for the relationship has ever been found. It would be more correct to say that the Bode formula represents a pattern in search of an explanation. Many theorists have tried to show that the Bode relationship follows from their model of the formation of the solar system, but so far no one has been completely successful in this attempt. However, it is hard to dismiss Bode's law as an insignificant accident — especially since a very similar relationship seems to exist for the satellite systems of Jupiter, Saturn, and Uranus.

As we observe the planets from Mercury out to Pluto, we notice that they seem to fall into two distinct groups. The planets closest to the sun are all relatively small, dense, and rocky in composition. They generally have relatively thin atmospheres, and tend to rotate rather slowly. These are the so-called terrestrial planets, named after their prototype, terra, or earth. This group includes Mercury, Venus, the earth and its moon, and Mars. The next four planets — Jupiter, Saturn, Uranus, and Neptune — are often called the Jovian planets, after the god Jove (Jupiter). They are far larger, but far less dense, than the terrestrial planets, being composed chiefly of hydrogen and helium, much of it in a liquid state. They have heavy, dense atmospheres, and rotate very rapidly for their size. These planets are also sometimes called the major planets, because of their size, or the solar planets, because their composition is much closer to that of the sun than the terrestrial planets. Pluto, which is small and dense, does not really fit into this group, though it is located beyond the orbit of Neptune.

The Jovian planets are larger in size and lower in density than the terrestrial planets; they also rotate faster.

The terrestrial planets contain almost no traces of hydrogen or helium, but are instead composed of oxygen, silicon, and iron. On the basis of composition, asteroids most likely belong to the terrestrial group, whereas comets are much more similar to the Jovian planets.

The contrast between terrestrial and Jovian planets is indeed striking and is something which must be explained by the origin of the solar system. Assuming that the terrestrial planets originated from the same cloud of primordial matter as did the sun and Jovian planets, why did the Jovians retain the light elements which were in all probability the primary constituent of this primordial cloud? An answer to this question must form the cornerstone of any theory of solar system origin.

One of the most remarkable differences between the terrestrial and Jovian planets appears when we compare the masses and compositions of their atmospheres. The Jovian planets have extensive atmospheres consisting of approximately two-thirds hydrogen and one-third helium. The larger terrestrial planets (earth, Venus, Mars) have smaller atmospheres rich in the elements carbon, nitrogen and oxygen. The smaller terrestrial bodies (Mercury, the moon) possess no atmosphere at all. The reason for such wide variations can be explained by two facts: the large mass difference between the planets, and their different surface temperatures, resulting largely from their varying distances from the sun. The more massive a planet, the greater its surface gravity is likely to be, and the more easily it will retain its atmospheric gases. Moreover, at the lower temperatures of the Jovian planets, the molecules move more slowly, and are thus less likely to attain the escape velocity necessary to be lost into space.

The terrestrial planets have solid crusts and relatively thin atmospheres; the Jovian planets are largely liquid and have very dense atmospheres.

The great distance from the sun of the Jovian planets may also account, at least in part, for their rapid rotation rates. The terrestrial planets have probably been slowed, over the eons, by the sun's tidal forces (Chapter 13). If we compare the effectiveness of the sun in slowing down the rotation of the various planets, however, we find that several of the planets—the earth and Mars, in particular—seem to be rotating more slowly than expected. This suggests that the solar tidal forces acting on them in the past may have been much greater than they are at present. One way to account for this is to assume that these planets were originally much larger than they are now. This hypothesis is consistent with the fact that if the earth is assumed to have been made from the same material as the sun, an original mass about the same as that of Jupiter would be needed to get the amounts of iron, oxygen, and silicon that the earth now contains. It seems likely, therefore, that the terrestrial planets were all originally gas giants like the Jovian planets. The higher temperature in the vicinity of the sun resulted in their losing most of their hydrogen and helium, leaving behind only the rocky cores that now make up these planets.

The terrestrial planets may originally have formed from large masses of gas, much of which was dissipated by the sun's heat.

Pluto has always remained a thorn in the side of astronomers who tried to tie the planets into two neat packages: the inner terrestrial and the outer Jovian. This is because Pluto is small and dense and without an atmosphere, much like a terrestrial planet, although it is located on the outermost fringes of the solar system along with the Jovian planets. A recent suggestion has attempted to resolve the difficulty. In this theory, Pluto in the beginning was not a planet at all but merely a satellite of Neptune. Neptune had two satellites, Pluto and Triton, with Pluto being closer to Neptune. During a chance near-collision between the two satellites, Triton's orbital motion was reversed, while Pluto was ejected into its own solar orbit, completely escaping the orbit of Neptune. Pluto has the most elliptical of all planetary orbits and its current orbit intersects that of Neptune periodically. Because of the differing inclinations of their orbital planes, however, the two bodies can never meet.

Pluto may once have been a satellite of Neptune.

Magnetic Fields: Dynamo Theory

Planetary magnetic fields shed additional light on the possible origins of the solar system. We know that the earth generates a magnetic field with a north and a south pole (Chapter 11). Space explorations of the past decade have revealed the somewhat surprising fact, however, that none of the other terrestrial planets has much of a magnetic field.

The mechanism that creates the earth's magnetic field is not fully understood, but it is generally believed that it can be accounted for, at least in broad outline, by the dynamo theory. Dynamo theory deals with the production and maintenance of magnetic fields through the motions of an electrically conductive fluid. The two properties needed for the creation of a planetary magnetic field are a fluid core, and a rapid rotation rate to keep the core in motion. In the case of the earth, both conditions are met. The earth rotates fairly rapidly, and possesses an internal fluid core of nickel-iron, or metallic silicates, both of which are good conductors of electricity. As a result, the earth generates the north-south magnetic field familiar to anyone who has ever used a compass.

The situation is quite different when we come to the other terrestrial planets. Mars rotates almost as fast as does the earth, but it apparently lacks a conductive fluid core. Its magnetic field is thus quite negligible compared to that of our own planet. Venus, on the other hand, does seem to have a molten conductive core, but it rotates so slowly that the strength of the magnetic field produced is very low. The same is true of Mercury and the moon, neither of which has a liquid core, or a rapid enough rotation rate to produce a magnetic field even if it did.

The earth and Jupiter both possess the conductive fluid cores and rapid rates of rotation necessary to produce a strong planetary magnetic field.

Jupiter, on the other hand, has a very powerful magnetic field, far stronger than the earth's. Both the conditions needed for the creation of such a field are present in Jupiter's case. The planet has a rapid rotation rate—less than 10 hours—and a conductive core of liquid metallic hydrogen. The magnetic properties of the other Jovian planets are not known with any certainty. We might expect, though, that they will have magnetic fields similar to that of Jupiter, if their internal pressures are great enough to compress hydrogen into its metallic state.

ORIGINS OF THE SOLAR SYSTEM

So far we have surveyed the evidence indicating that the solar system is indeed an organized system, whose regularities are far from accidental. The high degree of order inherent in the arrangement and motions of the planets, their satellites, and the lesser bodies of the system strongly suggest that the entire system probably had a common origin. Another fact pointing to the same conclusion is the extreme isolation of the system. The nearest star is over 60,000 times as far from our sun as is Pluto, the outermost planet.

A reasonable theory of the origin of the solar system ought to account for as many of the chief structural features of the system as possible. Among the most important questions to be answered are:

1. Why do the planets move in nearly circular orbits in the same plane, and in a common direction?
2. What accounts for the spacing of the orbits (Bode's law)?
3. Why is the angular momentum of the system distributed in a seemingly paradoxical manner, with less than 2 percent of it possessed by the sun, the most massive object in the solar system?
4. Why are there two planetary families, the terrestrial and the Jovian, differing so markedly in mass and composition?
5. Why are the rotation rates of the terrestrial planets so much lower than those of the Jovians?
6. Do the Jovian planets indeed possess atmospheres whose com-

position reflects that of the material from which the planets originally formed?

7. What accounts for the presence of an asteroid belt where we might expect to find another planet?
8. How do the surface features of the planets relate to their formation?
9. What is the origin of the lesser bodies of the system — the satellites and comets in particular?

There is also a great deal of interplanetary material in the solar system, ranging from the large asteroids to particles of atomic size. Any theory of origin and evolution must also account for this interplanetary debris. Finally, there has undoubtedly been considerable evolutionary change in the solar system over the 4 to 5 billion years that we believe it to have existed. We might expect an adequate theory of origin to help us understand the subsequent evolution of the system.

Theories of Origin: A Historical Survey

The earliest ideas about the solar system have been discussed in Chapter 12. It is worth noting that until the time of Newton, the problem of developing a satisfactory model of the solar system seemed more urgent than that of how the system came into existence. Most thinkers probably accepted the idea of direct divine creation of the system; the question for philosophers and scientists, therefore, was to fathom the order of the divinely-created system.

The first "scientific" theory of cosmogony, or solar system origin, was proposed by the French philosopher, scientist, and mathematician Rene Descartes in 1644. Descartes dealt not with the theological problem of the "creation" of the solar system, but rather with the question of how the system might have arisen from something which already existed. He proposed that the solar system was formed from some massive rotating disk of gas and dust in the heavens. He suggested that matter was gathered at the edges of this huge vortex and was formed into the sun, planets, satellites, and smaller celestial objects. Simple as it may seem, this picture of solar system formation bears much similarity to many of the modern ideas in this field.

During the next three centuries, a great many theories of various sorts were proposed to account for the origin of the solar system. The theories were so numerous precisely because the observational data to be explained were so few. Early telescopes could provide men with very little knowledge of the other bodies of the solar system, and the geology of the earth offered only a very incomplete picture of our planet's early history. There were only a handful of observations to be explained: the regular spacing of the planetary orbits (Bode's law), the direction of planetary spins and orbits, and the paradoxical distribution of angular momentum in the solar system. With few facts to be explained, there were few constraints on possible theories.

As astronomical knowledge increased, theories became more sophisticated in order to explain additional information about the arrangement and dynamics of the the planetary system. Most of the theories that emerged could be classified into one of three general categories:

1. theories in which material is extracted from the sun by a passing star and quickly forms planets (**encounter theories**)
2. theories in which the sun forms and then captures a cloud of interstellar material which becomes the planets (**consecutive formation theories**)
3. theories in which the sun and the planets form at the same time from the interstellar material (**simultaneous formation theories**)

With our present knowledge of the solar system, it appears that the third group of theories has the most merit. Those theories in which the sun and planets are formed simultaneously seem to satisfy most modern astronomers, although many important observations are still unexplained by any of the numerous variations on this theme. The other two groups of theories listed — those of a dualistic nature and those in which it is postulated that the planets formed after the sun — were each very popular at different times in the history of astronomy. It is worth taking a look at the theories which have evolved since the time of Descartes in order to understand why ideas which once seemed so plausible are now rejected.

The Encounter Hypothesis

A century passed until anybody thought to challenge Descartes' concept of solar system formation. In 1745, George Louis LeClerc de Buffon put forward a second theory: that a massive comet came close to the sun and collided with it, causing the ejection of material which later condensed to form the planets and their satellites. From our present knowledge of comets, this idea sounds ridiculous. In Buffon's time, however, comets were thought to be very massive objects, comparable in this respect to stars. Buffon's idea was the first of what later were to be called the dualistic theories of cosmogony, referring to the fact that two celestial bodies, the sun plus some foreign invader, were involved in a catastrophic event which created our solar system. Collision theories were put forth by many others; the idea was still alive into the early part of the present century, and for good reason. It was one way to explain why the largest part of the solar system's angular momentum resided with the planets. The energy necessary to account for this was supplied to the system from the outside in the form of a catastrophic collision.

Along these same lines, in 1916 and 1929, Jeans and Jeffreys proposed that tidal forces from the close approach of a star pulled a stream of gas from the sun to form the planets and satellites. As the star approached, massive tidal bulges were raised on the sun, and when the star got close enough, matter was ejected from the sun in the form of a long, cigar-shaped filament, tapered at the ends. The end of the filament followed the star as it sped past, and was thereby set into motion in a curved path about the sun. Some of the ejected material fell back onto the sun and started it rotating in the same direction as the revolution of the filament around it. The filament ultimately broke off into several separate pieces, which condensed to form the planets. Asteroids, comets and meteorites formed from the interplanetary debris.

These ideas now seem implausible. The density of stars in space is very low, and the probability of a chance encounter between two of them at a distance which could lead to tidal disruption is almost negligible. The average separation of stars in the sun's neighborhood approximates that of ping-pong balls 400 miles apart, and the chance of a collision between two such objects is minute. Moreover, from analysis of the motions of nearby stars, it seems likely that at least two of them may have planets of their own. On the basis of these findings, it has been estimated that between one and ten percent of the stars in our galaxy may have planets. There is no way so many planetary systems could be accounted for by chance encounters between stars.

Another objection to this theory is that it seems very unlikely that any material which presently makes up the planets could have come directly from the interior of the sun by a process involving tidal forces. For one thing, such material would not have had sufficient density or self-gravitation to condense in the presence of the sun's own tidal forces. Gas

Figure 14-2
The encounter hypothesis (Jeans-Jeffries version, 1917). A passing star pulls a long, tapered filament of material from the sun. As the star passes, this material is set into orbital motion about the sun. Eventually, planets condense — the largest in the middle, smaller ones at each end.

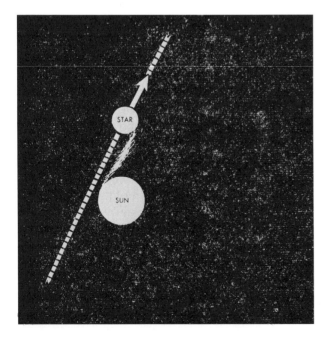

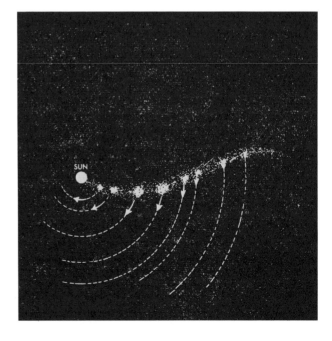

from the interior of the sun would have been so hot that it would have expanded and dissipated almost immediately, or fallen back onto the sun; it could not have condensed into planets.

Consecutive Formation Theories

Theories of this variety have had many proponents over the years, but currently seem unacceptable. The main premise of these theories is that the sun came into being first, and subsequently captured interstellar material in the form of a cloud. Though it is possible that a star may capture interstellar material, it is unlikely that this material would then condense and form planets in orbital motion in the same plane as that of the sun's rotation. In fact, there is no reason to believe that planets and satellites should suddenly form at all. Another drawback of this theory is the fact that it does nothing to explain why the sun rotates so slowly.

Some have tried to explain this by assuming that, instead of the sun capturing interstellar material, it somehow shed part of its mass to form the planets. This is similar to the colliding star theory, but it contains no mention of solar collision, only the assumption that somehow the sun was unstable enough to eject some of its mass. It is true, as we know, that most of the angular momentum of the solar system as a whole resides in the orbital momenta of its planets and not in the axial rotation of the sun. Let us assume that all of its angular momentum once resided in the sun. If this were true, the sun would rotate in a period of about 12 hours, instead of the 27 days it now takes to spin on its axis. So rapid a rotation should create a very flattened sun, but this distortion would still not be sufficient to cause the sun to throw off any of its mass. Moreover, the solar equator is inclined by 7° to the plane of the solar system, so that even if the sun had ever shed any mass from its equator to form the planets and satellites, it would have done so in the wrong direction.

The Nebular Hypothesis

Immanuel Kant and Pierre Simon de Laplace in the late century were the first to suggest that the sun and planets formed at the same time. According to this theory (which we present in a slightly updated version), the planets did not come from the sun at all, but were formed concurrently with a **protosun,** which then evolved to become the central star. The theory proposes that a huge disc-shaped cloud, or nebula of cold gas, was in slow rotation about a central axis, extending in space way beyond what are now the outermost planets. This large cloud of gas contracted under the mutual gravitation of its parts. As it contracted, it had to spin faster and faster in order to conserve angular momentum. Eventually, the rotational speed increased so much that the outer region of the cloud could no longer be contained by gravitational forces, causing a series of

Consecutive formation theories fail to explain how the planets condensed, or why they orbit in one plane about the sun.

rings to be shed from the main mass. As contraction proceeded, successive rings broke off at smaller distances from the center, and eventually formed into planets and satellites.

Kant and Laplace's theory, which came to be known as the **nebular hypothesis,** proved to be one of the most popular of all the theories. In fact, it stood practically unchallenged for more than a century, and despite modifications of some of its premises, it still provides the basis of current models.

There is a good deal of evidence which supports a nebular hypothesis. We have accumulated much meteoritic material on our recent space explorations. By radioactive dating, it has been determined that the ages of materials from places far removed from each other, such as meteorites in space and rocks on the surface of the moon, are approximately the same: 4.6 billion years. If such random samples are all of the same age then it seems logical to suppose that they were all formed as part of the same process. Also, the relatively small range of ages found in these samples leads us to believe that the entire process of formation took less than 100 million years.

In spite of all this, there were objections to several aspects of the Kant-Laplace theory. Some astronomers argue that the rings necessary for the formation of the planets would not have formed by themselves. The velocity necessary to detach such rings from a cloud of gas, they reason, would be far too great. Furthermore, even if the rings had formed, they would not have coalesced into planets, but would instead have dissipated into space. Moreover, the simple nebular hypothesis fails to explain the slow rotation rate of the sun. There is nothing in the theory that

Kant and Laplace proposed that the planets formed from the same contracting cloud of gas that gave birth to the sun.

Figure 14-3
The nebular hypothesis (Kant's version, 1755). Dense clots of dust and gas form in the contracting solar nebula. These grow by accretion to form the planets and their satellites, while the remainder of the nebula condenses to form the sun.

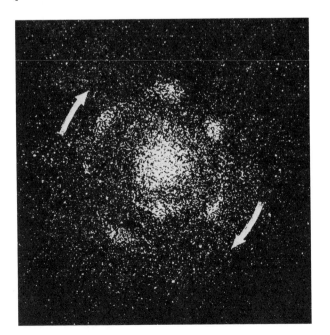

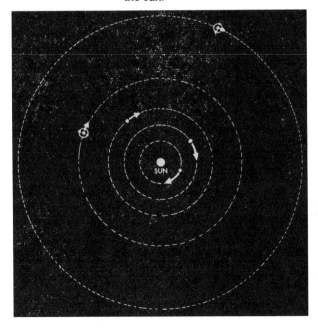

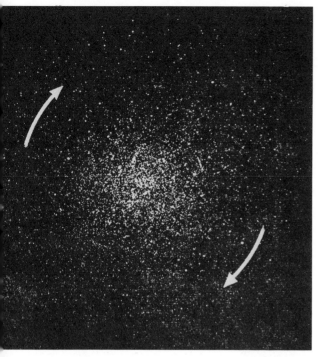 

Figure 14-4
The nebular hypothesis (Laplace's version, 1796). The rapidly rotating solar nebula sheds rings of hot gas as it contracts. The material left behind in the rings condenses to form the planets: the rest continues to shrink and form the sun.

would account for the angular momentum of the solar system being distributed so unequally.

Recent Theories

Most of the more recent hypotheses are proposed in the context of a nebular model of formation. in 1945, Weizäcker proposed that planets had been formed from the vortices of a turbulent mass of gas rotating about the sun. By a proper choice of turbulent eddies, he could approximate the positions for the planets predicted by Bode's law (Figure 14–5). A similar theory was proposed by Kuiper in 1950 in which he suggested that the planets and satellites formed from "protoplanets."

Two of the more recent nebular theories were proposed by F. Hoyle and W. H. McCrea in the last two decades. Both of these theories are significant in that they attempt to explain the problem of angular momentum. In Hoyle's version, the massive slowly rotating solar nebula begins to collapse under gravity until an unstable protosun is formed. Rings of ionized rings by magnetic forces. Gaseous matter in the rings condenses, some of the angular momentum of the protosun is transferred to these ionized rings by magnetic forces. Gaseous matter in the rings condenes, and the accretion of cosmic dust particles increases the mass of the planets as they form. Near the protosun, however, lighter elements are dissipated by the increasing radiation. The result is two separate groups of planets — the terrestrials close to the sun, and the Jovians farther out.

Figure 14-5
Weizsäcker's theory (1945). As a result of the turbulent motion of the gases in the solar nebula, vortices form in the equatorial plane. Accretion takes place where the flow patterns around the edges of the vortices intersect (heavy concentric circles). With a proper choice of initial conditions, this theory can be made to account, not only for the rotation and revolution of the planets, but also their spacing (Bode's law).

Another model which includes a role for magnetic forces was proposed in 1970 by Alfven and Arrhenius. They suggest that the protosun has a large magnetic field associated with it, so that interstellar gas and dust become concentrated and attracted to the surface of the protosun. In this process the gas becomes partially ionized, and magnetic fields which are set up cause it to spin around the sun in a circular pattern. Soon solids condense from this ionized gas, or plasma, and become narrowed into separate orbits, called jet streams. In time the streams of particles accrete to form planets, which develop sizeable magnetic fields. These magnetic fields in turn begin to control the motion of the plasma in their neighborhood. The result is a series of additional jet streams in orbit around the planets, which eventually become planetary satellites.

Summary: A Composite Model

Which features of the various modern theories seem to be the most promising in describing and explaining observable features of the solar system? As we have seen, current efforts in the field are invariably done

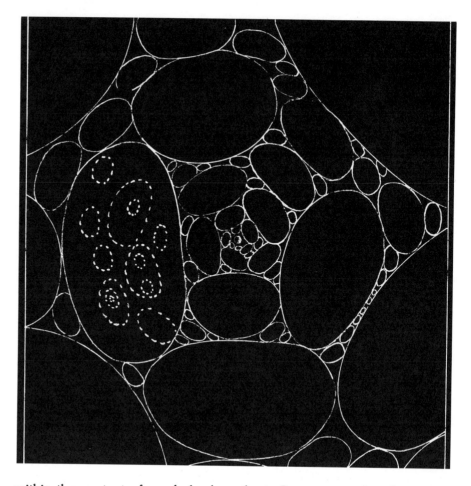

within the context of a nebular hypothesis. Let us start, therefore, with the nebular hypothesis, and attempt to construct a composite model of the origin of the solar system.

Nebular theories all begin with the massive heterogeneous cloud or nebula of interstellar gas and dust, extending out many times the distance of our own current planetary system. The gas consists of about 75 percent hydrogen and 25 percent helium, by mass, together with a small admixture of some of the heavier elements. The particles of the gas are in random motion, and turbulence is set up. As a result, there is small net rotation and angular momentum. Under the influence of its own gravitational field, the nebula gradually contracts, spinning faster and faster in the process. Perhaps at this point, there is some spin-off of gas and dust to form the comets at the outer edges of the nebula.

Contraction of the nebula continues, and the rotation rate increases to conserve angular momentum. Before long, the nebula is no longer the formless mass it was at the outset, but instead assumes an oblate spheroidal shape, with a very dense central bulge, which eventually becomes the protosun. At this point the density of the entire nebula has begun to increase, and the dust particles are able to aggregate and form large particles, which capture water, ammonia, methane and carbon dioxide. The protosun is soon able to radiate heat as its gravitational potential energy is converted into thermal energy. As a result, the large particles closer to the radiating protosun lose their volatile materials, while those farther away are shielded from the central radiation by the intervening nebula.

As this process continues particles accrete in large asteroid-like **planetesimals** which continue to grow in size owing to increasing gravitational attraction. As these planetesimals attract nebular material in the region of their orbits, they grow into **protoplanets.** Protoplanets acquire gaseous envelopes by gravitational attraction, and those closer to the center are slowed down by the tidal forces of the protosun. The area which will later divide the innermost or terrestrial planets from the outermost or Jovian planets is acted on by gravitational perturbations from the Jovian protoplanets, and planets are prevented from forming in this area. Alternatively, several planetesimals remain in this area and collide; in either case, an asteroid belt is created.

In addition to slowing down the rotations of the terrestrial planets, the protosun, which is now generating a great deal of radiant heat, is able to ionize the gases around these planets. The ionized gases couple magnetically to the sun, and transfer angular momentum away from the sun as they are lost into space. After this occurs, the protosun begins to ignite thermonuclear reactions and becomes a stable star.

Some of the protosun's angular momentum was probably transferred to the ionized gases of the young solar system by magnetic forces.

Evolution of the Solar System

The solar system that formed some 4.6 billion years ago has remained remarkably stable since then. This is not to say, however, that the system has been completely static since its birth, for it most definitely has not. We will briefly trace the later history of the system, and examine the changes which have occured since its formation.

From evidence which we have been able to gather during recent space explorations, we have been able to place the approximate age of the solar system at 4.6 billion years. Sources of samples such as moon rocks and meteorites suggest to us that formation took about 100 million years ago. There is evidence, also, that at the time the planets were formed, the solar wind was as much as 10 million times as intense as it is today. This estimate comes from the study of rocks in the crust of the earth and moon, which seem to contain a disproportionately high concentration of radioactive minerals. If the interiors of these bodies contained a comparable amount of radioactive substances, the heat generated would probably be enough to melt them right out to the crust. Nonradioactive elements in the crust were probably made radioactive as a result of bombardment by atomic and subatomic particles in the solar wind.

Radioactive elements heated the crusts of the terrestrial planets, causing the release of gases that formed the planetary atmospheres.

Study of ancient rocks also suggests that the formation of the planets probably took place under highly reducing conditions — that is, there was an abundance of hydrogen. It thus seems likely that the Jovian planets retain their original atmospheres — predominantly hydrogen and helium. The terrestrial planets probably acquired their present atmospheres by the slow emanation of gases from the interior of the planet, a process known as degasing or outgasing. This is possible when there is a slow buildup of heat, as might have occurred when long-lived radioactive elements began to decay. The production of heat from such decay, which is a very slow process, could have raised the internal temperature of such planets from several hundred to a few thousand K. Furthermore, heat production of this sort is proportional to the size of the planet. Thus, earth and Venus could have built up higher temperatures in this manner than Mercury or the moon.

The geologic record also bears out the fact that there was liquid water present 3.5 billion years ago. Metamorphosed sediments dating back this far are convincing evidence that the earth was at a temperature above the freezing point of water. This is particularly intriguing, for we

The Evolution of Atmospheres

The history of the terrestrial planetary atmospheres is a fascinating study in contrasts, with relatively small differences in initial conditions leading to widely dissimilar results. Mercury, as we pointed out in the previous chapter, is too small and too hot to hold much atmosphere of any sort. But why did Venus develop a "greenhouse" atmosphere of carbon dioxide, while the earth did not? A slight difference in surface temperature seems to have been responsible. Earth was cool enough for water to remain in a liquid form on its surface. The liquid water dissolved the carbon dioxide gas emerging from the planet's crust, keeping much of it out of the atmosphere. (It is strange to think of earth's primeval oceans consisting of something resembling rather flat club soda!) In time, the dissolved carbon dioxide reacted with calcium compounds and other materials in the earth's crust, producing such minerals as chalk and limestone. Thus the earth's carbon dioxide was safely fixed in solid form.

The surface temperature of Venus, however, was warm enough for most of the water produced by outgasing to remain in the form of water vapor. With little liquid water to intercept and dissolve it, carbon dioxide from the Venusian crust entered the atmosphere, where it began to trap the sun's radiation in the manner described in Chapter 13. As the planet's surface temperature rose due to this "greenhouse effect," less water could exist in liquid form, and more carbon dioxide could enter the atmosphere unimpeded. The cycle, in other words, reinforced itself, resulting in a "runaway" greenhouse effect that has made Venus perhaps the hottest body in the solar system. Mars, by contrast, is cold enough for both its water and its carbon dioxide to be frozen in the polar ice cap. With no free atmospheric carbon dioxide to trap the sun's heat, it seems destined to remain cold, unless there is a major shift in its climate for other reasons.

know that 3 to 4 billion years ago, the sun was 50 to 60 percent smaller than it is today. This means that the earth's surface temperature could not have been more than about 15° F, which is below the freezing point of water. To resolve this paradox, various scientists have suggested that either 1) the earth was 10 percent closer to the sun 3.5 billion years ago, or 2) there was a "greenhouse" effect present on earth, so that the sun's radiant heat was stored more efficiently by atmospheric gases such as carbon dioxide. It is quite possible that both factors were present.

We saw in Chapter 13 that the geological history of the moon suggests that it was subject to a fierce bombardment by massive meteoroids or asteroid-sized objects. These impacts created the huge basins we know as seas, and triggered the volcanic outpourings which filled them with lava. The episode of bombardment seems to have begun not long after the formation of the solar system, and ended rather abruptly some 4 billion years ago. Similar geological features have been found on other terrestrial bodies — the Hellas basin on Mars, the Caloris basin on Mercury, the recently discovered lava flow on Venus, and perhaps certain basins on earth, largely camouflaged by erosion and other geological processes. It seems likely that all of these scars were caused by collisions with debris left over from the formation of the solar system.

About 4 billion years ago, the bodies of the solar system seem to have undergone heavy bombardment by debris left over from their formation.

SUMMARY

The solar system exhibits a number of striking features that any theory of its origin would have to explain. For example, all the planets except Pluto orbit in nearly the same plane. Furthermore, they all orbit in the same direction — the direction of the sun's spin. Each successive planet's distance from the sun, moreover, is close to that prescribed for it by Bode's law. And the planets (with the exception of Pluto again) are divided into two groups: the terrestrial planets, which are closer to the sun, smaller, dense and rocky, have thin atmospheres, and spin slowly; and the Jovian planets, which are farther from the sun, larger, composed chiefly of gases, and spin rapidly.

Many different theories of the origin of the solar system have been advanced, but they all fall into three classes. Encounter theories assert that a star passing the sun pulled some material out of it, and the planets were formed from this solar substance. Consecutive formation theories see the planets as having coalesced from a cloud of interstellar material captured by the sun. Simultaneous formation theories picture the sun and planets as forming at the same time from the same material. With the evidence available to us today, it is the simultaneous formation theories that seem the most convincing.

According to the most recent encounter theory, a star swung close to the sun and sucked a long, cigar-shaped filament of material from the sun's tidal bulge. Following the star as it passed by, the filament was set into orbital motion around the sun and eventually condensed into the planets. What makes this theory seem implausible is that the distance between stars is so great that the probability of a collision is practically nil.

The crescent earth rises above the desolate lunar landscape to greet Apollo astronauts in orbit around the moon. Since the moon's period of revolution about the earth is synchronized with its period of rotation, it always keeps the same face turned toward the earth. As a result, an observer on the moon's surface would not see the earth rise and set. The earth would remain fixed in a given region of the sky, wobbling only slightly over a period of weeks or months. It would, however, go through the entire cycle of phases, from dark to full and back again, just as does the moon seen from earth. (NASA photograph.)

Skylab in orbit, 470 km above the earth. The experimental space station, 36 m in length, was launched in May, 1973. Three successive three-man crews have visited Skylab, the longest for 84 days, and more than 50 major research projects have been performed there. All of the photographs of the sun on the last three pages of this portfolio were taken by Skylab's cameras. The wing-like panels are banks of solar cells that provide power for the space station. (NASA photograph.)

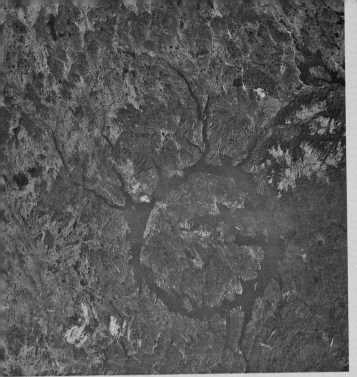

Above, the Manicouagan Reservoir, a fossil crater 60 km in diameter in Quebec, contrasted with Crater Schmidt on the moon. Lunar craters remain unchanged for millions of years, but those on earth undergo weathering, or are buried by tectonic activity. This crater, formed some 200 million years ago and long buried, was uncovered by glacial action. The photograph was taken from an earth satellite at an altitude of 920 km. The colors are not real, merely a device to facilitate interpretation. At right, Harrison H. Schmitt, now a U.S. Senator from New Mexico, stands beside the famous split rock on the lunar surface. A large boulder, probably dislodged by a meteor impact, apparently rolled down a nearby hillside at some time in the past and broke into five pieces, of which this is one. This chance find has enabled scientists to examine the inside of a large moon rock. As a result, it has been determined that two separate cataclysmic events occurred in this region about 4 billion years ago. From such evidence it may be possible to piece together a much fuller picture of the moon's history. (Lunar photos: NASA, courtesy Carl Zeiss, Inc.; satellite photo, EROS.)

At left, Mars, photographed from earth. Notice the white polar icecap, and the dark mottlings, which may be caused partly by dust storms in the Martian atmosphere. Earlier observers claimed to see an intricate network of geometrically straight lines on the planet's surface, but few Martian geological features are really visible from earth. Below, a Viking photograph of the barren, rocky Martian terrain. The landscape is not unlike that of deserts in the western United States, but the pink sky provides a distinctly unearthly touch. Viking's search for life on Mars has provided tantalizing but inconclusive results. (Lick Observatory; NASA.)

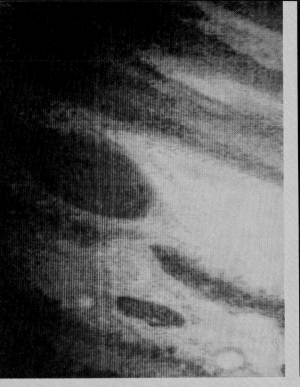

Above, left, Jupiter's Great Red Spot, photographed from a distance of 383,000 km by Pioneer 11 in 1974. This fascinating feature has brightened and faded at times, but remains much as it appeared when first discovered more than 200 years ago. It is believed to be a gigantic storm in the planet's dense, turbulent atmosphere, large enough to swallow the earth many times over. Above right, Jupiter's polar regions, not visible from earth, seen from the Pioneer 11 spacecraft. The bands and convective cells of the planet's complex weather patterns are clearly visible in both photographs. (NASA.)

At right, an image of Jupiter indicating relative amounts of absorption by methane in the planet's atmosphere. The methane absorption band lies at 887 nm, in the infrared portion of the spectrum. In this false-color image, white represents the least methane; red, orange, yellow, green, and blue indicate progressively more, with violet indicating the most. It is interesting to note the relative absence of this gas in the region of the Great Red Spot. (Photograph by Alan M. Goldberg and Dr. Thomas B. McCord; Copyright 1976, Massachusetts Institute of Technology.)

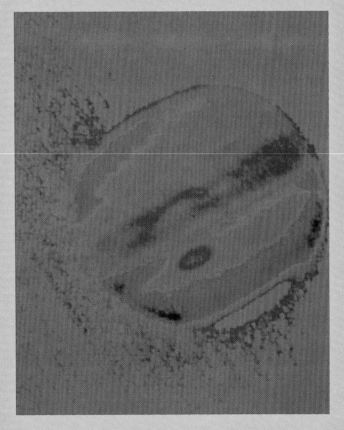

This photograph of Saturn with its rings conveys only a small measure of its beauty, seen through even a moderate sized telescope. The planet is flattened at the poles and faintly banded. The rings consist of small chunks of icy material; they are so tenuous that stars can often be seen through them. The outer regions of the rings revolve about the planet more slowly than the inner regions, in conformity with Kepler's laws for orbiting bodies. The gaps in the rings are due to the gravitational perturbations of Saturn's satellites, especially Mimas. We hope to get a closer look at this most beautiful object in the solar system during the Pioneer flyby in 1979. (New Mexico State University Observatory.)

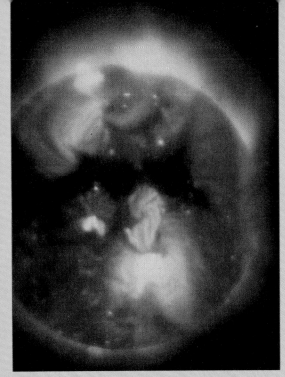

Left, an image of the sun from the X-ray telescope aboard Skylab. Notice the irregular outline of the sun in this photograph. Most of the solar X-rays seem to be produced in regions well above the photosphere. (Solar Physics Group, American Science and Engineering.)

Below, a false-color image of the sun from Skylab, taken just before the eclipse of June 30, 1975. The disc of the sun is artifically blocked out, allowing the solar corona to be seen. The different colors represent different degrees of brightness. Can you guess why the dark side of the moon appears to be faintly illuminated? (High Altitude Observatory and NASA.)

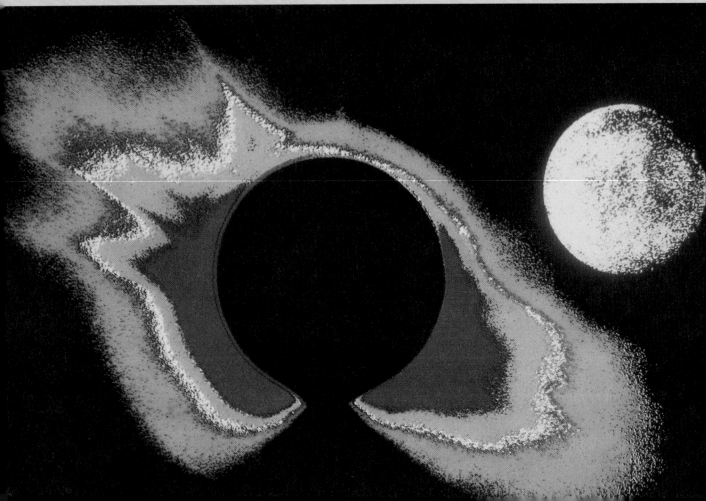

Above, an active area on the west limb of the sun, photographed with the ultraviolet spectroheliometer aboard Skylab on December 4, 1973. The false-color image shows loop-like structures extending as far as 125,000 km above the active region. Three different ultraviolet wavelengths are represented. Radiation from O IV (triply ionized oxygen) at 100,000 K is shown in shades of blue, while the emissions from O VI (5-times ionized oxygen) at about 250,000 K appear as shades of green. These two wavelengths outline the loops of magnetic force and clearly show their structure: relatively cool cores surrounded by sheaths of increasing temperature. The red tones, indicating radiation from Mg X (9-times ionized magnesium) at 1.5 million K, represent hot, dense coronal material over the active region. (Harvard College Observatory and NASA.)

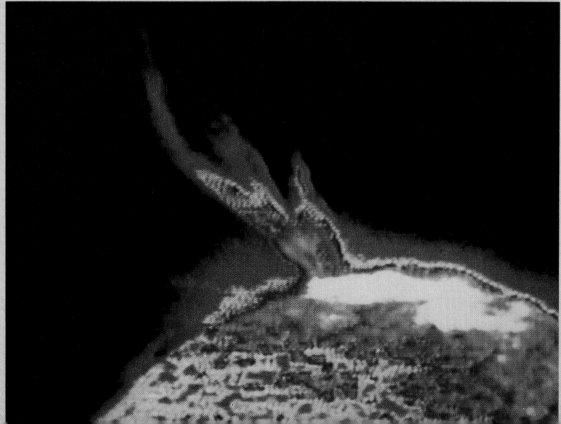

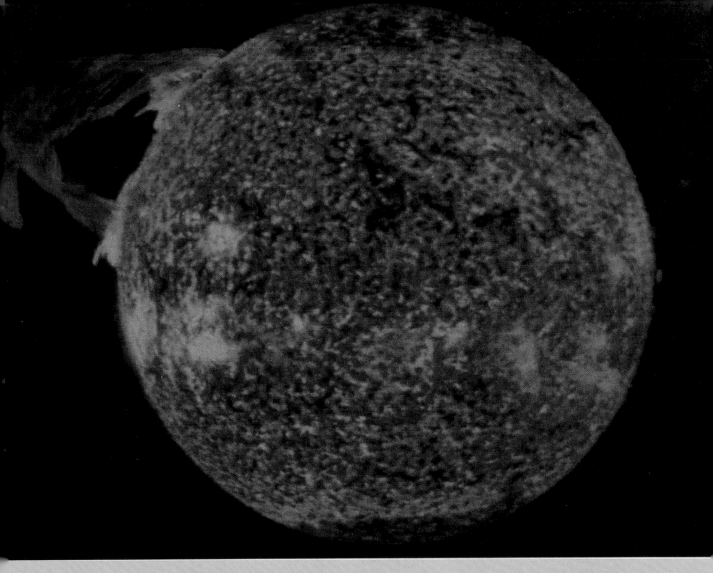

Preceding page (below): A large solar eruption, photographed by Skylab astronauts on August 21, 1973. The image was taken in the light of He II (singly-ionized helium) at 30.4 nm, using the extreme ultraviolet spectroheliograph aboard Skylab. When the picture was taken, just 90 minutes after the start of the eruption, clouds of helium had already been blown some 550,000 km from the solar surface, and eventually reached nearly a million km. The different colors represent different intensities of ultraviolet light. (NASA.)

Above: A spectacular eruptive prominence, photographed in the ultraviolet light of He II on December 19, 1973, from Skylab. The arch of gas spans a distance of nearly 600,000 km on the solar surface. The last previous picture of this region, taken 17 hours earlier, showed this feature as a huge quiescent prominence, hovering in the solar atmosphere. The intricate structure of the ionized gas is shaped by the local magnetic field; it suggests a twisted sheet of material in the process of unwinding itself. Notice the absence of super-granulation near the poles, and the active regions at the left of the disc. (Naval Research Laboratory and NASA.)

According to the simultaneous formation theories, sometimes called nebular theories, the solar system formed from an enormous cloud or nebula of interstellar gas and dust. This cloud must have been several times larger than the solar system is today. The particles in it were originally in turbulent motion, and the cloud had a small net rotation. Contracting under its own gravity, the cloud spun faster and faster. It flattened at the edges and developed a dense central bulge—the protosun, which soon began to radiate heat. The heat dissipated the lighter, more volatile elements in the vicinity of the protosun. Particles accreted into asteroid-like planetisimals, which kept growing, soon becoming proto-planets. The rotation of those nearer the protosun was slowed by tidal forces. Finally the protosun began to ignite thermonuclear reactions and became a stable star. rotation of those nearer the protosun was slowed by tidal forces. About 4.6 billion years ago, the protosun initiated fusion reactions and became a stable star evaporating lighter elements from the inner regions of the solar system.

Planetary atmospheres were formed by outgasing from their crusts. Mercury was too hot and small to hold its atmosphere; that of Venus created a runaway greenhouse effect, keeping the planet's surface temperature high, while that of Mars was largely frozen down. The huge lava-filled basins on the moon, Mars, Mercury, and Venus may have been created by volcanic outpourings triggered by collisions with massive asteroid-sized bodies left over from the formation of the solar system. This bombardment ended abruptly 4 billion years ago.

EXERCISES

1. How do you explain the fact that the sun has most of the mass of the solar system, but the planets have most of the angular momentum? What would happen to the sun if all the angular mementum of the planets were transferred to it?
2. Describe some simple relationships that exist between the periods of revolution and the periods of rotation of the planets.
3. Why do the planets all travel in nearly circular orbits?
4. Discuss the appropriateness of calling the Titius-Bode relationship a law. How does it compare with other physical laws, such as the inverse square law for light?
5. Give a brief, general comparison of the Jovian and terrestrial planets. What features do the members of each group have in common? In which group would you classify Pluto? Why?
6. Why does the earth have a magnetic field, while Mars and Venus do not?
7. What do you think is the reason the Jovian planets rotate far more rapidly than the terrestrial planets?
8. Pluto, the outermost planet, is very different from the neighboring Jovian planets. How would you explain this difference?
9. Describe briefly the three classes of theories that attempt to explain the origin of the solar system.
10. Which theory of solar system origin is most accepted today? What are the chief objections to the other theories?
11. How have the original Kant and Laplace theories been modified in the light of modern knowledge?

life in the universe

Viking soil sampler on Mars

"I can tell what the inhabitants of Venus are like; they resemble the Moors of Granada; a small black people, burned by the sun, full of wit and fire, always in love, writing verse, fond of music, arranging festivals, dances, and tournaments every day."

So wrote Bernard de Fontanelle, the imaginative seventeenth-century scientist and man of letters. Fontanelle's guess turned out to be wrong; he did not forsee the greenhouse effect, which makes the surface of Venus several hundred degrees hotter than Granada.

Today, we try to approach the question of extra-terrestrial life in a more sophisticated way than did earlier thinkers. Our starting point is one of the most basic assumptions in astronomy: that the laws of physics are constant throughout the universe. In a sense, this is necessary. If we give our imaginations free rein, we can imagine life forms anywhere and everywhere—crystalline creatures on the surface of Pluto, beings of pure energy in the heart of the sun. Yet if we go to the other extreme, and think only of beings exactly like us, such as Fontanelle's "Moors of Granada," we may have to conclude that they exist nowhere else in the cosmos. Scientists therefore try to formulate a definition of life that is based on our experience of terrestrial life, yet is broad enough to include the very different kinds of life forms we might some day encounter among the stars.

Unfortunately, even on earth, the distinction between living and non-living things is not easy to pin down, and biologists find different definitions useful in different contexts. One definition is based on the fact that many of the processes and activities of living things seem directed toward the maintenance of homeostasis. That is, living things try to stay internally the same in spite of environmental changes.

Reproduction provides an alternative way of defining life. Though some organisms which are clearly alive, such as mules and plant grafts, cannot reproduce, the cells of these organisms can . All living things contain the same reproductive chemical —deoxyribonucleic acid, or DNA—which stores, in a chemical "code," all the information necessary for producing a new organism. The offspring may be identical to the parent, or may combine genetic information from two parents. Occasionally the message is accidentally altered, and the offspring contains a slight change, or mutation, which makes it different from others of its kind. Usually this mutation is harmful, and the organism has less chance of surviving to reproduce itself. We can define living things, then, by saying that they reproduce, mutate, and pass on the beneficial mutations to their offspring.

This definition of life leaves at least one important question unsettled — must life be carbon-based? Only carbon seems capable of forming the long chains that make up the backbones of the basic chemicals of life: proteins and nucleic acids (such as DNA). We can easily imagine physical conditions different from those on earth, which might make a carbon chemistry impractical. But could life then evolve so as to cast some other element in this role? We can only speculate.

Another question that has not yet been fully answered is how life originated in the first place. One ancient theory held that organisms are spontaneously generated from non-living matter. Flies come from spoiled meat, for example, crocodiles from river mud, and fireflies, of course, from fire. But in the nineteenth century Louis Pasteur carefully disproved the theory of spontaneous generation by showing that no organisms will appear in sterilized solutions.

The most convincing theory of the origins of life developed so far combines the insights of geology, chemistry, and the theory of evolution. Its starting point is the fact that the earth's early atmosphere was not like that of today, but consisted primarily of methane, ammonia, water vapor, nitrogen and hydrogen. The Russian chemist A. I. Oparin and the British biologist J. B. S. Haldane independently proposed in the 1920s that such an atmosphere provided the raw materials for life, which appeared through a long, slow process of chemical evolution. According to Haldane, the action of ultraviolet light on these gases produced organic compounds, which accumulated in the primitive oceans. This idea received strong support after Stanley Miller conducted a historic experiment in 1953. Miller circulated a mixture of the gas past an electrical discharge, representing lightning, and then through water. By the end of a week the water was rich in organic compounds, including amino acids, which are the building blocks of proteins — the basic materials of living cells. Since then, similar experiments have yielded other important compounds that are fundamental to life.

Perhaps the way in which life first originated on earth is not the only process by which organisms may emerge, but it is the most useful model we have for exploring the possibility of life beyond earth. Such an inquiry begins most naturally in our own back yard, the solar system. It seems, however, that few, if any, of the planets are hospitable to life. Mercury has almost no atmosphere, and temperatures on the side facing the sun reach a scorching 600° F. Since the planet makes a complete rotation with respect to the sun every 176 days, no surface area can escape this heat. Venus is at least as hot as Mercury, due to the greenhouse effect produced by its dense layer of clouds. If these clouds contain significant amounts of water vapor, it might be possible for some hardy microorganisms to exist in suspension at certain altitudes, but the composition of the clouds is not clear at this time, and water vapor seems very scarce.

As every reader of science fiction knows, the most habitable alternative to earth in our solar system is Mars. The existence of volcanoes and lava on Mars suggests that its early atmosphere probably resembled the early atmosphere of earth, with methane and ammonia in contact with pools of warm water. If so, life could have begun as it did on earth, and perhaps some microorganisms were able to adapt through evolution and survive the drying up of the planet.

In July and September, 1976, two Viking spacecraft landed on Mars with instruments designed to test for the presence of such organisms. Each of the landers was equipped to perform three types of biological experiments on soil samples and to test

A nineteenth century fantasy about life on the moon

leased in waste products. The third experiment (gas exchange) monitored the level of carbon dioxide, and tested for hydrogen, oxygen, nitrogen and methane, in a chamber during incubation of a soil sample in varying amounts of concentrated nutrient solution. Presumably, the metabolic processes of Martian organisms would change these levels.

The results of the Viking tests were tantalizingly ambiguous. The pyrolitic release and labeled release experiments yielded positive results, which could not be repeated after heat sterilization of the sample. The gas exchange experiment produced a substantial amount of oxygen and showed a significant increase in the carbon dioxide level. Though the results of the first two tests were consistent with life processes, and are somewhat difficult to explain in terms of inorganic chemical reactions, a number of inorganic theories have been advanced which cannot be ruled out.

With the possible exception of Jupiter, the remaining planets seem quite inhospitable to life. Jupiter is apparently composed chiefly of liquid hydrogen, with no solid surface at all. However, it is not as cold as might be expected, due to heat from the planet's interior. Its dense, turbulent atmosphere contains hydrogen, methane, and ammonia, much as the earth's once did, and it is thought that energy may be readily available in the form of electrical discharges, similar to the lightning that accompanies earthly storms. All in all, conditions in the Jovian atmosphere are strikingly like those on the primitive earth.

What are the possibilities that life exists outside our solar system? We would expect life only in galaxies whose stars have evolved sufficiently to produce substantial amounts of the heavier elements, and not in galaxies such as the Magellanic Clouds, for example, which seem to contain only relatively young stars. Moreover, even the simplest living things consist of a wide variety of molecules which take part in many chemical reactions. Interstellar space is much too empty for the chance encounters of molecules to yield an organism.

We must consider, then, whether the formation of

for the presence of organic compounds in the soil. Since carbon dioxide and carbon monoxide are present in the Martian atmosphere, some organisms may be expected to assimilate at least one of them. Therefore the first experiment (pyrolytic release) tested for either photosynthesis or chemical fixation of these gases. The second experiment (labeled release) was designed to see if radioactively labeled carbon dioxide in a solution of nutrients would be incorporated by organisms into their own internal chemistry and later partially re-

planets around stars is very likely. There is little direct evidence that planets exist outside our solar system. However, there is some indirect evidence. We know that binary star systems are very common in the galaxy, and that the average separation between the stars of a binary pair is comparable to the distance between the sun and the large, Jovian planets. Jupiter itself is often described as a star that failed, having too small a mass to initiate nuclear fusion. It seems plausible, then, to imagine that many stars like our sun may have dark companions less massive than the smallest stars in binary systems—in other words, planets.

The most important evidence, however, comes from stellar rotations. Stars cooler than the F2 types have relatively low rotation rates. This is generally explained by postulating the transfer of angular momentum to a disc of protoplanetary material. It is thought that this process can only take place for cooler stars; around hotter stars, such a disc would soon be dissipated by stellar radiation, and would not last long enough to siphon off much angular momentum.

It is one of the fascinating consistencies of the universe that the types of stars which seem most likely to have planets are also the types of stars on whose planets we would most reasonably expect to find life. That is, only spectral types cooler than A shine long enough for life to evolve. Types O, B and A stay on the main sequence for a mere three billion years at most—hardly enough time for significant biological evolution. We must also eliminate the cooler K and M stars. Though they may have planets, these types are not hot enough to sustain life at expected planetary distances. The region around a star which is neither too hot nor too cold for life is called the ecosphere. Consideration of stellar lifetimes and ecospheres indicates that the cooler F, G and hotter K types are the most promising stars, which is consistent with the fact that our sun is a G2 type. Similarly, we must eliminate all but a few percent of the stars in binary systems. With rare exceptions, such as very close binary pairs, planetary orbits in these systems would be too irregular to be contained in an ecosphere.

If the occurence of planets outside our solar system is not rare, we may also ask whether the presence of oceans on the planets may be expected. The probability that life will emerge from chance combinations of organic compounds is highest in some kind of ocean, where organic molecules may be concentrated and yet have freedom to move

During the Viking probe, a special instrument collected soil samples rich in organic materials.

about. Of the substances which are abundant enough to be available for an ocean, only ammonia, methane, hydrogen sulfide and water exist as liquids at the temperatures and pressures which might be expected on a planetary surface of these water remains as the most likely substance to form oceans, and planets which are about the same size and temperature as the earth will probably have an ocean.

Since there is reason to believe that planets with oceans exist beyond our solar system, and that life forms may have evolved on some planets, we may also ask whether some of these forms possess the kind of intelligence that results in a technological civilization which is capable of communicating with us. At a conference in 1961 at the National Radio Astronomy Observatory in Green Bank, West Virginia, Frank Drake stated the problem in the form of an equation. The number of such civilizations (N) in our galaxy is:

$$N = R_* f_p n_e f_l f_i f_c L$$

This formulation has become a touchstone for discussion and investigation by exobiologists.

The number of advanced civilizations at a given time will of course be proportional to the number of stars in the galaxy. But the present rate of star formation is much lower than the rate of formation when the galaxy was young. Since G-type stars, which are most likely to support life, have life spans comparable to the age of the galaxy, we can use the average rate of star formation during the galactic lifetime, or R_*. This gives us an estimate of the number of stars whose civilizations, like ours, may have recently become technological. Most estimates of R_* lie between one and ten stars per year.

The fraction of these stars which have planets is f_p. If we accept the nebular hypothesis that planets condense out of the same cloud of dust and gas as their stars, and the argument that slow stellar rotations indicate that angular momentum has been taken up by planets, then f_p seems to be large. Even if we eliminate multiple star systems because their planetary orbits would be too irregular, we are left with a value of f_p of about 0.5. We must then consider n_e, or the expected number of planets in each system which are hospitable to life. In our own solar system we know of one such planet. Since the cooler M stars are more common than our G type, n_e may be less than one in most cases unless a greenhouse effect keeps the more distant planets warm. However, an n_e of three is at least conceivable for our system if we include Mars and Venus. Therefore we might adopt n_e equal to one as a reasonable estimate.

As we progress to the right of the equation the terms become harder to estimate and to verify experimentally. Though conditions may be appropriate on a planet this does not guarantee that life will emerge. Therefore we must take the fraction of suitable planets on which life actually appears (f_l). Scientists at the Green Bank conference took the optimistic view that Main Sequence stars from about F2 through K5 have sufficiently long lifetimes to make the appearance of life almost inevitable. They therefore adopted f_l equal to one. Similarly, they argued that the eventual emergence of intelligence is inevitable if life has sufficient evolutionary time. (It is interesting to note that the intelligence need not be limited to biological organisms. It might be embodied in artificial devices such as sophisticated computers or robots.) Accordingly, they assigned the value of one to f_i, the fraction of life-supporting planets inhabited by intelligent forms.

The value of N must be further restricted by the fraction of intelligent forms which have developed sufficient technology and interest to be capable of communicating with us, or f_c. This is extremely difficult to estimate. Studies of civilizations like ancient China and the Aztecs leave open the question of whether they would have developed a technology if they had not come into contact with other civilizations. Many civilizations have passed into oblivion on earth estimates of f_c are about 0.1.

Finally, we must ask how long a technological civilization may be expected to survive (L). Though we have just achieved the ability to exchange radio signals with extrasolar civilizations, we have also achieved the ability to annihilate ourselves with nuclear or biological weapons, or to pollute or overpopulate the environment until it will no longer sustain us. Indeed, the question "Is there intelligent life on earth?" was posted on Frank Drake's office door. The maximum lifetime of a civilization is limited by the lifetime of its star. For a G star, L is not more than about 10^{10} years. The great uncertainty in L, as well as in other terms of the equation, makes many scientists unwilling to commit themselves to a value of N. The Russian exobiologist I. S. Shklovskii and the American Carl Sagan optimistically set N at about 10^6, while others have suggested values as low as 10^4.

Although these values represent a large number of stars, the great distances between stars, even within our own galaxy, make it unlikely that one of these civilizations is closer to us than several hundred light years. It would seem, then, that there is a kind of "cosmic quarantine" which prevents us from visiting or being visited by extrasolar beings. The theory of relativity demonstrates that travel at speeds faster than light is impossible, and experimental evidence has verified this speed limit. Hundreds, or more likely thousands, of years would be required to reach another civilization. We can, however, trade radio and television messages at distances within our own galaxy. Even then, an exchange such as "Hello, is anyone out there?" followed by "Yes, we hear you," would take hundred of years. As Carl Sagan has pointed out, this is not exactly a snappy conversation. But the cultural consequences of contact with another civilization could be so great that a long period for reflection and consideration of the message might be an advantage. We would not want to be too hasty. Fears that such contact would be detrimental to our civilization therefore seem unfounded, and it has been suggested that we might gain valuable knowledge from such contact which could help us solve some of the problems which threaten our existence. Similarly, the concern's that we might harm another civilization are probably unwarranted since any civilization which could communicate with us would have to be more advanced than we are.

There is also the possibility that a civilization is already sending signals without waiting to see if anyone is listening. In 1959, Drake initiated Project Ozma at Green Bank, named after the princess of the land of Oz in the well-known children's books. He pointed a radio telescope at Tau Ceti and Epsilon Eridani and listened for intelligent transmissions at the 21 cm wavelength emitted by neutral hydrogen atom and is therefore important in radio astronomy. A civilization which is technologically advanced and interested in astronomy might reasonably be expected to have radio equipment that operates on this wavelength. The chances of receiving coded signals from these or any other two stars are vanishingly small (and Drake found nothing) but they may be increased by a systematic search of many stellar systems. Proponents of such an investigation point out that its total cost would represent only a small fraction of the present military budget of the United States, while the results could open a new era in human history.

Late in 1976, NASA announced the most ambitious project yet undertaken in the search for extraterrestrial life. One of its goals is the construction of a computerized radio analyzer that will be able to scan simultaneously a million different narrow wavelength bands between 21 cm (the wavelength of the radiation emitted by neutral hydrogen atoms) and 18 cm (the wavelength of the radiation emitted by the hydroxyl radical, OH). Since ordinary water, in chemical terms, consists of hydrogen plus hydroxyl, this region of the spectrum has been dubbed "the waterhole" by radio astronomers. It is a logical place for attempts at extraterrestrial communication. In the words of a recent NASA report, "different galactic species might meet there just as different terrestrial species have always met at certain more mundane waterholes." Another part of the project will involve new attempts to discover planets in orbit about nearby stars, by means of improved interferometry techniques, and eventually with telescopes placed in orbit above the atmosphere.

APPENDIX

SOME USEFUL MATHEMATICAL TOOLS

INDEX NOTATION

It is traditional and convenient to write $a \times a \times a$ as a^3, $a \times a \times a \times a$ as a^4, and so on. In general, we can say that a^n represents a multiplied by itself n times. We read this notation as "a raised to the nth power," or "a to the nth" for short. The number n is called the index. There are two commonly encountered cases with special names: a^2 is read as "a squared," and a^3 as "a cubed."

Index notation is very convenient for multiplying or dividing. $(a^n) \times (a^m) = (a \times a \times a \ldots n$ times$) \times (a \times a \times a \ldots m$ times$) = (a \times a \times a \ldots m + n$ times$) = a^{m+n}$. For example, $10^3 \times 10^4 = 10^7$. In other words, when we multiply powers of a number, all we need do is add the indices. Similarly, $a^m/a^n = a^{m-n}$. From these formulas it is easy to show that $a^{-n} = 1/a^n$, and $a^0 = 1$. Thus $5^{-3} = 1/5^3 = 1/125$.

The number which, when squared, gives a is called the square root of a. It is written $\sqrt{a}$. Since $\sqrt{a} \times \sqrt{a} = a$, we can write as $a^{1/2}$. (Notice that $a^{1/2} \times a^{1/2} = a^{1/2+1/2} = a^1 = a$.) Similarly, the cube root of a can be written as $a^{1/3}$. Since most fractional powers, such as $a^{4.7}$, cannot be computed without mathematical tables, a slide rule, or a scientific calculator, they are not used in this book. The special cases in which the index contains the fraction $1/2$, however, are fairly easy to deal with:

$P^2 \propto a^3$; $P \propto a^{3/2} = a^{1\ 1/2} = a\sqrt{a}$ (Kepler's third law, p. 116)

Lifetime $\propto m^{2.5} = m \times m \times \sqrt{m}$ (Stellar lifetimes, p. 145)

One other relationship is often useful to remember: if a is increased by a factor of z (that is, multiplied by z), then a^n is increased by a factor of z^n. For example:

$3^3 = 27$; $(3 \times 2)^3 = 3^3 \times 2^3 = 27 \times 8 = 216 = 6^3$

POWERS OF 10

Very large and very small numbers can be written in a very compact and convenient form by using index notation. The method makes use of the fact that our number system is based on 10, and that all the powers of 10 are easy to calculate:

$10^2 = 100$; $10^5 = 100,000$; $10^{-1} = 1/10 = .1$; $10^{-4} = 1/10,000 = .0001$, and so on. We can therefore write a number like 234,000,000,000,000 as 234×10^{12}, or 23.4×10^{13}, or 2.34×10^{14}, and so on. Similarly, .0000000812 can be written as 81.2×10^{-9}, or 8.12×10^{-8}, and so on. In practice, it is conventional to write the number with one figure to the left of the decimal point: 5.88×10^{12}, not 588×10^{10}. However, it is sometimes more convenient to leave the power of 10 at some constant value and let the number to the left of the decimal point vary, especially for comparisons. For example, the diameter of Mars is about 6800 km and that of Jupiter 143,000 km. It is easier to compare 6.8×10^3 with 143×10^3 than with 1.43×10^5.

In astronomy, we often encounter very large numbers: It is helpful to remember: $10^3 = 1$ thousand; $10^6 = 1$ million; $10^9 = 1$ billion; $10^{12} = 1$ trillion.

ORDER OF MAGNITUDE CALCULATIONS

In astronomy, as in all sciences, it is often useful to be able to get quick approximations without having to perform detailed calculations. Suppose, for example, we want to know, roughly, how far light can travel in 450 years. Will it be a few million kilometers? A few trillion kilometers? Index notation makes it much easier to make such estimates. We know that the speed of light is about 3×10^5 km/sec, and that there are about 3×10^7 sec in a year. The distance we are looking for, therefore, will be about $(3 \times 10^5) \times (3 \times 10^7) \times (4.5 \times 10^2)$ km. First, we perform the calculations using only the powers of 10, forgetting for a moment about the numbers to the left of the decimal point: $10^5 \times 10^7 \times 10^2 = 10^{14}$. Our answer, in other words, will be at least in the hundred-trillions. Then, we glance at the other numbers to see how many orders of magnitude they are likely to add. We can easily see that $3 \times 3 \times 4.5$ will be more than 10, but less than 100. Our answer, therefore, will be between 10^{15} and 10^{16} km; it will take the form $n \times 10^{15}$, where n is a number between 1 and 10.

PROPORTION

If four quantities are related in such a way that $a/b = c/d$, then we can also write $ad = bc$; $a = bc/d$; $b = ad/c$; $b/a = d/c$; $c/a = d/b$, and so on. Note that a can always be written as $a/1$.

If $y = ax$, y is said to be directly proportional to x. If x is doubled, so is y; if x is multiplied by 100, y will be also, and so on.

If $y = a/x$, y is said to be inversely proportional to x. If x is doubled, y will be halved; if x is multiplied by 100, y will be divided by 100, and so on.

CIRCLES

The circumference of a circle is 2π times its radius. π is a number that cannot be represented by a fraction, nor as a decimal with a finite number of places, although it can be calculated to any number of places we require. In practice, it can be taken as approximately $22/7$, or 3.1416

If r is the radius, the following formulas can be used:

Circumference of a sphere or circle = $2\pi r$; area of a circle = πr^2

Surface area of a sphere = $4\pi r^2$; volume of a sphere = $4/3 \pi r^3$

UNITS AND CONSTANTS

The system of units we are familiar with in everyday life is known as the English system. In it the unit of force (and, somewhat confusingly, of mass) is the pound, and the unit of length is the foot. Unfortunately, this is not a decimal system—that is, the various units are not related to each other by powers of 10 (i.e. 10, 100, 1000, etc.) A mile, for example, is 5280 ft, an inch is $1/12$ ft, and an ounce is $1/16$ lb. This makes it inconvenient to use. (How many inches are there in a mile? How many tons do 1,280 oz make?)

In scientific work, therefore, the metric system is used. Although these units seem strange at first, they are employed in most countries of the world, and the U.S. will be adopting them in the near future. In the metric system the unit of length is the centimeter (cm) or the meter (m), and the unit of mass the gram (g) or the kilogram (kg). The unit of time is the second, as in the English system. In this book, we have generally used the centimeter and the gram as our basic units (the cgs system), although meters and kilograms have occasionally been employed where they were more convenient, and English units have often been used in examples from everyday experience because of their familiarity. Here are some other metric units, with their English equivalents.

1 kilometer (km) = 1000 meters = 100,000 centimeters = 10^{12} nanometers

1 kilometer = .62 miles; 1 mile = 1.6 kilometers.

1 meter = 3.28 feet; 1 foot = .305 meters

1 centimeter = .394 inches; 1 inch = 2.54 centimeters

1 nanometer (nm) = 10^{-9} meters = 10^{-7} centimeters

1 kilogram = 1000 grams = 2.2 pounds; 1 pound = .45 kilograms

1 gram = .035 ounces; 1 ounce = 28.35 grams

TEMPERATURE SCALES

In English-speaking countries, the Fahrenheit temperature scale is in general use; in most of the rest of the world, the Celsius or centigrade scale is employed. In both, the zero-point has been chosen for practical convenience in everyday life. In scientific applications, however, we generally use the Kelvin, or absolute scale, in which the zero-point is the temperature of a body from which all the available thermal energy has been extracted. This temperature is known as absolute zero; no lower temperature is attainable. A degree on the Kelvin scale is equal to a Celsius degree; both are equal to ⁹⁄₅ (1.8) Fahrenheit degrees.

	F	C	K
Boiling point of water	212	100	373
Human body	98.6	37	310
Melting point of ice	32	0	273
Coldest recorded temperature on earth	−130	−90	183
Absolute zero	−459	−273	0

To convert Celsius to Kelvin temperatures, add 273. To convert Fahrenheit to Celsius temperatures, and vice versa, the following formulas can be used:

$$F = \tfrac{9}{5}C + 32; \quad C = \tfrac{5}{9}(F - 32)$$

MAGNITUDES OF ASTRONOMICAL OBJECTS

Astronomers talk about the brightness of objects in terms of their **magnitudes.** The magnitude scale is a refinement of a system first used by Hipparchus. He classified all the visible stars into six classes. The brightest stars he called stars of the first magnitude and the faintest that he could see he called stars of the sixth magnitude. (This choice of six classes has been justified by modern studies of perception. It is now known that a much finer division does not result in a greater accuracy of classification.)

In 1856 the rather imprecise magnitude system of Hipparchus was put on a firmer foundation by Norman R Pogson. Using the methods of stellar photometry that had been developed, he noted that a first magnitude star gave us 100 times more light than a sixth magnitude star. Further research showed that without much readjustment, the old magnitude scale could be made quite regular by noting that each magnitude was 2.512 times fainter than the preceding magnitude. A star of magnitude 4, for example, is 2.512 times fainter than a star of magnitude 3, which in turn, is 2.512 times fainter than a star of magnitude 2. A star of magnitude 6

Differences in Magnitudes	0.0	0.5	0.75	1.0	1.5	2.0	2.5	3.0	3.5	4.0	5.0	6.0	7.0	8.0	9.0	10.0
Ratio of Light	1	1.6	2	2.5	4	6.3	10	16	25	40	100	250	6301	6600	4000	10,000

is therefore $2.512 \times 2.512 \times 2.512 \times 2.512 \times 2.512 = 100$ times fainter than a star of magnitude 1.

After this precise system was set up it was noticed that some stars, such as Canopus, Sirius, and Vega, are too bright to be classed with the other first magnitude stars. Vega, for example, is 2.512 times brighter than a first magnitude star and is therefore classified as a star of magnitude 0.0. A star that is 2.512 times brighter than Vega would have a magnitude -1.0. When telescopes became available stars fainter than magnitude 6.0 could be seen. The magnitude scale was therefore extended. A star 2.512 times fainter than a sixth magnitude star is called a star of magnitude 7.0 and a star that is 2.512 times fainter yet is called a star of magnitude 8.0, and so on.

A precise meaning can also be given to fractional magnitudes. As a simple example, consider a star of magnitude 4.5. Such a star is as many times brighter than a star of magnitude 5.0 as it is fainter than a star of magnitude 4.0. But a star of magnitude 4.0 is 2.512 times brighter than a star of magnitude 5.0. Everything works out correctly if the star of magnitude 4.5 is $\sqrt{2.512}$ ($= 1.585$) times fainter than a star of magnitude 4.0 and just as many times brighter than a star of magnitude 5.0.

Magnitudes are a way of talking about brightness. Astronomers who have become used to the magnitude system like to talk about *luminosity* in a similar manner. This has resulted in the definition of a new quantity called **absolute magnitude.** The absolute magnitude of an astronomical object is the magnitude that it would have if it were at 10 parsecs. (Whenever confusion is likely to arise the ordinary magnitude is called **apparent magnitude.** To illustrate this new concept, consider a star of apparent magnitude 12.5 at a distance of 100 parsecs. If brought to 10 parsecs it would be ten times closer and therefore 100 times brighter. Its magnitude would be 7.5 (five magnitudes *less* than 12.5) and this is its absolute magnitude.

THE U B V SYSTEM

After the brightness of a star has been measured, the next step is often to determine its color. One standard way of doing this has been to photograph it with a blue-sensitive plate, and then with a plate and filter combination that has the spectral sensitivity of the human eye. The first photograph indicates the amount of blue light received from the star; the magnitude determined from the photograph is called the **photographic magnitude.** The magnitude determined from the second photograph indicates the amount of yellow light received from the star. It is called the **photovisual magnitude,** because it corresponds to the magnitude scale one could set up by visual observations. A measure of the colour of the star, called the **Color Index,** is defined as the photgraphic magnitude minus the photovisual magnitude. The magnitude scales are adjusted to give a CI of 0.0 for A type stars. Very blue stars have a CI of -0.6, and the reddest stars have a CI of about $+2.0$.

Nowadays photoelectric devices are used to measure the light received from stars. The most common of these devices is the photomultiplier tube. In such a tube photons from a star dislodge electrons from a specially prepared plate. These electrons are accelerated and made to strike other plates. Each electron causes three to five times as many secondary electrons to be dislodged. After this multiplicative process has occurred a few times, an amplified flow of current results. This amplified current is easily detected and analysed. Photomultipliers enable us to measure brightness to accuracies of $1/100$ of a magnitude.

H.L.Johnson and W.W.Morgan have made very precise measurements using the 1P21 photomultiplier tube. They used three filters, each of which admits about 199 nm of the spectrum. The U filter admits all the near ultraviolet light from the atmospheric cutoff — about 300 nm — to the visible boundary — about 400 nm. The B filter admits the range from about 400 nm to a little beyond 500 nm; it corresponds to the response of blue-sensitive photographic plates. The V filter admits a range of wavelengths from 500 nm to a little beyond 600 nm. This filter coresponds to the response of the human eye. (Since the 1P21 does not respond to longer wavelengths it is not possible to examine red or infrared bands.)

THE U B V system has some interesting points. A color index may be defined by subtracting the V magnitude from the B. $B - V$ corresponds to the ordinary CI. But third color can be used to make up another color index, $U - B$. This other color index can then be compared with the color index $B - V$.

Now, a color index tells us something about the temperatures of stars assuming that they are black bodies. The comparision of the two color indices tells us something about the way in which stars differ from black bodies. In a metal rich star, for example, the blue region of the spectrum is filled with absorption lines. In such a star the B magnitude is larger than normal, and therefore $U - B$ is reduced in comparision with $B - V$, which is increased. The U B V system therefore enables us to estimate metal abundances.

The third color is present for a more compelling reason. If a color index is determined in two different systems, using different photocells and filters, it is possible to compare results only if there is a third color. The added information obtained with the third filter has been so intriguing that that the tendency has been to develop multi-color systems. In the last few years 6 and 8 – colour systems have been developed in conjunction with photelectric devices that have responses extending out into the infrared.

THE MOVING CLUSTER METHOD

We have seen in Chapter 2 that the tangential velocity of a star can be determined if we can measure its proper motion, and know its distance.

Conversely, knowing the tangential velocity of any star whose proper motion can be measured will enable us to find its distance. Unfortunately, it is often impossible to determine the tangential velocities of stars. There are some exceptional cases, however, that are very valuable to astronomers.

The stars of a cluster generally move through space together—that is, they move along nearly parallel paths. Because of the effects of perspective, these paths will seem to converge when plotted on the celestial sphere. (The phenomenon is similar to the apparent convergence of trees or poles along a railroad track, discussed in Chapter 2.) The point toward which they appear to converge is called the vanishing point. From its location, it is possible to deduce the direction in which the stars are really moving. Knowing this direction, and also knowing the radial component of the stars' velocities from their spectra, we can find their tangential velocity. The tangential velocity, together with the observed proper motion, gives us the distance to the cluster.

This method has been used to determine the distance to the Hyades and a few other clusters. Since the Hyades form the basis of the cluster-fitting method on which every other scale determination of the universe is based the method of moving clusters plays an important role in astronomy.

THE SUN

Quantity	Value	How determined
Mean distance from earth	1 AU 149,600,000 km	Radar reflection from planets
Angular diameter from earth	32'	Direct measurement
True diameter	1,392,000 km (109 earth diameters)	Direct measurement
Mass	1.99×10^{33} g (330 earth masses)	Acceleration of earth and planetary orbits
Density	1.4 g / cm³ (average) 150 g / cm³ at core 10^{-7} g/cm³ at photosphere	Mass / Volume
Temperature	1.5×10^7 K at core 5800 K at surface	Luminosity and radius; spectrum
Solar constant (energy received by earth)	1.37×10^6 erg/sec/cm²	High altitude spacecraft
Luminosity	3.86×10^{33} erg/sec 5×10^{23} horsepower	Solar constant and geometry
Spectral type	G2	Spectrum
Apparent magnitude	−26.8	Photometer
Absolute visual magnitude	4.71	Apparent magnitude and AU
Period of rotation	25 days at equator 27½ days at 45°N and S 31 days at pole	Movement of sunspots; Doppler shift in spectra of photosphere

CELESTIAL COORDINATES

In Chapter 2, we mentioned that the celestial equator and celestial poles are the projections of the earth's equator and poles onto the celestial sphere. Astronomers use a coordinate system based on the celestial equator and celestial poles just as we use, on earth, a system based on the equator and poles. The two coordinates used in the celestial coordinate system are **declination** (abbreviated δ) and **right ascension** (abbreviated α). Declination is the angular distance of an object north or south of the celestial equator. Conventionally, northern objects have a positive declination and southern ones a negative declination; except for this, it corresponds to the way in which latitude is used on earth. The other coordinate, right ascension, corresponds to longitude on earth. We define right ascension by taking a series of circles, running north and south, passing through the celestial poles. These circles are fixed on the celestial sphere, and correspond to the meridians of longitude on earth. The zero point of this coordinate system is taken to be the vernal equinox, where the sun lies on the first day of spring. Right ascension is measured eastward in hours, 24 hours making a complete circle. Each hour is, therefore, equal to 15° and each minute is equal to 15′. Coordinates of stars in this system are not affected by the position of the observer, or the day of the day.

Several other coordinate systems are also in use. One employs the plane of the ecliptic as its "equator." Another uses the equatorial plane of the galaxy, coinciding roughly with the visible MilkyWay. For most ordinary purposes, however, the system of declination and right ascension is the most common and useful.

THE BRIGHTEST STARS

Star	Constellation	α hr min	δ ° ′	Apparent magnitude	Distance (pc)	Proper Motion (″/yr)	Absolute magnitude	Spectral type
*Sirius	Canis Major	6 44	−16 41	−1.5	2.7	1.32	+1.4	A1 MS
Canopus	Carina	6 24	−52 41	−0.7	55	0.02	−3.1	FO Supergiant
*Alpha Centauri		14 38	−60 44	−0.3	1.3	3.68	+4.4	G2 MS
Arcturus	Bootes	14 15	+19 19	−0.1	11	2.28	−0.3	K2 Giant
Vega	Lyra	18 36	+38 46	0.0	8.1	0.34	+0.5	A0 MS
*Capella	Auriga	5 15	+45 52	0.0	14	0.44	−0.7	G2 Giant
*Rigel	Orion	5 13	−8 14	0.1	250	0.00	−6.8	B8 Supergiant
*Procyon	Canis Minor	7 38	+5 17	0.3	3.5	1.25	+2.7	F5 MS-subgiant
Achernar	Eridanus	1 38	−57 22	0.5	20	0.10	−1.0	B5 MS
*Beta Centauri		14 02	−60 15	0.6	90	0.04	−4.1	B1 Giant
Altair	Aquila	19 50	+8 48	0.8	5.1	0.66	+2.2	A7 MS-subgiant
Betelgeuse	Orion	5 54	+7 24	0.8 var	150	0.03	−5.5	M2 Supergiant
*Aldebaran	Taurus	4 34	+16 28	0.9	16	0.20	−0.2	K2 Giant
*Alpha Crucis		12 25	−63 00	0.9	120	0.04	−4.0	B1 Subgiant
Spica	Virgo	13 24	−11 01	1.0 var	80	0.05	−3.6	B1 MS
*Antares	Scorpio	16 28	−26 22	1.0 var	120	0.03	−4.5	M1 Supergiant
Pollux	Gemini	7 44	+28 05	1.2	12	0.62	+0.8	K0 Giant
Fomalhaut	Piscis Austrinus	22 56	−29 45	1.2	7	0.37	+2.0	A3 MS
Deneb	Cygnus	20 41	+45 11	1.3	430	0.00	−6.9	A2 Supergiant
Beta Crucis		12 46	−59 33	1.3	150	0.05	−4.6	B0 Subgiant

MS = Main Sequence; * = Multiple system; var = Variable

ASTRONOMICAL CONSTANTS

AU	astronomical unit	1.496×10^8 km
ly	light-year	9.46×10^{12} km
pc	parsec	3.086×10^{13} km = 3.26 ly = 206,265 AU
	sidereal year	365.256 days = 3.156×10^7 sec
$r_\oplus$	radius of earth	63.78 km
$m_\oplus$	mass of earth	5.98×10^{27} g
$r_\odot$	radius of sun	6.960×10^5
$m_\odot$	mass of sun	1.989×10^{33} g
$L_\odot$	luminosity of sun	3.90×10^{33} erg/sec

PHYSICAL CONSTANTS

c	velocity of light	2.998×10^{10} cm/sec
	mass of hydrogen atom	1.67×10^{-24} g
	radius of hydrogen atom (ground state)	.053 nm
	mass of electron	9.11×10^{-28} g
h	Planck's constant	6.626×10^{-27} erg sec
G	constant of gravitation	6.668×10^{-8} dyne cm²/g²
	constant in Wien's law	.290 cm/deg
	Stefan-Boltzmann constant	5.67×10^{-5} erg/cm² deg⁴ sec

NOTES FOR OBSERVERS

The objects described below are among the easiest to observe with a small telescope (40–150 mm aperture), and/or among the more interesting to observe with a moderate telescope (150–300 mm). Most of them are also beautiful.

(a) Nebulas, Clusters, Galaxies

Object	Popular name	α	δ	Nature	Description
M 1	Crab nebula	5ʰ 32ᵐ	+21° 59′	Supernova remnant	Appears as an elongated patch in a moderate telescope.
M 42	Great nebula in Orion	5ʰ 33ᵐ	−5° 25′	Emission nebula	Vast cloud of gas with embedded stars. Hazy patch to naked eye; spectacular in small–moderate telescope.
NGC 7662		23ʰ 23ᵐ	+42° 12′	Planetary nebula	Unusually bright
M 27	Dumbbell nebula	19ʰ 58ᵐ	+22° 35′	Planetary nebula	Elongated patch of light; needs moderate telescope for a good view.
M 57	Ring nebula	18ʰ 52ᵐ	+32° 58′	Planetary nebula	Bright, ring-shaped nebula.
M 97	Owl nebula	11ʰ 12ᵐ	+55° 17′	Planetary nebula	Very large nebula; needs a dark night.
M 36		5ʰ 33ᵐ	+34° 07′		Three fine clusters in Auriga. M 37 is the easiest,
M 37		5ʰ 49ᵐ	+32° 33′	Open clusters	with about 100 stars visible in a 150 mm
M 38		5ʰ 25ᵐ	+35° 48′		telescope.
M 45	Pleiades	3ʰ 45ᵐ	+23° 57′	Open cluster	Six stars visible to naked eye; about 130 in a 150 mm telescope.
M 44	Praesepe (beehive)	8ʰ 37ᵐ	+20° 10′	Open cluster	Large, scattered cluster, almost a naked eye object —use low power.
h + χ	Persei	2ʰ 17ᵐ	+56° 55′	Open clusters	The famous double cluster.
M 103		1ʰ 30ᵐ	+60° 26′	Open cluster	About 1° diameter
NGC 2244		7ʰ 39ᵐ	+4° 54′	Open cluster	Beautiful; visible to naked eye.
NGC 663		1ʰ 42ᵐ	+61° 00′	Open cluster	Visible with small telescope.
NGC 457		1ʰ 16ᵐ	+58° 03′	Open cluster	Tight cluster, < 1/3°
M 2		20ʰ 31ᵐ	−1° 03′		Very attractive clusters, needing > 120 mm for
M 3		13ʰ 40ᵐ	+28° 38′	Globular clusters	adequate resolution. In smaller instruments, hazy
M 5		15ʰ 16ᵐ	+2° 16′		spots.
M 13	Hercules cluster	16ʰ 40ᵐ	+36° 30′	Globular cluster	Brightest in northern skies.
M 22		18ʰ 33ᵐ	−23° 57′	Globular cluster	Very bright, but in southern sky.
M 92		17ʰ 16ᵐ	+33° 11′	Globular cluster	
M 31	Great nebula in Andromeda	0ʰ 41ᵐ	+41° 0′	Galaxy (spiral)	Brightest external galaxy, a hazy spot to naked eye. Bright oval patch in binoculars or telescope. Needs a clear dark night for best view.
M 51	Whirlpool nebula	13ʰ 28ᵐ	+47° 27′	Galaxy (spiral)	The spiral arms require a large telescope.
M 81	Great spiral in Ursa Major	9ʰ 53ᵐ	+69° 18′	Galaxy (spiral)	Bright galaxy; easy object for telescopes > 100 mm.
M 82		9ʰ 54ᵐ	+69° 56′	Galaxy (irregular)	A silverfish-like image near M 81.
M 87		12ʰ 28ᵐ	+12° 40′	Galaxy (elliptical)	Very bright elliptical radio galaxy.

(b) Double stars

Object	Popular name	α	δ	Nature	Description
——	Mizar	13ʰ 22ᵐ	+55° 11′	——	Famous double, middle star in handle of Big Dipper.
——	ε Lyrae	18ʰ 43ᵐ	+38° 44′	——	Double double.
——	Albireo	19ʰ 29ᵐ	+27° 51′	——	Beautiful bright double in Cygnus; colors orange and blue.
——	σ Orionis	5ʰ 30ᵐ	−2° 38′	——	Spectacular multiple system with components of different colors.
——	α Herculis	17ʰ 12ᵐ	+14° 27′	——	Marked color contrast, yellow and blue

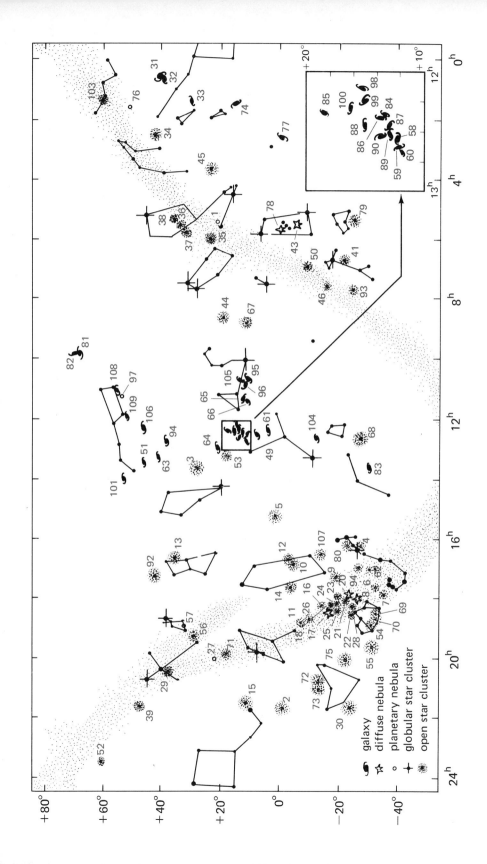

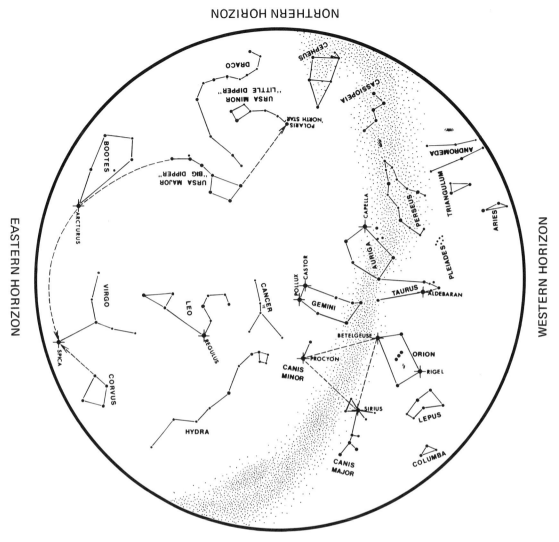

NORTHERN HORIZON

EASTERN HORIZON

WESTERN HORIZON

SOUTHERN HORIZON

THE NIGHT SKY IN MARCH

Latitude of chart is 34°N, but it is practical throughout the continental United States.

To use: Hold chart vertically and turn it so the direction you are facing shows at the bottom.

Chart time (Local Standard):

10 p.m. First of month
9 p.m. Middle of month
8 p.m. Last of month

Star Chart from GRIFFITH OBSERVER monthly magazine

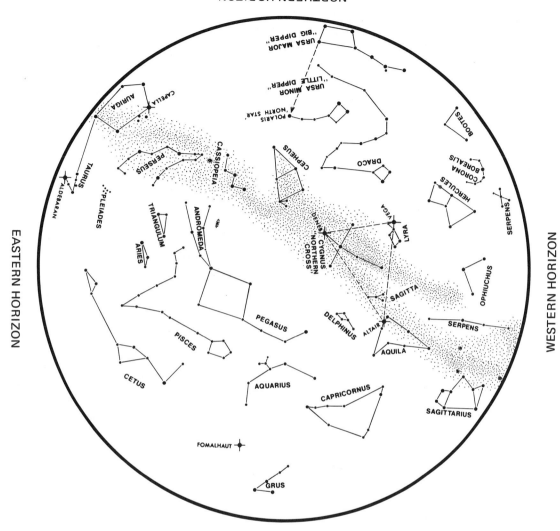

NORTHERN HORIZON

EASTERN HORIZON

WESTERN HORIZON

SOUTHERN HORIZON

THE NIGHT SKY IN OCTOBER

Latitude of chart is 34°N, but it is practical throughout the continental United States.

To use: Hold chart vertically and turn it so the direction you are facing shows at the bottom.

Chart time (Local Standard):

10 p.m. First of month
9 p.m. Middle of month
8 p.m. Last of month

Star Chart from GRIFFITH OBSERVER monthly magazine

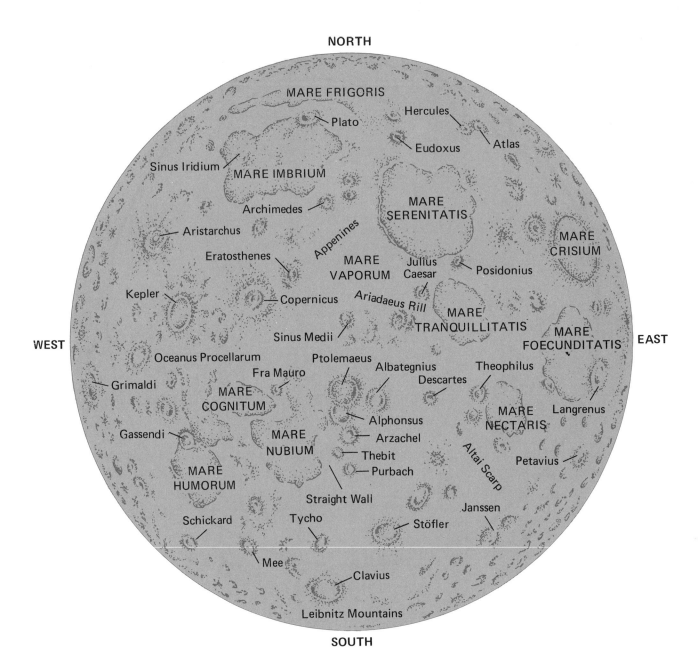

NORTH

MARE FRIGORIS

Plato
Hercules
Eudoxus
Atlas

Sinus Iridium
MARE IMBRIUM

Archimedes
MARE SERENITATIS

Aristarchus
MARE CRISIUM

Eratosthenes
Appenines

MARE VAPORUM
Julius Caesar
Posidonius

Kepler
Copernicus
Ariadaeus Rill
MARE TRANQUILLITATIS

WEST
Sinus Medii
MARE FOECUNDITATIS
EAST

Grimaldi
Oceanus Procellarum
Fra Mauro
Ptolemaeus
Albategnius
Descartes
Theophilus

MARE COGNITUM
Alphonsus
MARE NECTARIS
Langrenus

Gassendi
MARE NUBIUM
Arzachel
Thebit
Purbach

MARE HUMORUM
Altai Scarp
Petavius

Straight Wall
Janssen

Schickard
Tycho
Stöfler

Mee

Clavius

Leibnitz Mountains

SOUTH

SATELLITES OF PLANETS

Planet	Satellite	Discovered	Mean Distance from Planet (km)	Sidereal Period (Days)	Diameter of Satellite (km)	Mass (Planet = 1)
Earth	Moon	—	384,404	27.322	3476	0.0123
Mars	Phobos	1877	9,380	0.319	25	(2.7×10^{-8})
	Deimos	1877	23,500	1.262	13	4.8×10^{-9}
Jupiter	V	1892	180,500	0.498	(150)	(2×10^{-9})
	I Io	1610	421,600	1.769	3640	4×10^{-5}
	II Europa	1610	670,800	3.551	3100	2.5×10^{-5}
	III Ganymede	1610	1,070,000	7.155	5270	8×10^{-5}
	IV Callisto	1610	1,882,000	16.689	5000	5×10^{-5}
	VI	1904	11,470,000	250.57	(120)	(8×10^{-10})
	VII	1905	11,800,000	259.65	(40)	(4×10^{-11})
	X	1938	11,850,000	263.55	(10)	(1×10^{-12})
	XIII	1974	12,400,000	282.0	(8)	(5×10^{-13})
	XII	1951	21,200,000	631.1	(10)	(7×10^{-13})
	XI	1938	22,600,000	692.5	(15)	(2×10^{-12})
	VIII	1908	23,500,000	738.9	(25)	(8×10^{-12})
	IX	1914	23,700,000	758	(15)	(2×10^{-12})
Saturn	Janus	1966	157,500	0.749	(350)	(3×10^{-8})
	Mimas	1789	185,400	0.942	(500)	6.6×10^{-8}
	Enceladus	1789	237,900	1.370	(500)	1.5×10^{-7}
	Tethys	1684	294,500	1.888	(1000)	1.1×10^{-6}
	Dione	1684	377,200	2.737	(1000)	2×10^{-6}
	Rhea	1672	526,700	4.518	1600	3×10^{-6}
	Titan	1655	1,221,000	15.945	5800	2.5×10^{-4}
	Hyperion	1848	1,479,300	21.277	(400)	2×10^{-7}
	Iapetus	1671	3,558,400	79.331	(1200)	4×10^{-6}
	Phoebe	1898	12,945,500	550.45	(300)	5×10^{-8}
Uranus	Miranda	1948	123,000	1.414	550	1×10^{-6}
	Ariel	1851	191,700	2.520	1500	1.5×10^{-5}
	Umbriel	1851	267,000	4.144	1000	6×10^{-6}
	Titania	1787	438,000	8.706	1800	5×10^{-5}
	Oberon	1787	585,960	13.463	1600	3×10^{-5}
Neptune	Triton	1846	353,400	5.877	6000	3×10^{-3}
	Nereid	1949	5,560,000	359.881	500	(10^{-6})

( ) = estimated

CREDITS

GLOSSARY

Absorption lines: dark lines that are produced when a continuous spectrum is passed through the vapor of an element.

Acceleration: the rate of change of velocity.

Alpha particles: the nuclei of helium atoms, consisting of two protons and two neutrons.

Asteroids: large pieces of matter, smaller than planets, primarily orbiting between Mars and Jupiter. Originally, they were called minor planets.

Astronomic binary: a star system which includes an unseen component, whose presence is detected solely by its gravitational influence on the visible star.

Astronomical unit: the distance of the Sun from the Earth, about 140 million km.

Atoms: the smallest possible subdivision of the chemical elements.

Barycenter (Center of Mass): the common point about which two gravitationally bound bodies orbit.

Binary star: two closely spaced stars that are physically related.

Black holes: collapsed stars whose surface gravitation is so intense that not even light can escape from it. Never observed, but theoretically possible.

Blink microscope (blink comparator): a device which allows an observer to view two different photos of the same area of the sky in rapid succession.

Boyle's Law: a law stating that the pressure of a gas is proportional to its density if the temperature is unaltered.

Bright line spectrum: a spectrum consisting solely of a few bright lines separated by dark space in which only certain wavelengths are being radiated.

C-N cycle: one of the two chief routes by which hydrogen can be converted into helium, employing the nuclei of carbon and nitrogen as intermediaries. The carbon nucleus undergoes various changes, in the course of which four hydrogen nuclei emerge as a single nucleus of helium.

Celestial equator: a circle on the celestial sphere, halfway between the poles, corresponding to the Earth's equator.

Celestial poles: unmoving points on the celestial sphere corresponding to the Earth's north and south poles.

Celestial sphere: in ancient astronomy, an enormous sphere encircling the Earth on which the stars were fixed.

Centripetal force: a force directed inward toward the center of a circular orbit.

Cepheid variables: yellow supergiant stars whose luminosities vary in regular cycles.

The Chandrasekhar Limit: the greatest mass that a white dwarf can have, roughly 1.4 solar masses.

Charles' Law: a law stating that the pressure of a gas is directly proportional to its temperature, provided its density is constant.

Chromosphere: a layer of the sun's atmosphere, the "sphere of color," difficult to observe without special instruments because of the intensely bright light of the photosphere.

Circumpolar stars: stars close enough to the north "celestial pole" to give the appearance of never setting.

Comets: made up of dust, rock fragments, ice, and frozen gases, comets become fluorescent as they approach the sun. The heat of the sun melts the frozen gases, which glow.

Conduction: the process by which heat spreads directly by contact.

Conjunction: the state in which a planet appears nearest another object on the celestial sphere.

Consecutive formation theories: theories of the origin of the solar system in which the sun forms and then captures a cloud of interstellar material, which becomes the planets.

Constellations: pictures suggested by the patterns of groups of stars.

Continental drift: the slow movement of the Earth's land masses; the drawing apart of continents.

Continuous creation of matter: a theory stating that as the universe expands, it is slowly filled with new matter.

Continuous spectrum: light containing all wavelengths; all the colors of the rainbow blending imperceptibly one into the other.

Convection: a method of heat transfer that depends on the motion of a gas or liquid.

Corona: part of the Sun's atmosphere, an envelope of extraordinarily hot gases; so hot that all atoms are ionized. Observable only during a total solar eclipse.

Coronagraph: a modification of the ordinary telescope, which produces an artificial total solar eclipse, facilitating study of the corona and chromosphere.

Cosmological principle: a principle stating that all observers everywhere should see the universe in essentially the same form.

Dark line spectra: spectra representing wavelengths of light not present in sunlight.

Deferent: in the geocentric theory of the universe, a large circle whose center is located at the center of the Earth.

Degenerate electrons: electrons trapped in an artificially high energy state.

Density: the amount of mass per unit of volume contained in any object.

Diffraction: the bending or spreading of waves around the edges of objects.

Dispersion: the separation of different wavelengths of light.

Doppler effect: an apparent change in the wavelength of radiation from a source, due to its relative motion toward or away from the observer.

Dynamic stage: the first of two stages in the progress of a protostar toward becoming a star, in which the gravitational contraction takes place unopposed to any significant degree by the pressure within the protostar.

Eccentricity: a measure of the elongation of an ellipse; a determination of the shape of an ellipse.

Eccentrics: circular orbits with the Earth slightly off center.

Eclipsing binaries: a binary star system in which each star passes in front of the other at regular intervals, obstructing its light.

Ecliptic: the apparent path of the sun against the background of the fixed stars.

Electromagnetic waves: disturbances of the electrical and magnetic properties of space.

Electron: a particle of an atom, having a negative electric charge and a very small mass. It is stable.

Ellipse: a closed curve, a close relative of a circle.

Elongation: the angle between a planet and the sun as seen from the Earth.

Emission nebulas: highly visible masses of interstellar gases.

Encounter theories: theories of the origin of the solar system in which material is extracted from the sun by a passing star and quickly forms planets.

Energy levels: the level of energy in an atom, represented by electron configurations.

Epicycle: in the geocentric theory of the universe, a small circle whose center is a point on the deferent. A planet is attached to the rim of the epicycle.

Equant: in Ptolemaic theory, a point from which the motion of a planet would appear uniform, but located neither at the Earth, nor at the center of the planetary orbit.

Equations of stellar structures: four mathematical equations summarizing the relationships that must exist among the physical variables at each point within a star.

Equinox: literally meaning "equal night," the term given to the times of the year when day and night are equal everywhere on Earth. The vernal (spring) equinox occurs about March 21, and the autumnal equinox about September 23.

Feedback: the principle by which the rate of energy release is adjusted so that it always balances the rate of energy which is radiated out.

Fireball: an extraordinarily bright meteor, resulting from the entrance of a relatively slow-moving meteoroid into the Earth's atmosphere.

Fission: the breaking apart of the nuclei of heavy atoms into lighter nuclei.

Flash spectrum: the brief visibility of the chromosphere during a total solar eclipse.

Foci: two points within an ellipse, analogous to the center of a circle.

Forbidden lines: lines in the emission spectra of nebulas; not observed on Earth because the energy states that give rise to them are metastable.

Fusion: the combination of light nuclei to form a heavier nucleus.

Galactic plane: the center of the galactic disc, denser than the rest of the disc, containing the spiral arms.

Geocentric theory: the so-called Ptolemaic theory, in which the Earth is a fixed body around which the other bodies of our system revolve.

Geodesic: the shortest distance between two points on a mathematically derived surface, in this case a sphere.

Globular clusters: groups of halo stars that seem to travel in elongated orbits about the center of the galaxy.

Globules: large blobs of about 120 solar measures, formed of dense clouds of dust and gas. As they collapse, they become still denser and tend to break up into smaller blobs with masses equal to those of stars. These smaller blobs are the protostars.

Gravitation: the mutual attraction of two orbiting bodies.

Gravitational potential energy: the possible energy available from the pull of the force of gravity.

H-R Diagram: the graph of stars in the solar neighborhood, plotting the luminosity of each star against its temperature.

H II regions: clouds of ionized hydrogen surrounding massive short-lived blue giants of spectral types O and B.

Heliocentric theory: a theory associated with Aristarchos in which the sun is the unmoving center of the solar system.

Helium flash: a runaway burning of helium in a star that has a core made up of degenerative gas.

Hertzsprung gap: the empty region on the H-R diagram in which only the massive stars that use the C-N cycle have started to become red giants.

High-velocity stars: stars in the neighborhood of our sun which have higher velocities in respect to the sun and other stars in the neighborhood as a result of travelling in randomly oriented, highly elongated orbits.

Hubble's Law: the "law of red shifts," stating that, with a few exceptions in our local group, all galaxies are receding from us, and the farther away they are, the faster they are receding.

Hydrostatic equilibrium: the condition in which the balance between gas pressure and gravity is maintained within a star. Gas pressure increases as the center of the star is approached. Gas pressure must balance the inward force of gravity at every point.

Inertia: the resistance of objects to any change in their motion.

Inferior conjunction: the position of an inferior planet between the sun and the Earth when in conjunction with the sun.

Inferior planets: Mercury and Venus, with orbits smaller than the Earth's.

Inverse square law: a law formulated by Isaac Newton stating that the brightness of a luminous body varies inversely as the square of its distance from the observer.

Ionization: the escape of an electron from the attractive force of the nucleus, as a result of the electron gaining energy.

Ionization energy: the energy necessary to strip an electron from an atom.

Jovian planets: Jupiter, Saturn, Uranus, and Neptune are larger, but less dense than the terrestrial planets, have heavy dense atmospheres, and rotate very rapidly for their size.

Kepler's first law: orbits are ellipses, with the sun located at one focus.

Kepler's second law: the law of areas: a radius drawn from the sun to an orbiting planet will sweep out equal areas in equal time, no matter where the planet is in orbit.

Kepler's third law: the law of period: the square of the period of revolution, *P,* of any planet is proportional to the cube of its orbit's semimajor axis, *a.*

Kinetic theory of gases: gases consist of extremely small, perfectly elastic particles constantly in motion. The particles bounce off each other and the wall of their container, but do not react in any other way.

Kirkwood gaps: empty regions in the asteroid belt.

Law of inertia: formulated by Galileo, this law states that bodies at rest tend to stay at rest, but that in the absence of a resisting force, a body that is set in motion will continue in the original state of motion forever.

Leap year: a year which is divisible by four and in which one extra day is added to the month of February.

Light year: the distance travelled by light through a vacuum in one year (9.5×10^{12} km.)

Luminosity: the rate of radiation of electromagnetic energy of a star.

Main sequence: in the H-R diagram, a band of stars stretching from the upper left (hot, luminous stars) to the lower right (cool, dim stars). Most of the stars in the diagram lie in this band.

Maria (singular: Mare): seas, or lowlands, on the moon's surface.

Mascons: high-mass areas scattered at random below the moon's surface.

Mass: the mass of a body is a measure of its inertia—its resistance to any change in its velocity.

Mass-luminosity relationship: the relationship between the mass and luminosity of a star; the luminosity of a star is usually roughly proportional to the 3.5 power of its mass.

Metastable states: energy states from which all possible downward transitions are forbidden; that is, from which an atom is unlikely to leave by radiation.

Meteor: a meteoroid which enters our atmosphere, is heated to incandescence by friction with the air, and burns up within a few seconds.

Meteor shower: an intense bombardment of meteors, resulting from an encounter with a large number of meteoroids travelling together in almost identical orbits.

Meteor trail: the streak of light created by a meteor during its passage.

Meteorite: a meteor that survives the passage through our atmosphere and strikes the Earth's surface.

Meteoroids: fairly small fragments of rocky or metallic material that travel through interplanetary space.

Molecule bands: groups of many closely spaced absorption lines produced by molecules.

Molecules: atoms chemically bound together.

Monochromatic filter: similar to the spectroheliograph, this instrument allows the observer to view the solar disc and its atmosphere in the light of a single line all at once, rather than slice by slice.

Nebula: masses of glowing interstellar gas.

Nebular hypothesis: the idea that the planets and the sun all evolved concurrently from a rapidly rotating solar nebulus, or disc-shaped cloud of gas.

Neutron: a neutral particle of an atom, with no electric charge.

Neutron star: a star that derives most of its pressure support from degenerate neutrons.

Newton's First Law: a body at rest will remain at rest, and a body in motion will remain in motion with constant speed in a straight line, unless disturbed by an outside force.

Newton's Second Law: a law stating that when a force acts upon a body, the resulting acceleration is directly proportional to, and in the same direction as, the force, but inversely proportional to the body's mass.

Newton's Third Law: Whenever an object exerts force on another object, the second object exerts an equal but opposite force on the first. For every action, there is an equal and opposite reaction.

Nuclear reaction: changes in the nucleus of an atom, releasing mass-energy.

Nucleus: in an atom, the central mass containing protons and neutrons, surrounded by shells of electrons that are bound to it by electrical attraction.

Non-thermal X-rays: X-rays that are not a part of a star's normal black-body spectrum.

Opacity: the ability of matter to absorb radiation.

Optics: the study of the laws regarding light and its properties.

Outgasing: the expulsion of water, carbon dioxide, and nitrogen from the Earth's crust early in its history.

Parallax: a change in the apparent direction of a distant object because of the observer's motion.

Parsec: the distance at which a star will appear to have an annual parallax of $1''$ of arc ($1/3600°$).

Perfect absorber: an ideal body that absorbs all the electromagnetic radiation of every wavelength that falls upon it.

Perfect Cosmological Principle: the principle that states that all observers should see the universe in the same way, no matter where or when they observe it.

Perfect radiator: an ideal body that radiates electro-magnetic energy with the maximum possible efficiency; non-existent in nature.

Phases of the moon: the various stages of the moon as it orbits the Earth.

Photons: the quanta of light that travel through space.

Photosphere: the visible layer of the sun, 500 km. thick; a part of the sun's atmosphere.

Plages: bright, granulated clouds of gas found in the chromosphere, usually near sunspots.

Planck's constant: the elementary quantum of action; its dimensions are energy multiplied by time, forming the units of a quantity called action.

Planetary nebulas: shells of gas and dust expanding away from the surface of red giant stars.

Planetesimals: large, asteroid-like accretions of particles.

Population I stars: stars of the galactic plane, especially of the spiral arms, that are relatively young; apparently mixed with Population II stars in the nuclear bulge.

Population II stars: stars that are found in the galactic halo, all very old.

Primary minimum: a large dip in the light curve resulting from the eclipse of a hot star by a cool star in a binary system.

Prominences: streams of relatively dense hydrogen that jut out into the corona from the photosphere.

Proper motion: the change in position of a star in the sky with time.

Proton: a particle of an atom carrying a positive electrical charge equal in magnitude to that of the electron. A proton is 1836 times as massive as an electron. It is stable.

Proton-proton chain: one of the chief routes by which hydrogen can be converted into helium, involving the progressive buildup first of heavy hydrogen, then of light helium, and finally of normal helium.

Protoplanets: planetesimals, on their way to becoming planets, which have grown by attracting nebular material in the region of their orbits.

Protostar: a collapsing cloud of gas and dust on its way to becoming a star.

Protosun: in the nebular hypothesis of the forming of the solar system, a cloud of cold gas that contracted under mutual gravitation to form the sun.

Pulsating, or oscillating, universe: a theory stating that the universe goes through a series of expansions and collapses, beginning in a hot, dense state, expanding to the limit that its force of gravity allows, then collapsing to a hot, dense state, only to start all over again.

Quantum hypothesis: an idea formulated by Max Planck that light can be radiated or absorbed only in tiny individual packages of energy called *quanta.*

Radial velocity: the motion of a star toward or away from the earth.

Radiation: a process whereby heat energy is carried through space in the form of electromagnetic waves.

Radio galaxies: galaxies that generate more radio energy than they do visible light.

Red giants: cool, red stars of exceptionally large size.

Reflecting telescope (reflector): a telescope using a mirror as the principal image-forming component, instead of a lens.

Reflection: the return of light rays by a smooth surface.

Refracting telescope (refractor): a telescope that is constructed entirely with lenses.

Refraction: the bending of light rays.

Resolving power (resolution): the ability of a telescope to record fine detail.

Retrograde motion: an apparent occasional westward motion, or reversal of the usual eastward motion, of the planets in respect to the stars.

Roche limit: a) the boundary of the region in which a star's gravitational force is dominant; b) the minimum distance at which a satellite can orbit its parent planet without being torn apart by tidal forces.

RR Lyrae stars: short-period variable stars whose luminosity varies in regular cycles of less than twenty-four hours.

Russell-Vogt Theorem: a theory stating that for a star in hydrostatic and thermal equilibrium living off nuclear energy, the equations of stellar structure have only one solution which depends on just two factors: the distribution of the chemical elements within the star, and its mass.

Scattering: the dispersal of starlight in random directions by large particles of dust.

Schwarzschild Radius: the radius which an object of a given mass must reach to become a black hole. The radius is also known as the *event horizon.*

Secondary minimum: a shallow dip in the light curve resulting from the eclipse of a cool star by a hot star in a binary system.

Seyfert galaxies: unusual spiral galaxies with small but very brilliaht nuclei that are evidently the scenes of very violent activity; discovered in 1943 by Carl Seyfert.

Sidereal day: a ''star's day,'' the time between appearances of a star in the sky, 23 hours and 56 minutes.

Simultaneous formation theories: theories of the origin of the solar system in which the sun and the planets form at the same time from interstellar material.

Solar flare: a sudden brightening in the region above a sunspot due to the excitation of a mass of gas by a huge outpouring of radiation.

Solstices, summer and winter: points at which the distance of the sun from the equator is the greatest. They fall about June 21 and December 21 each year.

Space-time: the treatment of space and time as a single entity, according to Einstein's general theory of relativity.

Spectograph: an instrument that separates and classifies various wavelengths of light.

Spectroheliograph: a device that enables the observer to screen out all wavelengths of the solar spectrum except a single narrow desired wavelength band.

Spectroscopic binary: a star system which can be identified as double or triple only by recurrent Doppler shifts.

Spectrum binary: a star system which can be identified as double only by the different spectral types of its components. That is, the composite spectrum of the two stars may contain features not ordinarily found together.

Spicules: numerous jets of very bright gas that shoot thousands of kilometers above the sun's disc from the chromosphere.

Steady State theory: a theory in which the Earth is infinitely old, never began, and will never end.

Stephan-Boltzman Law: a law which states that the total energy per unit area emitted each second by a perfect radiator, summed up over all wavelengths, is proportional to the fourth power of its temperature.

Superior conjunction: the position of a planet beyond the sun when it is in conjunction with the sun.

Superior planets: planets with orbits larger than the Earth's: Mars, Jupiter, Saturn, Uranus, Neptune, and Pluto.

Subgiant: the less massive of two stars in a binary system in which the more massive component is a Main Sequence star.

Sunspots: Irregular regions of the photosphere that appear dark because they are cooler than their surroundings. Sunspots have an extremely strong magnetic field.

Supergiants: the largest of the cool, red stars known as red giants.

Supernovas: seemingly new stars of extraordinary brightness which represent the explosion of an entire star.

Synchronous rotation: the rotation of a satellite in exact relation to its revolution about its parent; that is, one rotation for each revolution.

Synodic period: the period of time, in geocentric theory, which the moon takes to travel around the celestial sphere; about 29.5 days.

Tangential velocity: the motion of a star across our line of sight.

Terrestrial planets: including Mercury, Venus, Mars, the Earth and its moon, these planets are small, dense, and rocky in composition, have relatively thin atmospheres, and rotate rather slowly.

Thermal equilibrium: a state in which the flow of heat into each layer of a star equals the flow of heat out.

Thermal radiation: the emission of light from a body in relation to the temperature of the body.

Thermal stage: the second of two stages in the progress of a protostar toward becoming a star, during which contraction proceeds slowly, made possible only by the loss of energy from the star's surface, because the pressure inside the protostar has built up and hydrostatic equilibrium has been reached.

Triple-Alpha process: the combination of three helium nuclei, also called alpha particles, to form a carbon nucleus.

Uncertainty principle: a principle stating that it is impossible to observe a system without disrupting it by a certain irreducible minimum amount.

Velocity: as distinguished from *speed,* velocity is the distance travelled each second, plus the direction travelled.

Visible spectrum: different wavelengths of light perceived by the eye that we call color: red, orange, yellow, green, blue, and violet.

Visual binaries: double stars that can be observed telescopically.

Wave (light wave, sound wave, etc.): a continously moving disturbance usually meaning *periodic wave,* a series of disturbances that repeat themselves regularly.

Wave frequency: the number of wavelength crest and trough cycles recorded by a stationary observer each second; measured in cycles per second, or *hertz* 9HZ).

Wavelength: the distance between one crest of a wave and the next, or one trough of a wave and the next, usually represented by the Greek letter *lambda.*

White dwarfs: a class of stars that are unusually small, with radii about 1/00 of that of the sun, and which are white, indicating that they are very hot.

Wien's Law: a law stating that the wavelength at which maximum radiation is emitted is inversely proportional to the temperature of the radiation source.

Wolf-Rayet stars: very hot and luminous stars with surface temperatures of about 100,000° K, whose spectra exhibit broad emission lines.

Zero Age Main Sequence (ZAMS): a narrow band stretching diagonally across the H-R diagram containing stars of different mass which have begun using nuclear energy.

Zodiac: twelve constellations through which the sun appears to move annually.

Zone of Avoidance: the region near the disc of our own galaxy where few, if any, galaxies can be observed. Galaxies which may be in this zone are probably obscured by the dense dust clouds in the Milky Way.

INDEX

77 78 79 80 9 8 7 6 5 4 3 2 1